# Complex Phenomena in Nanoscale Systems

edited by

Giulio Casati
Center for Linear and Complex Systems, Como, Italy

and

Davron Matrasulov
Heat Physics Department, Uzbek Academy of Sciences
Tashkent, Uzbekistan

AF834564

Springer

Published in cooperation with NATO Public Diplomacy Division

Proceedings of the NATO Advanced Research Workshop on
Recent Advances in Nonlinear Dynamics and Complex System Physics:
From Natural to Social Sciences and Security
Tashkent, Uzbekistan
12 – 17 October 2008

Library of Congress Control Number: 2009929397

ISBN 978-90-481-3119-8 (PB)
ISBN 978-90-481-3118-1 (HB)
ISBN 978-90-481-3120-4 (eBook)

---

Published by Springer,
P.O. Box 17, 3300 AA Dordrecht, The Netherlands.

*www.springer.com*

*Printed on acid-free paper*

---

All Rights Reserved
© Springer Science + Business Media B.V. 2009
No part of this work may be reproduced, stored in a retrieval system, or transmitted in any form or by any means, electronic, mechanical, photocopying, microfilming, recording or otherwise, without written permission from the Publisher, with the exception of any material supplied specifically for the purpose of being entered and executed on a computer system, for exclusive use by the purchaser of the work.

## PREFACE

The objective of this NATO Advanced Research Workshop on "Recent Advances in Nonlinear Dynamics and Complex System Physics" was to bring together researchers working in different areas of nonlinear dynamics and its application to various topics such as nanoscale physics, cold atoms, Bose-Einstein condensates, quantum information, non-equilibrium systems and econophysics. Most of the talks were concentrated on quantum transport and nonlinear phenomena in nanostructures and cold atom systems, including quantum dots, graphene and trapped cold atoms. Special attention was brought to possible practical applications in nano-scale physics and engineering. A special feature of the meeting was the fact that most of the speakers and participants were young researchers who presented interesting talks in different areas. Over a total of 55 talks, 28 were 50-min talks by well known scientists while the other talks were half-hour ones, mostly by young researchers.

Panel discussions have attracted a broad audience of theoretical and experimental physicists interested in a deeper understanding of various particular aspects and recent progress on the topics of the meeting. A round table discussion has been organized on the final day of the conference in order to comment on the conference talks and to present concluding remarks.

The workshop has been wonderfully and efficiently organized by the local committee chaired by P.K. Khabibullaev and with K. Nakamura, A.A. Saidov, Kh.Yu. Rakhimov, U. Salomov, G. Milibaeva as members. A group of PhD students from the Heat Physics Department helped for several and important organizational matters. Our special thanks go to Olga Karpova for her help with the secretarial duties before, during and after the conference. We would like to thanks also Khakim Butanov, Nurmukhammad Iskandarov and Hamid Yusupov for their valuable assistance in the preparation of the proceedings of the conference.

Finally, we wish to thank NATO Science for Peace and Security Program for the financial support. Additional support was provided by the Heat Physics Department of the Uzbek Academy of Sciences and Physical Society of Uzbekistan.

Como, Italy — Giulio Casati
Tashkent, Uzbekistan — Davron Matrasulov

# Complex Phenomena in Nanoscale Systems

# NATO Science for Peace and Security Series

This Series presents the results of scientific meetings supported under the NATO Programme: Science for Peace and Security (SPS).

The NATO SPS Programme supports meetings in the following Key Priority areas: (1) Defence Against Terrorism; (2) Countering other Threats to Security and (3) NATO, Partner and Mediterranean Dialogue Country Priorities. The types of meeting supported are generally "Advanced Study Institutes" and "Advanced Research Workshops". The NATO SPS Series collects together the results of these meetings. The meetings are co-organized by scientists from NATO countries and scientists from NATO's "Partner" or "Mediterranean Dialogue" countries. The observations and recommendations made at the meetings, as well as the contents of the volumes in the Series, reflect those of participants and contributors only; they should not necessarily be regarded as reflecting NATO views or policy.

**Advanced Study Institutes (ASI)** are high-level tutorial courses intended to convey the latest developments in a subject to an advanced-level audience

**Advanced Research Workshops (ARW)** are expert meetings where an intense but informal exchange of views at the frontiers of a subject aims at identifying directions for future action

Following a transformation of the programme in 2006 the Series has been re-named and re-organised. Recent volumes on topics not related to security, which result from meetings supported under the programme earlier, may be found in the NATO Science Series.

The Series is published by IOS Press, Amsterdam, and Springer, Dordrecht, in conjunction with the NATO Public Diplomacy Division.

**Sub-Series**

| | | |
|---|---|---|
| A. | Chemistry and Biology | Springer |
| B. | Physics and Biophysics | Springer |
| C. | Environmental Security | Springer |
| D. | Information and Communication Security | IOS Press |
| E. | Human and Societal Dynamics | IOS Press |

http://www.nato.int/science
http://www.springer.com
http://www.iospress.nl

**Series B: Physics and Biophysics**

**CONTENTS**

## LIST OF CONTRIBUTORS

F.Kh. Abdullaev
Physical-Technical Institute of the Academy of Sciences, G. Mavlyanov 2-b, Tashkent 100084, Uzbekistan

Y. Alhassid
Center for Theoretical Physics, Yale University, New Haven, Connecticut 06520, U.S.A.

Sh. Ataev
Heat Physics Department of the Uzbek Academy of Sciences, 28 Katartal Street, 100135 Tashkent, Uzbekistan

Atabek E. Atamuratov
Urganch State University, Urganch, Uzbekistan
atabek@ursu.uzpak.uz

Sanat K. Avazbaev
Heat Physics Department of the Uzbek Academy of Sciences, 28 Katartal Street, 100135 Tashkent, Uzbekistan
sanat_avazbaev@yahoo.com

H.L. Calvo
Facultad de Matemática, Astronomía y Física and Instituto de Física (CONICET), Universidad Nacional de Córdoba, Ciudad Universitaria, 5000 Córdoba, Argentina

M.D. Croitoru,
Departement Fysica, Universiteit Antwerpen, Groenenborgerlaan 171, 2020 Antwerpen, Belgium

D. Delande
Laboratoire Kastler Brossel, Universit¢u Pierre et Marie Curie, 4, place Jussieu 75252 Paris Cedex 05, France

R.M. Galimzyanov
Physical-Technical Institute of the Academy of Sciences, G. Mavlyanov 2-b, Tashkent 100084, Uzbekistan

Bertrand Georgeot
Université de Toulouse; UPS; Laboratoire de Physique Théorique (IRSAMC); CNRS; LPT (IRSAMC), Toulouse, France
georgeot@irsamc.ups-tlse.fr

O. Giraud
Université de Toulouse; UPS; Laboratoire de Physique Théorique (IRSAMC); CNRS; LPT (IRSAMC), Toulouse, France

Martina Hentschel
Max-Planck Institut für Physik Komplexer Systeme, Nöthnitzer Street 38, 01187 Dresden, Germany
martina@mpipks-dresden.mpg.de

N.E. Iskandarov
Heat Physics Department of the Uzbek Academy of Sciences, 28 Katartal Street, 100135 Tashkent, Uzbekistan

Khayotullo N. Ismatullaev
Physical-Technical Institute of the Academy of Sciences, G. Mavlyanov 2-b, Tashkent 100084, Uzbekistan
hayoti@yahoo.com

Rodolfo A. Jalabert
Institut de Physique et Chimie des Matériaux de Strasbourg, Strasbourg, France
jalabert@ipcms.u-strasbg.fr

Bekir S. Kandemir
Department of Physics, Faculty of Sciences, Ankara University, Ankara, Turkey
Bekir.S.Kandemir@science.ankara.edu.tr

F.N.R. Koch
Physikalisches Institut, Universität Heidelberg, Philosophenweg 12, 69120 Heidelberg, Germany

B. Kubala
Department of Physics, Arnold Sommerfeld Center for Theoretical Physics, Center for NanoScience, Ludwig-Maximilians-Universität, Theresienstrasse 37, 80333 Munich, Germany

Kazue Kudo
Ochadai Academic Production, Ochanomizu University, Tokyo, Japan
kudo.kazue@ocha.ac.jp

V.I. Kuvshinov
Joint Institute for Power and Nuclear Research, Krasina Streeet 99, Minsk 220109, Belarus

A.V. Kuzmin
Joint Institute for Power and Nuclear Research, Krasina Streeet 99, Minsk 220109, Belarus

T.-Y. Kwon
Max-Planck Institut für Physik Komplexer Systeme, Nöthnitzer Street 38, 01187 Dresden, Germany

Florian Lenz
Physikalisches Institut, Universität Heidelberg, Philosophenweg 12, 69120 Heidelberg, Germany
lenz@physi.uni-heidelberg.de

M. Ludwig
Department of Physics, Arnold Sommerfeld Center for Theoretical Physics, Center for NanoScience, Ludwig-Maximilians-Universität, Theresienstrasse 37, 80333 Munich, Germany

Florian Marquardt
Department of Physics, Arnold Sommerfeld Center for Theoretical Physics, Center for NanoScience, Ludwig-Maximilians-Universität, Theresienstrasse 37, 80333 Munich, Germany
Florian.Marquardt@physik.lmu.de

J. Martin
Université de Toulouse; UPS; Laboratoire de Physique Théorique (IRSAMC); CNRS; LPT (IRSAMC), Toulouse, France

Shumpei Masuda
Department of Applied Mathematics and Physics, Graduate School of Informatics, Kyoto University, Kyoto, Japan
syunpei710@yahoo.co.jp

Davron Matrasulov
Heat Physics Department of the Uzbek Academy of Sciences, 28 Katartal Street, 100135 Tashkent, Uzbekistan
dmatrasu@phys.ualberta.ca

Tomoshige Miyaguchi
Department of Applied Physics, Osaka City University, Japan
tomo@a-phys.eng.osaka-cu.ac.jp

Takaaki Monnai
Department of Applied Physics, Waseda University, Tokyo, Japan
monnai@aoni.waseda.jp

Tania S. Monteiro
Department of Physics and Astronomy, University College London, Gower Street, London WC1E 6BT, UK
t.monteiro@theory.phys.ucl.ac.uk

Katsuhiro Nakamura
Heat Physics Department of the Uzbek Academy of Sciences, 28 Katartal Street, 100135 Tashkent, Uzbekistan
nakamura@a-phys.eng.osaka-cu.ac.jp

Z. Narzikulov
Institute of Nuclear Physics, Tashkent 102132, Uzbekistan

B.L. Oksengendler
Institute of Polymer Chemistry and Physics AS of Uzbekistan, 7B Kadiriy Street, 100128 Tashkent, Uzbekistan

H.M. Pastawski
Facultad de Matemática, Astronomía y Física and Instituto de Física (CONICET), Universidad Nacional de Córdoba, Ciudad Universitaria, 5000 Córdoba, Argentina

Francois M. Peeters
Departement Fysica, Universiteit Antwerpen, Groenenborgerlaan 171, 2020 Antwerpen, Belgium
francois.peeters@ua.ac.be

C. Petri
Physikalisches Institut, Universität Heidelberg, Philosophenweg 12, 69120 Heidelberg, Germany

Vadzim A. Piatrou
Joint Institute for Power and Nuclear Research, Krasina Streeet 99, Minsk 220109, Belarus
PiatrouVadzim@tut.by

Abdulla Rakhimov
Institute of Nuclear Physics, Tashkent 102132, Uzbekistan
rakhimovabd@yandex.ru

A. Rancon
Department of Physics and Astronomy, University College London, Gower Street, London WC1E 6BT, UK

V.G. Romanovski
CAMTP – Center for Applied Mathematics and Theoretical Physics, University of Maribor, Krekova 2, SI-2000 Maribor, Slovenia

Marko Robnik
CAMTP – Center for Applied Mathematics and Theoretical Physics, University of Maribor, Krekova 2, SI-2000 Maribor, Slovenia
robnik@uni-mb.si

J. Ruostekoski
Department of Physics and Astronomy, University College London, Gower Street, London WC1E 6BT, UK

Almas F. Sadreev
Kirensky Institute of Physics, 660036 Krasnoyarsk, Russia
almsa@ifm.liu.se

Uktam R. Salomov
Heat Physics Department of the Uzbek Academy of Sciences, 28 Katartal Street, 100135 Tashkent, Uzbekistan
salukrah@yahoo.com

P. Schmelcher
Physikalisches Institut, Universität Heidelberg, Philosophenweg 12, 69120 Heidelberg, Germany

Sebastian Schmidt
Institute for Theoretical Physics, ETH Zurich, Zurich, Switzerland
schmidts@phys.ethz.ch

A.A. Shanenko
Departement Fysica, Universiteit Antwerpen, Groenenborgerlaan 171, 2020 Antwerpen, Belgium

Zarif A. Sobirov
Heat Physics Department of the Uzbek Academy of Sciences, 28 Katartal Street, 100135 Tashkent, Uzbekistan
sobirovzar@yahoo.com

Nigora N. Turaeva
Institute of Polymer Chemistry and Physics AS of Uzbekistan, 7B Kadiriy Street, 100128 Tashkent, Uzbekistan
nturaeva@sarkor.com

Hiroki Tutu
Department of Applied Analysis and Complex Dynamical Systems, Graduate School of Informatics, Kyoto University
tutu@acs.i.kyoto-u.ac.jp

Denis Ullmo
Université Paris-Sud, LPTMS, Orsay Cedex, France
denis.ullmo@u-psud.fr

H. Yusupov
Heat Physics Department of the Uzbek Academy of Sciences, 28 Katartal Street, 100135 Tashkent, Uzbekistan

A. Zakhidov
UTD NanoTech Institute, BE26, Richardson TX 75083-0688

# TUNING THE SUPERCONDUCTING PROPERTIES OF NANOMATERIALS

M.D. Croitoru, A.A. Shanenko, and F.M. Peeters*
*Departement Fysica, Universiteit Antwerpen, Groenenborgerlaan 171, 2020 Antwerpen, Belgium; francois.peeters@ua.ac.be*

**Abstract.** Electron confinement and its effect on the superconducting-to-normal phase transition driven by a magnetic field and/or a current is studied in nanowires. Our investigation is based on a self-consistent numerical solution of the Bogoliubov–de Gennes equations. We find that in a parallel magnetic field and/or in the presence of a supercurrent the transition from the superconducting to the normal phase occurs as a cascade of discontinuous jumps in the superconducting order parameter for diameters $D < 10 \div 15\,\text{nm}$ at $T = 0$. The critical magnetic field exhibits quantum-size oscillations with pronounced resonant enhancements as a function of the wire radius.

**Key words:** Quantum confinement; Bogoliubov–de Gennes equations; Nano-superconductivity

## 1. Introduction

Our understanding of the superconducting properties in nanostructures has experienced a notable development in the last few years (von Delft 2001; Han and Crespi 2004; Grigorenko et al. 2008; Gladilin et al. 2004; Shanenko and Croitoru 2006; Shanenko et al. 2006a). This has been partly propelled by the recent developments in fabrication techniques of high-quality nanoscopic metallic structures (Guo et al. 2004; Özer et al. 2006, 2007; Eom et al. 2006; Zgirski et al. 2005; Tian et al. 2005; Jankovič et al. 2006; Gournis et al. 2008; Altmore et al. 2006). For example, in Zgirski et al. (2005) it was shown that the application of low energy (~1 keV) $Ar^+$ ion sputtering allows one to reduce an Al or Sn nanowire width down to 10 nm. The electron mean free path in such structures was estimated to be about or larger than the specimen thickness and, therefore, such nanowires can be considered to be in the clean limit for the transverse (confined) electron motion. That means that nonmagnetic impurities can only influence the electron motion parallel to the nanowire. Within this method the resistance $R(T)$ of the same nanowire before and after sputtering sessions was obtained. The authors reported a clear, gradual increase of the superconductor critical temperature with decreasing wire

G. Casati and D. Matrasulov (eds.), *Complex Phenomena in Nanoscale Systems*,
NATO Science for Peace and Security Series B: Physics and Biophysics,
© Springer Science+Business Media B.V. 2009

thickness. This is the opposite behavior as found earlier for sufficiently disordered nano-samples where a suppression of the superconducting temperature with decreasing wire width is observed.

Studies of superconducting properties of quasi low-dimensional structures (films, wires, dots, etc.) have a long history. Most of these studies referred to strongly disordered or granular structures (Altshuler et al. 1980; Altshuler 1985; Lee 1985; Skvortsov and Feigel'man 2005; Oreg and Finkelstein 1999). The main purpose of those earlier investigations was to explain the suppression of superconductivity in such high-resistive structures. Many of these studies have ignored the effect of confinement of the itinerant carriers, because in such strongly disordered structure this effect is not of primary importance.

In Shanenko et al. (2006a) it was found that recent experimental observations of the width-dependent increase of the superconducting-transition temperature of high quality nanowires (Zgirski et al. 2005; Tian et al. 2005; Savolainen et al. 2004) is a manifestation of size-dependent superconducting resonances as a consequence of quantum confinement. Such a size resonance superconducting quantity occurs when an electron subband passes through the Fermi surface (Shanenko and Croitoru 2006; Shanenko et al. 2006a; Blatt and Thompson 1963; Shanenko et al. 2006b). Effects of nanowire topology on superconductivity have been investigated and it was shown that quantum-size oscillations are accompanied by quantum-shape variations in all superconducting properties (Croitoru et al. 2007). These theoretical works not only helped to clarify the physics behind the observed results, but they also show the way towards possible tuning of the superconducting characteristics by changing the cross section of the structures.

One can expect that quantum confinement should also influence the superconducting-to-normal phase transition driven by a magnetic field or a supercurrent. Indeed, according to the Ginzburg–Landau theory, the critical magnetic field is expected to increase as $1/D$ in the Meissner state, with $D$ the diameter of the mesoscopic wire. Moreover, the superconducting-to-normal phase transition in a magnetic field is of second order for such mesoscopic wires while being of first order in bulk type I superconductors. In this paper we present an overview of our recent theoretical results on the critical magnetic field and on the critical current in superconducting nanowires.

## 2. Theoretical Model and Numerical Approach

To study the interplay of superconductivity and quantum confinement in high-quality metallic nanowires, we investigate a superconducting nanocylinder with diameter $D$ and length $L$ in the clean limit. In the presence of quantum confinement the translational invariance in the confined directions is broken and the superconducting order parameter depends on the position $\Delta = \Delta(\mathbf{r})$. To investigate equilibrium superconductivity in this case the use of the Bogoliubov–de Gennes (BdG) equations (Bogoliubov 1959; de Gennes 1966) is required. These equations can be written as

$$\varepsilon_i|u_i\rangle = \widehat{H}_e|u_i\rangle + \widehat{\Delta}|v_i\rangle, \tag{1}$$

$$\varepsilon_i|v_i\rangle = \widehat{\Delta}^*|u_i\rangle - \widehat{H}_e^*|v_i\rangle, \tag{2}$$

where $\varepsilon_i$ stands for the Bogoliubov-quasiparticle (bogolon) energy, $|u_i\rangle$ and $|v_i\rangle$ are the particle-like and hole-like ket vectors and the single-electron Hamiltonian, shifted by the chemical potential $\mu$, is given by

$$\widehat{H}_e(\mathbf{r}) = \frac{\widehat{\mathbf{P}}^2}{2m_e} + V(\mathbf{r}) - \mu, \tag{3}$$

with $\widehat{\mathbf{P}} = \hat{\mathbf{p}} + m_e\mathbf{v}_s$. For the sake of simplicity, the confining interaction is taken as: $V(\mathbf{r}) = V_B\,\theta(R-\rho)$ with the barrier potential $V_B \to \infty$ ($R = D/2$ and $\rho$ is the transverse coordinate from the cylindrical set $\rho$, $\varphi$, $z$). In (1) and (2), $\widehat{\Delta}$ stands for the gap operator whose matrix element are given by $\langle\mathbf{r}|\widehat{\Delta}|\mathbf{r}'\rangle = \Delta(\mathbf{r})\delta(\mathbf{r}-\mathbf{r}')$, with $\delta(x)$ the Dirac $\delta$-function. The BdG equations should be solved in a self-consistent manner together with the relation

$$\widehat{\Delta} = g\sum_i(1-2f_i)\hat{\delta}_i, \tag{4}$$

with $i = (j, m, k)$ due to the cylindrical symmetry and $\langle\mathbf{r}|\widehat{\delta}_i|\mathbf{r}'\rangle = \langle\mathbf{r}|u_i\rangle\langle v_i|\mathbf{r}'\rangle\,\delta(\mathbf{r}-\mathbf{r}')$,where $g$ is the coupling constant [for a delta-function effective electron–electron interaction $-g\delta(\mathbf{r}_1 - \mathbf{r}_2)$], and $f_i = 1/(e^{\beta E_i} + 1)$ is the Fermi distribution for the bogolons. The sum in (4) runs over the states with the single-electron energy

$$\xi_i = \left[\langle u_i|\widehat{H}_e|_{v_s=0}|u_i\rangle + \langle v_i|\widehat{H}_e^*|_{v_s=0}|v_i\rangle\right] \in [-\hbar\omega_D, \hbar\omega_D], \tag{5}$$

with $\omega_D$ the Debye frequency and $v_s = |\mathbf{v}_s|$. Equation (5) introduces the cut-off over the single-electron momentum $\mathbf{p}$, to remedy the delta-function approximation for the effective electron–electron interaction. Such an approximation neglects the complex structure of the Fourier transform of the pair interaction. The problem is cured by the cutoff in the canonical-momentum space. In addition, notice that there are two branches for a solution of the BdG equations: $(i,+)$ and $(i,-)$ (see de Gennes 1966; Swidzinsky 1982). For them we have $E_{i,+}|_{v_s=0} > 0$ and $E_{i,-}|_{v_s=0} < 0$. The sum in (4) should be taken over the physical states [the $(i,+)$ branch], i.e. $E_i = E_{i,+}$. It is of importance to remark that some of $E_i$ can become negative at $v_s \neq 0$. This is a signature of the reconstruction of the ground state due to the current- or magnetic field-induced depairing of electrons.

For a given mean electron density $n_e$ the chemical potential $\mu$ is determined from

$$n_e = \frac{2}{\pi R^2 L} \sum_i [f_i \langle u_i | u_i \rangle + (1 - f_i) \langle v_i | v_i \rangle]. \tag{6}$$

Expanding $|u_i\rangle$ and $|v_i\rangle$ in terms of the eigenfunctions of $\widehat{H}_e$, one can convert (1) and (2) into a matrix equation. Then, the numerical problem is solved by means of diagonalizing the relevant matrix and invoking iterations, in order to account for the self-consistency relation given by (4).

## 3. Quantum-Size Oscillations and Resonances

The physics behind the quantum-size superconducting resonances is as follows. The superconducting order parameter is not simply the wave function of an ordinary bound state of two fermions but the wave function of a bound fermion pair in a medium (Bogoliubov 1959; Gor'kov 1958). In the homogeneous case the Fourier transform of the Cooper-pair wave function is suppressed for wavenumbers less than the Fermi one due to the presence of the Fermi sea (Cooper 1956). Therefore, the Fourier transform of the Cooper-pair wave function appears to be essentially nonzero only in the vicinity of the Fermi wavenumber. Generally, the superconducting order parameter strongly depends on $N_D$, the number of single-electron states (per spin projection) situated in the Debye window around the Fermi level ($\xi_i \in [-\hbar\omega_D, \hbar\omega_D]$). More precisely, the mean energy density of these states taken per unit volume $n_D = N_D/(2\hbar\omega_D V)$ is the key quantity. In the presence of quantum confinement the band of single-electron states in a clean nano-sample is split up in a series of subbands. While the specimen thickness increases (decreases), these subbands move down (up) in energy. Note that the position of the bottom of any subband scales as $1/D^2$, with $D$ the specimen thickness. Each time when the bottom of a parabolic subband passes through the Fermi surface, the density $n_D$ increases abruptly. As an example in Fig. 1a, b, single-electron subbands are schematically plotted vs. the wave vector of the quasi-free electron motion along the nanowire. The single-electron states located in the Debye window (making a contribution to the superconducting characteristics) are highlighted by the broken lines. In Fig. 1a the bottoms of all subbands 1, 2 and 3 are situated outside the Debye window, and we are in the off-resonance regime. However, when increasing the thickness of the sample, the bottom of subband 3 moves down so that it enters the Debye window

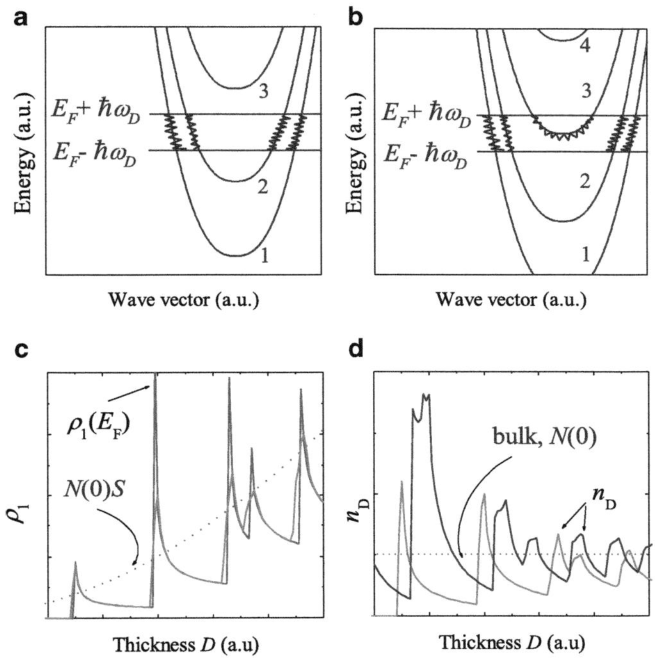


*Figure 1.* (**a**) Single-electron subbands vs. the wave vector for motion parallel to the nanowire for the off-resonant case. (**b**) The same as in panel (**a**) but for the resonant situation. (**c**) The $1D$ energy density of states at the Fermi level $\rho_1(E_F)$ vs. the wire thickness $D$, when the $E_F = E_{F,bulk}$. The red curve is the density of states per unit length, the green curve is the density of states averaged over the Debye window. (**d**) The mean energy density of states in the Debye window $n_D$ vs. $d$: the green curve is for a constant Fermi level, the blue curve results from taking account of a change in $E_F$ for narrow nanofilms

(as shown in Fig. 1b), the mean density of states $n_D$ increases and a superconducting resonance develops. This leads to a sequence of peaks in $n_D$ as a function of the sample thickness $d$ (see Fig. 1d) and, as a consequence, any superconducting quantity exhibits quantum-size oscillations with remarkable resonant enhancements. Such superconducting resonances are significant in nanoscale samples but smoothed out with an increase in $D$ (see Fig. 1d), when $n_D$ slowly approaches its bulk limit $N(0) = mk_F/(2\pi^2\hbar^2)$, with $k_F$ the bulk Fermi wavenumber.

In the case of a nanowire the density of states per unit length is given by

$$\rho_1(E) = \frac{\sqrt{m/2}}{\pi\hbar} \sum_{j=0}^{\infty} \theta\left(E - E_j\right) / \sqrt{E - E_j}, \tag{7}$$

where $\theta(x)$ is the Heaviside function. Here $E_j$ is the bottom of the $j$-subband due to the quantization of the electron motion in the confined direction. The red curve in Fig. 1c displays schematically quantum-size oscillations of the nanowire density of states per unit length when the Fermi level is pinned to its bulk value, while the green one is the same quantity but averaged over the Debye window. The blue and green curves in Fig. 1d correspond to the averaged (over the Debye window) density of states per volume $n_D \sim \rho_1(E_F)/S$, when the Fermi level fluctuates to keep the electron density of the system constant (blue curve) and when the Fermi level is constant (green curve).

Notice that on the average $n_D$ is near its bulk value. However, as shown below, the superconducting characteristics are enhanced at the resonant points as compared to their bulk values and these resonances are correlated with the resonances in the single-particle density of states. The reason is that the single-electron wave functions are no longer three-dimensional plane waves, and this change plays a significant role on the nanoscale.

## 4. Results and Discussion

Below we investigate how the quantum size resonances influence the critical magnetic field and the critical supercurrent in superconducting nanowires. We restrict ourselves to Al nanowires whose material parameters are: $\hbar\omega_D/k_B = 375\,\mathrm{K}$, $gN(0) = 0.18$ (de Gennes 1966). To investigate the quantum-size variations of physical properties of nanoscale systems within the parabolic band approximation (based on the use of the band mass $m_e$), one should use an effective Fermi level rather than the true Fermi level (for more details, see Shanenko et al. 2006b; Wei and Chou 2002). In the present work $E_F = 0.9\,\mathrm{eV}$. This value of $E_F$ for Al is justified from the good agreement with the experimental data found in Shanenko et al. (2006a). In the present work we restrict ourselves to a cylinder nanowire with diameter $D$ and length $L = 5\,\mu\mathrm{m} \gg \lambda_F$.

### 4.1. PARALLEL CRITICAL MAGNETIC FIELD IN NANOWIRES

The formation of single-electron subbands as a consequence of quantum confinement strongly influences the superconducting-to-normal phase transition driven by a magnetic field. According to the Ginzburg–Landau theory this transition is of second order for mesoscopic wires, while being of first order in bulk type I superconductors. This critical magnetic field increases as $1/D$ in mesoscopic wires (Silin 1951; Lutes 1957). Recent calculations within the BdG formalism for wires with $D = 20$–$200\,\mathrm{nm}$ (Han and Crespi 2004) has confirmed the GL conclusion about the second order transition, which is in agreement with recent experimental data for Sn (Tian et al. 2005; Jankovič et al. 2006) and Zn (Kurtz et al. 2007) nanowires with $D \gtrsim 20\,\mathrm{nm}$.

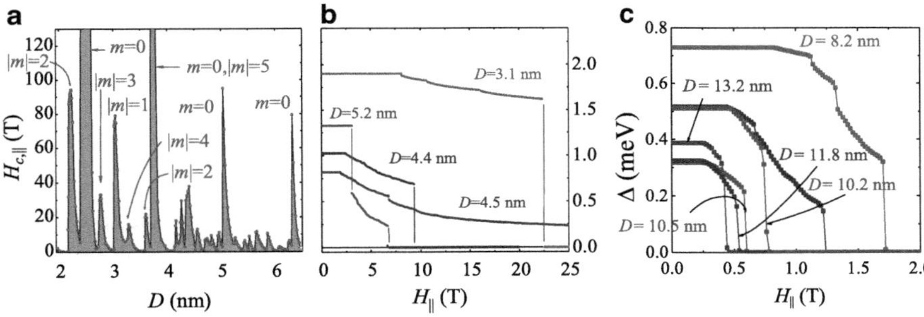


*Figure 2.* (**a**) Critical parallel magnetic field $H_{c,\|}$ vs. the nanowire diameter $D$ and (**b**) spatially averaged order parameter $\overline{\Delta}$ as function of $H_\|$ for the resonant diameters $D = 3.1$, 4.4, 4.5, and 5.2 nm. (**c**) Spatially averaged order parameter vs. $H$ for the resonant diameters $D = 8.2$, 10.2, 10.3, 10.5, 11.8, and 13.2 nm

However, the situation changes significantly for narrower wires. In Fig. 2a, we show over calculated critical field $H_{c,\|}$ for zero temperature as a function of the nanowire diameter $D$. At the resonances $H_{c,\|}$ exhibits huge enhancements as compared to the bulk critical magnetic field ($H_{c,bulk} \simeq 0.01$ T). According to Shanenko and Croitoru (2006) at a resonance the main contribution to the superconducting quantities comes from the subband whose bottom passes through the Fermi surface. Due to quantum size oscillations in the pair-condensation energy, we get corresponding oscillations in the critical magnetic field whose resonances, therefore, can be labeled by the quantum numbers of the subband $(j, m)$ that passes through the Fermi level. From this figure we notice, that the resonances in $H_{c,\|}$ dependent strongly on the diameter $D$ and the azimuthal quantum number $m$ of the subband.

The states with large $m$ are strongly influenced by the magnetic field and, so the resonances in $H_{c,\|}$ governed by large $m$ are, as a rule, less pronounced. In contrast, the resonances controlled by $m = 0$ have a weak magnetic field dependence. Figure 2b shows several typical examples of how the spatially averaged order parameter $\overline{\Delta}$, calculated at the resonance diameters, depends on the external parallel magnetic field. We find the very interesting result that the superconducting-to-normal phase transition driven by a magnetic field parallel to the nanowire occurs as a cascade of discontinuous jumps in the mean value of the order parameter (see Shanenko et al. 2008).

The physics underlying this cascade behavior is as follows. Within Anderson's approximate solution of the BdG equations (Anderson 1959), the excitation energies $E_{jmk}$ can be approximated by the following expression

$$E_{jmk} = \sqrt{\xi^2_{jmk} + |\Delta_{jm}|^2} - \mu_B m H_\|, \tag{8}$$

where $\xi_{jmk}$ is given by (5), $\mu_B$ stands for the Bohr magneton, and

$$\Delta_{jm} = \int_0^R \mathrm{d}\rho\rho\Delta(\rho)\left[\left|u_{j,m,k}(\rho)\right|^2 + \left|v_{j,m,k}(\rho)\right|^2\right] \tag{9}$$

is the averaged value of the order parameter as seen by *jmk*-quasiparticles. From this expression it is clear, that the quasi-particle states with $m > 0$ move down in energy with increasing magnetic field $H_{\|}$, and that the larger the relevant azimuthal quantum number $m$, the faster the decrease in the excitation energy. To be more specific, let us consider a nanowire with diameter $D = 1.94$ nm. In Fig. 3a the quasiparticle energies as a function of wave vector $k$ are given for three cases. The left panel displays the quasi-particle energy dispersion in the absence of a magnetic field $H_{\|} = 0$, the middle panel shows

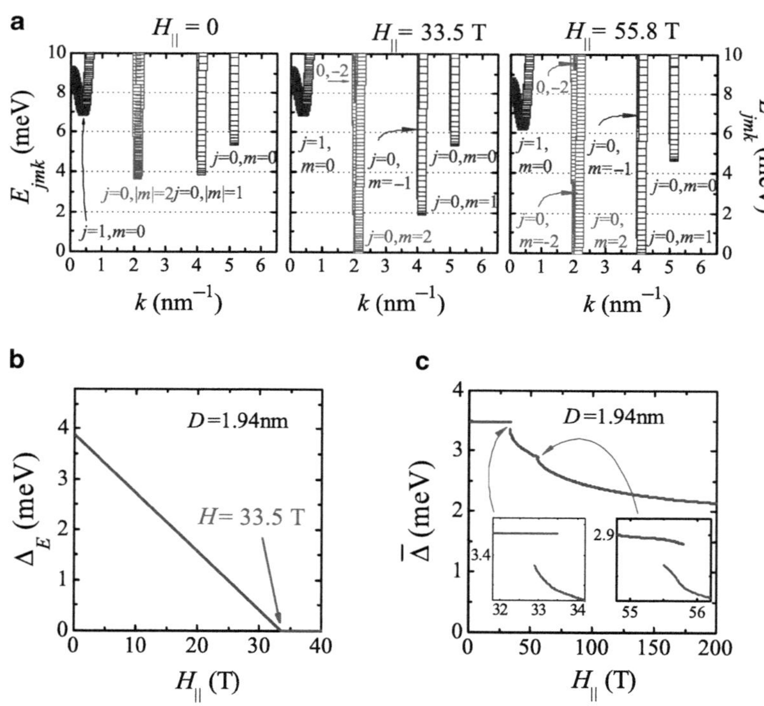


*Figure 3.* (**a**) The quasi-particle energies $E_{jmk}$ vs. $k$ for the four relevant branches $(j, m) = (0,0); (0,1); (0,2)$ and $(1,0)$ at $H = 0$, $H = 33.5$ T and $H = 55.8$ T for the resonant diameter $D = 1.94$ nm. (**b**) Superconducting energy gap vs. $H$, and (**c**) the mean order parameter vs. $H$

this quantity for the magnetic field $H_{\|} = 33.5\,\mathrm{T}$, at which the first small jump occurs in $\overline{\Delta}$, and the right panel shows the energy dispersion for the magnetic field $H_{\|} = 55.8\,\mathrm{T}$, at which the second jump occurs in $\overline{\Delta}$. Comparing this figure with Fig. 3c one can see, that each time when a quasi-particle branch specified by $j, m > 0$ touches the Fermi surface, a discontinuous jump in $\overline{\Delta}$ occurs. Or, introducing the energy gap for the $(jm)$ quasi-particle branch $\Delta_E^{(jm)}$, a jump in $\overline{\Delta}$ appears when one of the energy gap $\Delta_E^{(jm)} = 0$. The insets of Fig. 3c show details of such jumps in the mean order parameter with clear signatures of a hysteretic behavior, characteristic for a first-order phase transition. For $m = 0$ the second term in (8) is switched off and, the mean order parameter slowly decreases with magnetic field $H_{\|}$, exhibiting only a sequence of weak discontinuous jumps, as seen from Fig. 3c. If one takes a look at Fig. 3b, where we show the total excitation energy gap, defined as $\Delta_E = \min \Delta_E^{(jm)}$, as a function of applied field, one can see that the energy gap decays linearly with the magnetic field, which reflects the linear dependence of the excitation energies of the quasi-particles on the external parallel magnetic field. Figure 3b illustrates that for the magnetic field $H_{\|} > 33.5\,\mathrm{T}$ a gapless superconductivity is realized. When an energy branch that controls a resonant enhancement approaches the Fermi surface, $\overline{\Delta}$ jumps down to zero and the superconducting solution disappears. Other quasi-particle branches are less important due to the smaller density of states and are responsible for small, sometimes almost insignificant, jumps.

Signatures of the cascade behavior in the superconducting-to-normal phase transitions driven by a magnetic field can be observed even for larger diameters up to 15 nm. For instance, Fig. 2c displays the mean order parameter as a function of the parallel magnetic field for several resonant thicknesses (8.2, 10.2, 10.3, 10.5, 11.8, and 13.2 nm). It is remarkable that only jumps to zero in mean order parameter are clearly seen for large diameters: a cascade of preceding small jumps has nearly degenerated into a continuous curve. When the thickness of the wire becomes larger than 20 nm the superconducting-to-normal phase transition driven by a magnetic field becomes of the second order in agreement with Han and Crespi (2004).

In our approach we have neglected Pauli paramagnetism and included only the orbital effect. This is justified when the paramagnetic Pauli limiting field is larger than the orbital value of $H_{c,\|}$. Pauli paramagnetism is important for those resonances that are governed by states with $m = 0$, and it can lead to corrections to the resonances governed by $m = 1$ or $m = 2$. However, most of the resonant enhancements for $D > 5\,\mathrm{nm}$ are produced by the states with $m > 2$. Thus, our numerical results are not very sensitive to the spin-magnetic interaction for $D > 5\,\mathrm{nm}$.

### 4.2. CURRENT-CARRYING STATE IN NANOWIRES

Now we investigate the destruction of the superconducting condensate by a supercurrent. One can expect that quantum confinement should also have a strong impact on the transition from a superconducting to a normal state but now driven by a supercurrent.

Figure 4a, b display the spatially averaged order parameter $\overline{\Delta}$ and the superconducting energy gap $\Delta_E$, calculated at zero temperature for three resonant wire sizes $D = 4.2, 4.8$, and 5.3 nm, as a function of the normalized superfluid velocity $v_s = \hbar q/m$. The normalization is performed with respect to the Landau bulk depairing velocity $v_{\mathrm{L,bulk}} = \Delta_{\mathrm{bulk}}/\hbar k_{\mathrm{F,bulk}}$ (Landau 1941). The resonant enhancement of the superconducting energy gap at these three diameters is illustrated in the inset of the right figure. From the left figure one can see, that the destruction of the superconducting state occurs in the same way as in the presence of a magnetic field, namely, as a cascade of discontinuous jumps in the mean order parameter. The hysteretic transition near $v_s = 54\, v_{\mathrm{L,bulk}}$ for the resonant thickness $D = 4.2$ nm is enlarged in the inset of Fig. 4a.

The superconductor energy gap $\Delta_E = \Delta_E\,(v_s/v_{\mathrm{L,bulk}})$ shown in Fig. 4b vs. the superfluid velocity exhibits an almost linear dependence. We can understand this dependence from the Bogoliubov equations within the Anderson approximation. Within this approximation, in the presence of supercurrent flow the quasi-particle energies in a cylindrical nanowire are given by the following dispersion relation

$$E^q_{jmk} = \sqrt{\left(\xi_{jmk} + \frac{\hbar^2 q^2}{2m}\right)^2 + \left|\Delta_{jm}\right|^2} + E^q_D\,(k), \tag{10}$$

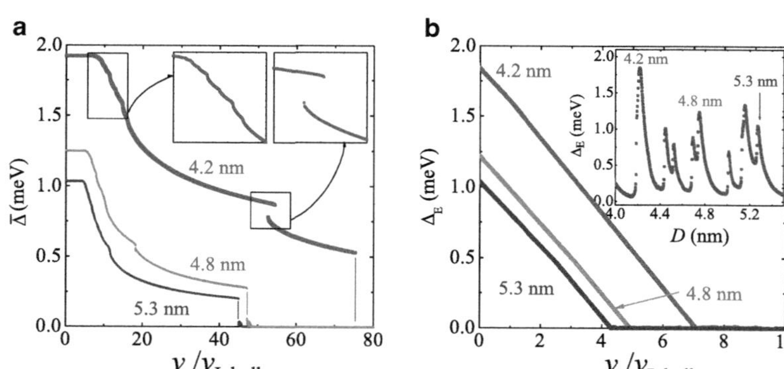


*Figure 4.* (**a**) The mean value of the order parameter $\overline{\Delta}$ (the insets are enlargements of the parts of the curves) and (**b**) the superconducting energy gap $\Delta_E$ at $T = 0$ vs. the normalized superfluid velocity (the inset shows the superconductivity energy gap vs. the nanowire diameter)

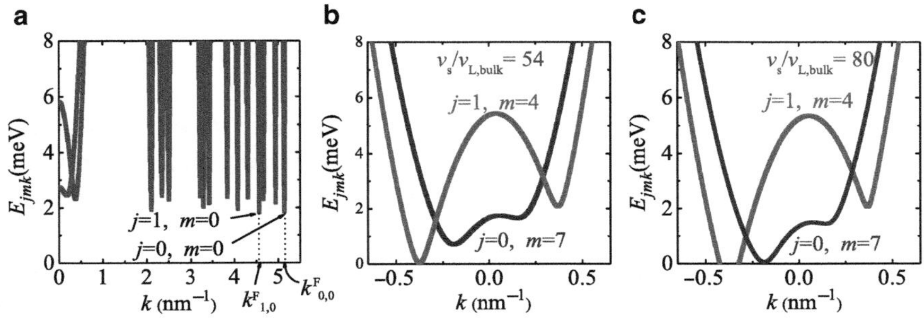


*Figure 5.* (**a**) Dispersion relation for the superconducting nanowire with $D = 4.2$ nm (only $k > 0$ are shown) and for $v_s/v_{L,\mathrm{bulk}} = 0$. Dispersion relation for the superconducting nanowire with $D = 4.2$ nm and for (**b**) $v_s/v_{L,\mathrm{bulk}} = 55$ and (**c**) $v_s/v_{L,\mathrm{bulk}} = 80$. Only the resonant subbands are presented

where the energy gap $\Delta_{jm}$ is introduced for each quasi-particle branch. Due to the Doppler shift term $E_D^q(k) = \hbar^2 kq/m$, part of the quasi-particle energies ($k < 0$) moves down with increasing superfluid velocity. Moreover, since the relevant states of each single-electron subband (within the Debye window) occupy only a narrow domain in the $1D$ $k$-space, we may assume that each subband has its own Doppler shift $E_D^q\left(k_{jm}^F\right)$, where $k_{jm}^F = \pm\sqrt{2m\left(\mu - \xi_{jmk=0}^{q=0}\right)}$. This approximation is certainly valid for those subbands, whose bottoms situate far from the Debye window. As seen from Fig. 5a, where the dispersion relation for the quasiparticles in a nanowire with $D = 4.2$ nm is plotted, at $v_s = 0$ the superconducting gap $\Delta_E$ is defined by the smallest quasi-particle branch gap $\Delta_E^{10}$. Hence at small values of the superfluid velocity the slope of the function $\Delta_E(v_s)$ is determined by the wavevector $k_{10}^F$. However, $k_{10}^F < k_{00}^F$ (see Fig. 5a) and at a certain value of the superfluid velocity, when $\Delta_E^{10}(v_s) = \Delta_E^{00}(v_s)$, a kink occurs in the plot of $\Delta_E(v_s)$. After that the slope of the function $\Delta_E(v_s)$ is determined by the wave-vector $k_{00}^F$.

The physics behind these jumps in the mean order parameter is the same as in the previous case of the magnetic field. However, the mechanism of the destruction of the superconducting condensate by a supercurrent possesses some peculiarities. Figure 5b, c display the dispersion relation for small wave vectors in case of a nanowire with $D = 4.2$ nm, when $v_s = 54\, v_{L,\mathrm{bulk}}$ and $v_s = 80\, v_{L,\mathrm{bulk}}$, respectively. As seen from these figures, in the current-carrying state the whole band structure of the superconducting wire is tilted with respect to the $k = 0$ point due to the superfluid flow. This tilting reflects the displacement of the entire distribution of electrons, including pairs, by an amount $q$ in momentum space. Each time when a quasi-particle branch touches the Fermi surface, a discontinuous jump in the mean value of the order parameter $\overline{\Delta} = \overline{\Delta}(v_s)$ curve occurs. This can be inferred by noticing that the snapshots of Fig. 5b, c correspond to the superfluid velocities, at which the two last discontinuities in $\overline{\Delta}$ occur (see Fig. 4). Figure 5b, c illustrate the branches of quasiparticles with the largest contribution to the order parameter, responsible for these discontinuities. When a branch controlling the resonant enhancement of the order parameter (in this case $\Delta_{j=0,|m|=7}$) approaches the Fermi surface we get the largest depairing transition. As seen from (10) the resonant subband has the smallest Doppler shift coefficient $k^F_{j,m}$, since it situates close to the point $k = 0$. Other quasi-particle branches produce only small kinks in the $\overline{\Delta} = \overline{\Delta}(v_s)$ curve due to a smaller density of states at the Fermi surface (see inset of Fig. 4a). Moreover, these branches have a large value for $k^F_{j,m}$, so they are more sensitive to nanowire imperfections. Consequently, the kinks in the $\overline{\Delta}$-curve obtained for a nanowire with uniform cross-section would be smeared out, if one takes into account such imperfections.

We identify the maximum supercurrent in a nanowire as the critical current density. In bulk samples depairing limits the maximum supercurrent to only about 1% above the current at the Landau depairing velocity (Swidzinsky 1982). Figure 6a shows the calculated critical current density as a function of the nanowire diameter. This result shows that quantum-size effects play a very important role in the size-dependent increase of the critical current. Real samples exhibit inevitable cross-section fluctuations that will smooth those quantum-size oscillations in the critical current. However, the average increase of the order parameter and, hence, of the critical current, is due to the resonance controlling quasi-particle branch. These states are characterized by small values of the vector $k$ and, consequently, they are

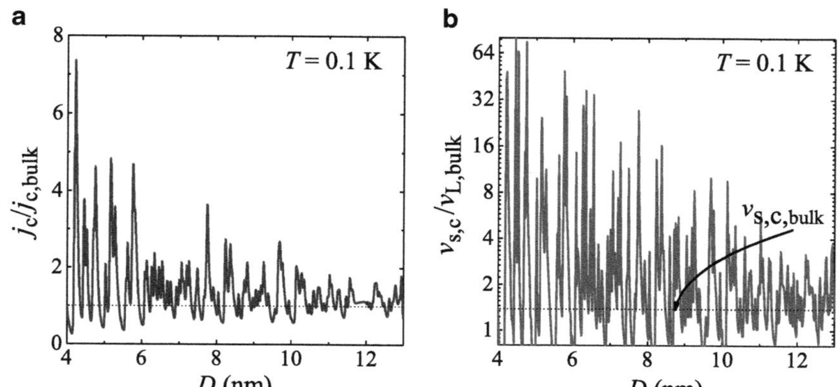


*Figure 6.* (**a**) Critical current and (**b**) critical superfluid velocity vs. the nanowire diameter calculated for $T = 0.1$ K

insensitive to nanowire imperfections. So the critical current at the resonant diameters are rather stable with respect to such fluctuations. For anti-resonant diameters the contributions to the order parameter from different subbands in the Debye window are more or less equal and the vectors $k$ are characterized by sufficiently larger values. Therefore, such states are more sensitive to nanowire imperfections, so the anti-resonant peaks seen in the figure would be smeared out if one takes into account these imperfections. Therefore, this results in an overall enhancement of the critical current density with decreasing wire thickness. Figure 6b displays the critical superfluid velocity $v_{s,c}$ as a function of the wire diameter. This quantity defined as the velocity at which complete destruction of the condensate occurs exhibits the same trend.

## 5. Conclusion

Quantum confinement plays a corner-stone role in high-quality metallic superconductors with nanoscale dimensions. Interplay of quantum confinement and superconductivity results in important qualitative changes in the superconductor characteristics. Based on a numerical self-consistent solution of the Bogoliubov–de Gennes equations, we showed that at zero temperature the superconducting-to- normal phase transition driven by a parallel magnetic field or by a supercurrent occurs as a cascade of discontinuous jumps in the order parameter. At the same time the critical magnetic field exhibits quantum-size oscillations with pronounced resonant enhancements. Therefore, nano-samples allow one to tune its superconducting characteristics (like critical magnetic field or supercurrent) by varying e.g. the nanowire radius.

## Acknowledgements

This work was supported by the Flemish Science Foundation (FWO-Vl), Scientific research fund of University of Antwerp (BOF UA 2008), Interuniversity Attraction Poles Programme – Belgian State – Belgian Science Policy (IAP) and the ESF-AQDJJ and ESF-VORTEX networks.

## References

Altmore F., Chang A.M., Melloch M.R., Hong Y. and Tu C.W.: 2006. *Phys. Rev. Lett.*, **97**, 017001

Altshuler B.L.: 1985. *JETP Lett.*, **41**, 648

Altshuler B.L., Aronov A.G. and Lee P.A.: 1980. *Phys. Rev. Lett.*, **44**, 1288

Anderson P.W.: 1959. *J. Phys. Chem. Solids*, **11**, 26

Blatt J.M. and Thompson C.J.: 1963. *Phys. Rev. Lett.*, **10**, 332

Bogoliubov N.N.: 1959. *Sov. Phys. Usp.*, **67**, 549

Cooper L.N.: 1956. *Phys. Rev.*, **104**, 1189
Croitoru M.D., Shanenko A.A. and Peeters F.M.: 2007. *Phys. Rev. B*, **76**, 024511
von Delft J.: 2001. *Ann. Phys. (Leipzig)*, **10**, 1
Eom D., Qin S., Chou M.-Y. and Shih C.K.: 2006. *Phys. Rev. Lett.*, **96**, 027005
de Gennes P.G.: 1996. *Superconductivity of Metals and Alloys*, (W.A. Benjamin, New York)
Gladilin V.N., Fomin V.M. and Devreese J.T.: 2004. *Phys. Rev. B*, **70**, 144506
Gor'kov L.P.: 1958. *Sov. Phys. JETP*, **7**, 505
Gournis D., Trikalitis P.N. et al: 2008. *Nano Lett.*, **8**, 3060
Grigorenko I., Zhu J.-X. and Balatsky A.: 2008. *J. Phys.: Condens. Matter*, **20**, 195204
Guo Y., Zhang Y.-F. et al: 2004. *Science*, **306**, 1915
Han J.E. and Crespi V.H.: 2004. *Phys. Rev. B*, **69**, 214526
Jankovič L., Gournis D. et al: 2006. *Nano Lett.*, **6**, 1131
Kurtz J.S., Johnson R.R. et al: 2007. *Phys. Rev. Lett.*, **98**, 247001
Landau L.: 1941. *J. Phys. USSR*, **5**, 71
Lee P.A. and Stone A.D.: 1985. *Phys. Rev. Lett.*, **55**, 1622
Lutes O.S.: 1957. *Phys. Rev.*, **105**, 1451
Oreg Y. and Finkelstein A.M.: 1999. *Phys. Rev. Lett.*, **83**, 191
Özer M.M., Thompson J.R. and Weitering H.H.: 2006. *Nature Phys.*, **2**, 173;
Özer M.M., Jia Y. et al: 2007. *Science*, **316**, 1594
Savolainen M., Touboltsev V. et al: 2004. *Appl. Phys. A*, **79**, 1769
Shanenko A.A. and Croitoru M.D.: 2006. *Phys. Rev. B*, **73**, 012510
Shanenko A.A., Croitoru M.D., Zgirski M., Peeters F.M. and Arutyunov K.: 2006a. *Phys. Rev. B*, **74**, 052502
Shanenko A.A., Croitoru M.D. and Peeters F.M.: 2006b. *Europhys. Lett.*, **76**, 498; newblock 2007, *ibid.*: *Phys. Rev. B*, **75**, 014519
Shanenko A.A., Croitoru M.D. and Peeters F.M.: 2008. *Phys. Rev. B*, **78**, 024505
Silin V.P.: 1951. *Zh. Eksp. Teor. Fiz.*, **21**, 1330
Skvortsov M.A. and Feigel'man M.V.: 2005. *Phys. Rev. Lett.*, **95**, 057002
Swidzinsky A.V.: 1982. *Spatially Inhomogeneous Problems in the Theory of Superconductivity*, (Nauka, Moscow)
Tian M.L., Wang J.G. et al: 2005. *Phys. Rev. B*, **71**, 104521
Wei C.M. and Chou M.Y.: 2002. *Phys. Rev. B*, **66**, 233408
Zgirski M., Riikonen K.-P., Touboltsev V. and Arutyunov K.: 2005. *Nano Lett.*, **5**, 1029

# OPTICAL MICROCAVITIES OF SPIRAL SHAPE: FROM QUANTUM CHAOS TO DIRECTED LASER EMISSION

M. Hentschel* and T.-Y. Kwon
*Max-Planck Institut für Physik Komplexer Systeme, Nöthnitzer Street 38, 01187 Dresden, Germany; martina@mpipks-dresden.mpg.de*

**Abstract.** Optical microcavities are open billiards for light in which electromagnetic waves can, however, be confined by total internal reflection at dielectric boundaries. These resonators enrich the class of model systems in the field of quantum chaos and are an ideal testing ground for the correspondence of ray and wave dynamics that, typically, is taken for granted. Using phase-space methods we show that this assumption has to be corrected towards the long-wavelength limit. We also discuss the issue of achieving directional emission from optical microcavity lasers, highly desired concerning applications in photonic devices, with a focus on cavities of spiral shape.

**Key words:** Optical microcavity; Billiard; Quantum chaos

## 1. Introduction

Research in the field of quantum chaos has closely accompanied the developments in mesoscopic physics that started about 20 years ago when micrometer-scale samples with very high-mobility for electrons in two dimensions, so-called quantum dots, were fabricated in semiconductor heterostructures for the first time (Sohn et al. 1997). Very soon, mesoscopic systems for light instead of electrons – optical microcavites and microlasers – were also fabricated (Vahala 2004). Up to now they are a topic of intensive research since, on the one hand, they extent the field of quantum chaos to intrinsically open systems, and on the other hand, they provide the opportunity to build micro*lasers* that emit light in just one direction.

Let us highlight these two points in some more detail. First, one has to realize an intrinsic difference between electronic and optical mesoscopic and nanoscopic systems, namely the mechanism that confines the electrons and light, respectively. Electrons carry a charge, and they are easily manipulated and confined on quantum dots by means of gate voltages. Light, in contrast, has no charge, and the confinement mechanism is indeed very different from that of electrons: It is the principle of total internal reflection at the optically thinner medium.

G. Casati and D. Matrasulov (eds.), *Complex Phenomena in Nanoscale Systems*,
NATO Science for Peace and Security Series B: Physics and Biophysics,

© Springer Science+Business Media B.V. 2009

The presence of light refraction (for angles of incidence below the critical angle $\chi_c = \arctan 1/n$, where $n$ is the refractive index of the cavity and we assume air ($n_0 = 1$) outside) and evanescent waves (even in the regime of total internal reflection) makes optical cavities intrinsically open systems. Note furthermore that electrons are described by the Schrödinger equation whereas light is governed by the Helmholtz equation (that follows, of course, from Maxwell's equations). Remarkably, these two equations are formally equivalent in two dimensions (up to polarization-dependent boundary conditions for the Helmholtz equation) which motivates a direct comparison between the resulting billiards for electrons and light, respectively. Consequently, optical microcavities enrich the class of model systems in the field of quantum chaos.

Another aspect that drives optical microsystems into the focus of interest is to a large extent application motivated. With the miniaturisation of optical (and electronic) devices, the fabrication limits of lasers based on Fabry–Perot resonators are reached. Since the need for highly unidirectional, coherent light sources of course nonetheless remains, alternative realizations and construction schemes have to be identified. The starting point have been microdisk resonators (McCall et al. 1992) where very high $Q$-factors were realized. They lack, however, the property of directional emission due to their rotational invariance. It was shown in the last years that deformation of the ideal disk can enhance the directionality of the light emission. Examples where this was demonstrated are the bow-tie mode in microlasers of quadrupolar shape (Gmachl et al. 1998), annular resonators with carefully tuned frequencies and geometry parameters (Wiersig and Hentschel 2006), or microcavities of Limacon shape (Wiersig and Hentschel 2008) where it was shown that engineering the farfield output characteristics via the so-called unstable manifold of the system is a promising and very robust approach that was, meanwhile, also confirmed experimentally (Yan et al. 2009).

One of the earlier, yet still popular attempts to achieve directional emission was to employ microcavities and microlasers of spiral shape (Chern et al. 2003; Chern et al. 2004; Ben-Messaoud and Zyss 2005; Fujii et al. 2005; Tulek and Vardeny 2007; Kim et al. 2008; Audet et al. 2007). These systems are characterized by an offset in the system's curved boundary, a so-called notch, cf. Fig. 1. Its size is parametrized by the geometry parameter $\epsilon$ and given as $\epsilon R_0$ where $R_0$ is the minimal radius of the cavity. Naively, one can think of the motivation to generate directional emission from spiral microlasers in a way as to excite so-called whispering-gallery modes (WGMs, i.e. modes that travel close to system boundary) as they are known from circular cavities. The idea was then that WGMs that travel clockwise and therefore do hit the notch, must leave the cavity through the notch, and hence

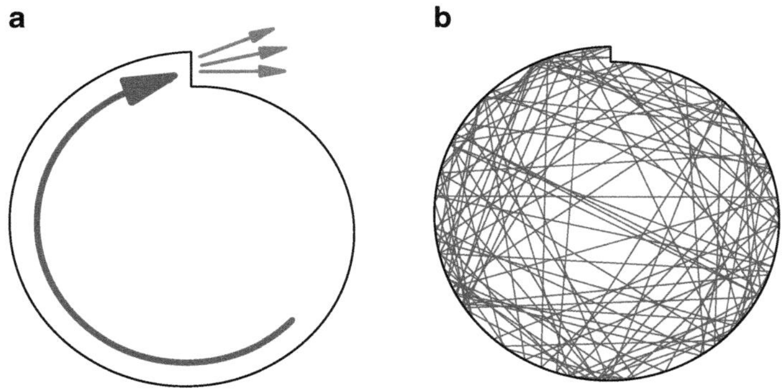


*Figure 1.* (**a**) Shape of the spiral microcavity with a schematically illustrated whispering–gallery mode that travels clockwise and would, therefore, hit the notch. If such modes could be realized in microlasers, the hope is that they would lead to directional emission originating from the notch region as indicated by the smaller arrows. (**b**) Typical ray trajectory in the spiral reflecting the characteristic chaotic dynamics of rays in such a geometry ($\epsilon = 0.1$). Clearly, the trajectory does not favor WGM-type orbits. It results in a structureless phase space (not shown) without regular islands. According to the principle of ray-wave correspondence, no directional emission can be expected in such a situation

in a unidirectional way. This is illustrated in Fig. 1a. We shall see below to what extent this idea holds and what the true mechanism behind directional emission from spiral mircolasers is.

A classic and practically approved way to study, and predict, the far-field characteristics of optical microcavities are ray-simulations. The principle behind is, of course, ray-wave correspondence and there are numerous examples where ray model simulations agreed very well with both experimental results and wave simulations (Lee et al. 2007; Schwefel et al. 2004; Shinohara and Harayama 2007; Tanaka et al. 2007; Hentschel and Vojta 2001; Wiersig et al. 2008). The correspondence holds not only in the semiclassical limit but, despite known semiclassical corrections to the ray picture (Goos and Hänchen 1947; Schomerus and Hentschel 2006), even down to wavelengths comparable to the system size (Wiersig and Hentschel 2008). This finding is to a certain extent surprising, and, at the moment, attributed to the importance of the so-called unstable manifold (Lee et al. 2005). Note, however, that (slight) deviations from ray-wave correspondence have been observed in several systems including spiral cavities (Unterhinninghofen et al. 2008; Lee et al. 2004; Altmann et al. 2008).

Motivated by these successes, we start this paper with a discussion of spiral cavities in terms of the ray picture and compare our far-field results with those of wave simulations. We then turn to a comparison with experimental results and report both experiments supporting our findings of *no* directional far-field emission and those experiments where directional emission was observed. The origin of these differences is discussed, and we explain the mechanism to be used in order to achieve directional emission from spiral microlasers that we revealed performing wave simulations of the active cavity (Hentschel and Kwon 2009).

## 2. Far-Field Characteristics of Spiral Microcavities: Ray and Wave Simulation Results

*Ray simulation results* for a sample trajectory are shown in Fig. 1b. The chaotic character of the orbit is clearly visible. It represents itself also in the far-field emission characteristics that is provided in Fig. 2 for spiral microcavities of both TM (transverse magnetic, i.e. electric field perpendicular to the cavity plane) and TE (transverse electric) polarization and for various geometries as well as two different refractive indices *n*. The emission occurs in a number of directions, and no directionality of the far-field pattern can be expected based on ray-picture simulations. Rather, emission occurs in a "spiky" fashion into a number of directions that sensitively depend on the geometry and reflect the chaotic character of the ray dynamics. This is especially apparent in the TM case. For TE polarization, one of the spikes seems to be singled out to be a preferred radiation direction. The reason behind is the existence of the Brewster angle: It corresponds to the angle of incidence where the reflection coefficient drops to zero, just before the regime of total internal reflection is reached. This results in a sort of "all-or-nothing" escape characteristics in the TE case (that would strictly speaking correspond to a step-like Fresnel law) that leads to the better radiations characteristics that was confirmed in many examples (Wiersig and Hentschel 2008; Gmachl et al. 2002).

*Wave simulation results* fully confirm the ray-picture based findings, see Figs. 3 and 4. In particular, the wave calculations reveal the origin of the far-field spikes. The configuration space presentation in the upper part of Fig. 4 suggests a WGM-type character of the resonances with elements reminiscent of quasiscar modes (Lee et al. 2004). At certain points along the cavity boundary the confinement of the modes by total internal reflection is violated, resulting in leakage. Light emerging from those regions forms the spikes visible in the far-field. We point out that the way that the light leaves the cavity (its "sense of rotation") clearly indicates that the supported WGMs travel *counter*clockwise – that is, opposite to the direction needed to realize emission of WGMs from the notch. The calculations based on the boundary elements method (Wiersig 2003) presented here (for TM polarization

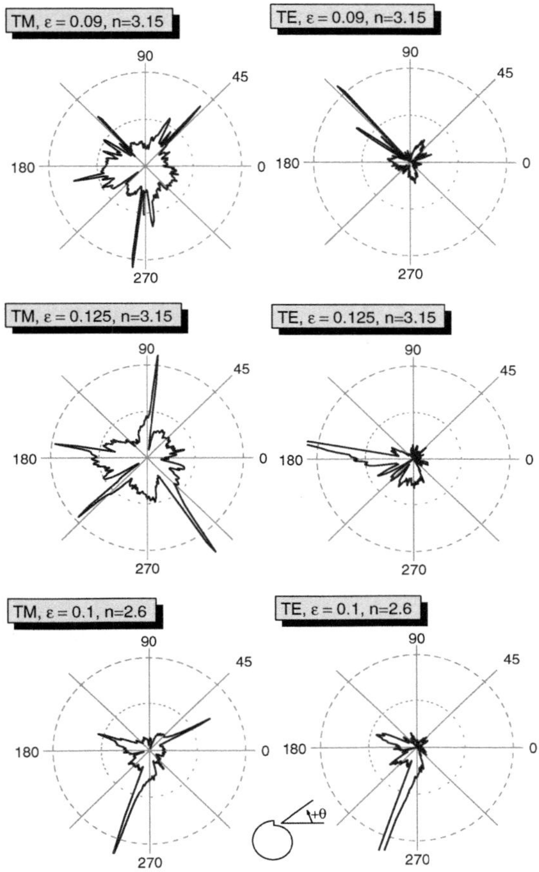


*Figure 2.* Far-field pattern of spiral microcavities from ray simulations of a Fresnel billiard based on the unstable manifold. Compared are the two possible polarizations of light (TM, left panels and TE, right panels) for different geometries (characterized by the parameter $\epsilon$) and refractive indices $n = 3.15$ corresponding to the one used in Audet et al. (2007) and $n = 2.6$ as in Chern et al. (2003). No directional emission can be expected based on ray simulations results, although directionality is improved in the TE case due to the existence of the Brewster angle. The upper two panels on the left correspond, qualitatively, very well to the experimental results reported in Audet et al. (2007). The inset at the bottom indicates how the far-field angle is measured

and $n = 3.2$) show that the high-$Q$ modes that can be expected to start lasing near the threshold (cf. their characteristics in Fig. 3) are dominantly counterclockwise modes that actually *do not see and do not hit the notch.*

The agreement between ray and wave simulation is of special importance because of the possibility of diffraction effects related to the sharp corners associated with the notch. Diffraction effects were made responsible for the

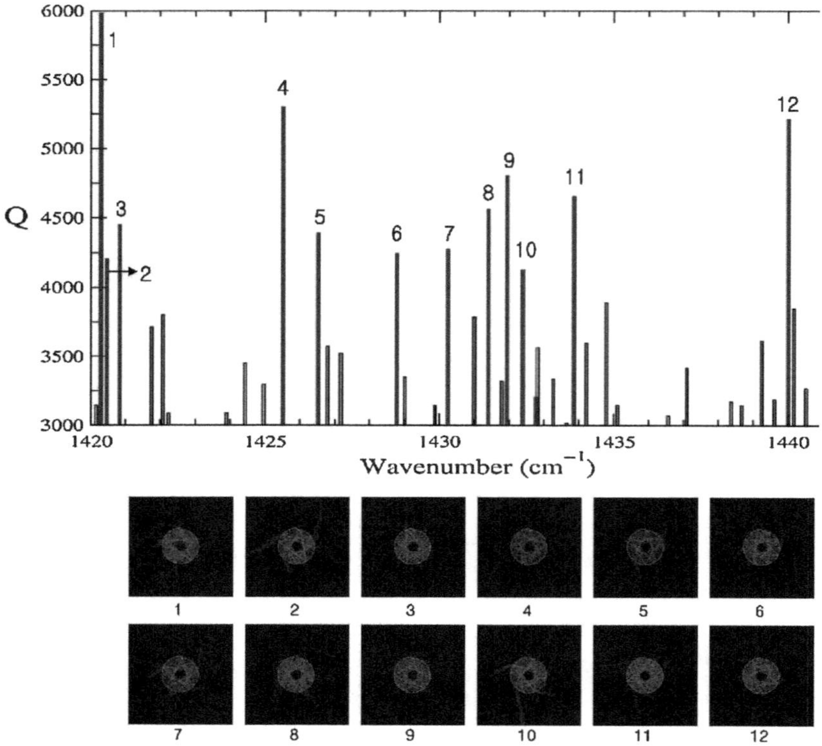


*Figure 3.* Resonances of a spiral microcavity ($\epsilon = 0.909$, TM polarization) as computed with the boundary element method. Resonances are characterized by a complex wave number $\Omega$ from which the $Q$-factor is derived as $Q = \mathrm{Re}\,\Omega/(2\,\mathrm{Im}\,\Omega)$. The upper panel shows $Q$ vs. the wavenumber $\mathrm{Re}\Omega$ for wavenumbers that were realized in the experiments in Audet et al. (2007). The lower panel shows a number of high-$Q$ modes in configuration space

directional emission from microcavities in Chern et al. (2003). In the present study we find, however, no supporting material for this statement. Rather, the ray-wave agreement (where the wave model naturally contains diffraction effects whereas the ray model does not) indicates that diffraction effects are, at least, not crucial (and in particular not responsible for the directional output, see below).

*Experimental results* reported in Audet et al. (2007) agree well with ray and wave simulation as we already mentioned. These experiments were performed with quantum cascade lasers that were uniformly pumped and used TM-polarized light (both conditions cannot be changed for technical reason). Most other experiments were performed with TE-polarized light and/or applied non-uniform pumping along the cavity boundary (Chern et al. 2003; Chern et al. 2004; Ben-Messaoud and Zyss 2005; Fujii et al. 2005; Tulek and

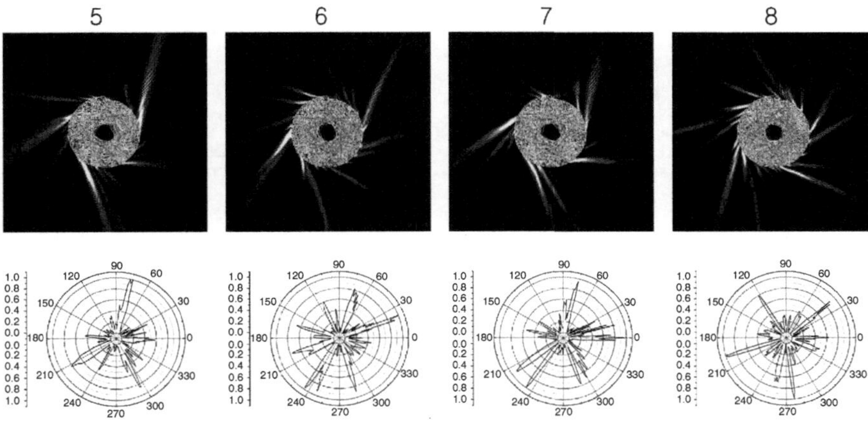


*Figure 4.* Resonances of a spiral microcavity (as in Fig. 3) in configuration space (upper panels). Note that the openness of the optical cavity leads to leakage of light in form of pronounced spikes that leave clear signatures in the far-field patterns. The resulting far-field patterns are shown in polar representation as in Fig. 2 in the lower panels. They correspond qualitatively very well to the ray simulation results shown in the upper two panels on the left in Fig. 2 and to the experimental findings reported in Audet et al. (2007)

Vardeny 2007; Kim et al. 2008) and the importance of the pumping scheme was realized already in Chern et al. (2003). It is therefore straightforward to assume that the emission properties, for the case of non-uniform pumping, are closely related to the pumping scheme applied. We investigate this issue in detail in the following section.

## 3. Towards Directional Emission from Spiral Microlasers

In order to study the characteristics of non-uniformly pumped devices, the spatial distribution of the active material has to be taken into account. This is possible within the Schrödinger-Bloch model (Harayama et al. 2005), the state-of-the-art instrument to describe active microcavities (Kwon et al. 2006). The results are shown in Fig. 5 where the far-field and configuration space patterns of uniformly and boundary pumped spiral microlasers are compared. The advantage of a boundary pumping scheme, leading to a dominant emission into a 45° far-field direction (measured from the notch) is clearly visible (Hentschel and Kwon 2009).

We have found that the directionality depends on the resonator geometry, i.e. on the size of the notch. The best performance is achieved when the notch length is about twice the cavity wavelength (Hentschel and Kwon 2009). In other words, the geometry must neither be too close to the disk (which would capture all light in WGMs and lead to uniform emission from

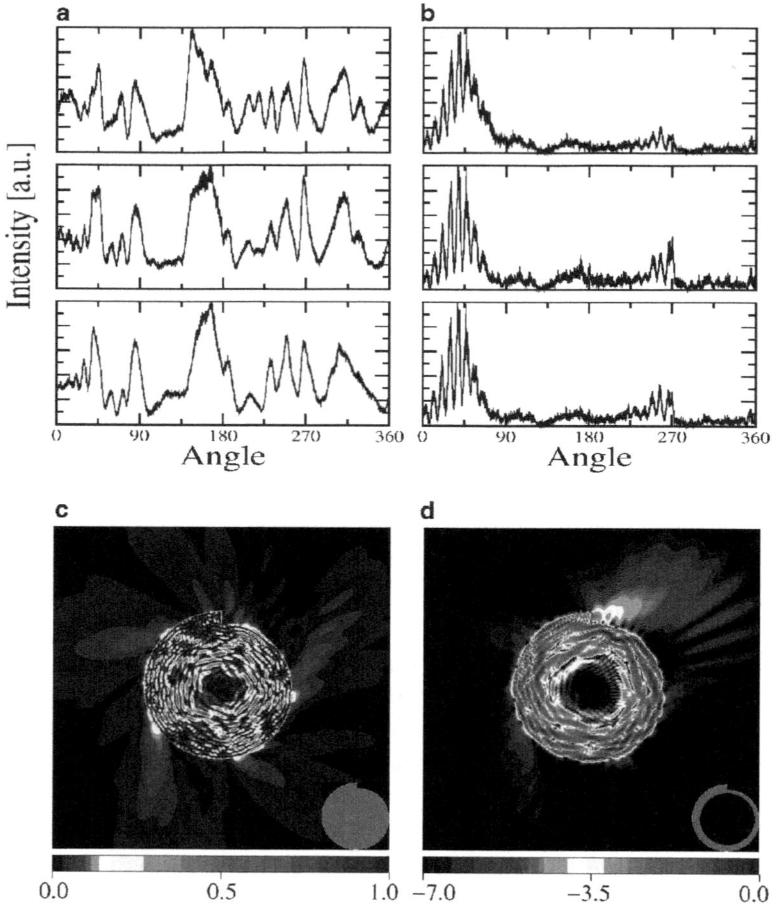


*Figure 5.* Comparison of uniformly pumped (left panels) and selectively boundary-pumped spiral microcavities. The pumped area is indicated in red in the insets in panels (**c**) and (**d**). The upper panels (**a**) and (**b**) show the far-field characteristics for increasing pumping strengths close to the lasing threshold, the far-field angle is measured as indicated in Fig. 2. Emission in a direction of approximately 45° is clearly visible in the boundary-pumped scheme. These far-field patterns were obtained by time-averaging after passing through a transient regime. The configuration space plots in (**c**) and (**d**) nicely illustrate the differences in the cavity wave patterns that, eventually, cause the very different far-field characteristics

evanescent waves) nor too far away from the disk geometry (in which case the system becomes too open and amplification cannot balance the radiation losses). Concerning the pumping scheme, pumping as close to the boundary as possible turned out to be the most favorable, but pumping a ring of about one tenth of the cavity radius gives still very reasonable results (Hentschel and Kwon 2009).

But what is the reason that makes the boundary pumping scheme so successful? In short (and we would like to refer the reader to Hentschel and Kwon 2009 for details), the boundary pumping makes possible the existence of WGMs of the type sketched in Fig. 1a. Recall that these *clockwise* traveling modes were missing in the simulations of the passive cavity, cf. Figs. 3 and 4 in the sense that their $Q$-factor was too low to allow for a lasing activity. The boundary pumping changes precisely this feature: It supports (pumps) especially those clockwise traveling WGMs and effectively increases their $Q$-factor. It turns out, however, that the true mechanism is somewhat more complicated. It involves an interaction that is reminiscent of mode beating between these (boundary-pumping supported) clockwise traveling WGMs and the "normal" counterclockwise traveling WGMs that we know from the passive cavity. This interaction is made possible by the boundary pumping scheme, otherwise the difference in the $Q$-factors of the counterpropagating WGMs would be too large. As a result, the light intensity inside the cavity flows between the clockwise and counterclockwise propagating whispering-gallery type states. This also implies that the light emission occurs in a pulsed fashion (Hentschel and Kwon 2009), and Fig. 5d shows a snapshot with the clockwise traveling state.

## 4. Conclusions

Optical microcavities and microlasers receive a strong interest as model systems for quantum chaos in open systems that are rather easily fabricated in the laboratory and as future optical devices with unidirectional emission properties. We have focused here on spiral microcavities where the desired directional far-field characteristics had been realized in a number of experiments and was now also theoretically understood for the case of TM-polarized light (Hentschel and Kwon 2009). It is crucial to pump the microlaser only along its spiral-shaped boundary and to use resonators with a certain geometry (notch size about twice the wavelength).

For the uniformly pumped case and TM polarization we showed that ray and wave simulations agree well with experimental results (Audet et al. 2007) and show no signatures of a directional emission characteristics. This confirms the big versatility of the simple ray model in yet another example. What remains open at this point is a deeper study of the ray-wave-experiment correspondence for TE-polarized spiral microcavities. We already mentioned that the unidirectionality is generally improved for TE devices due to the existence of the Brewster angle. To what extent the Brewster angle may influence boundary-pumping schemes and the resulting conditions for unidirectional emission will be the subject of further studies.

## Acknowledgements

M.H. thanks the German Research Foundation (DFG) for support through the Emmy-Noether Program and through the Research Unit (Forschergruppe) FG 760. T.-Y. Kwon was partly supported by the "Korea Research Foundation Grant" funded by the Korean government (MOEHRD) (contract number KRF-2006-352-C00022).

## References

Altmann E.G., Del Magno G. and Hentschel M.: 2008. *Europhys. Lett.*, **84**, 10008
Audet R., Belkin M.A. et al: 2007. *Appl. Phys. Lett.*, **91**, 131106
Ben-Messaoud T. and Zyss J.: 2005. *Appl. Phys. Lett.*, **86**, 241110
Chern G.D. et al: 2003. *Appl. Phys. Lett.*, **83**, 1710
Chern G.D. et al: 2004. *Appl. Phys. Lett.*, **84**, 2485
Fujii A., Nishimura T. et al: 2005. *Jpn. J. Appl. Phys.*, **44**, L1091
Gmachl C., Capasso F. et al: 1998. *Science*, **280**, 1556
Gmachl C., Narimanov E.E., Capasso F., Baillargeon J.N. and Cho A.Y.: 2002. *Opt. Lett.*, **27**, 824
Goos F. and Hänchen H.: 1947. *Ann. Phys.*, **1**, 333
Harayama T., Sunada S. and Ikeda K.: 2005. *Phys. Rev. A*, **72**, 013803
Hentschel M. and Kwon T.-Y.: 2009. *Opt. Lett.*, **34**, 163
Hentschel M. and Vojta M.: 2001. *Opt. Lett.*, **26**, 1764
Kim Ch.-M., Cho J., Lee J. et al: 2008. *Appl. Phys. Lett.*, **92**, 131110
Kwon T.-Y., Lee S.-Y., Kurdoglyan M.S., Rim S., Kim Ch.-M. and Park Y.-J.: 2006. *Opt. Lett.*, **31**, 1250
Lee S.-Y. et al: 2004. *Phys. Rev. Lett.*, **93**, 164102
Lee S.-Y., Ryu J.-W., Kwon T.-Y., Rim S. and Kim C.-M.: 2005. *Phys. Rev. A*, **72**, 061801(R)
Lee S.-B., Yang J. et al: 2007. *Phys. Rev. A*, **75**, 011802
McCall S.L., Levi A.F.J. et al: 1992. *Appl. Phys. Lett.*, **60**, 289
Schomerus H. and Hentschel M.: 2006. *Phys. Rev. Lett.*, **96**, 243903
Schwefel H.G.L., Rex N.B., Tureci H.E., Chang R.K., and Stone A.D.: 2004. *J. Opt. Soc. Am. B*, **21**, 923
Shinohara S. and Harayama T.: 2007. *Phys. Rev. E*, **75**, 036216
Sohn L.L., Schön G. and Kouwenhowen L.P.: 1997. *Mesoscopic Electron Transport*, Kluwer, Dordrecht
Tanaka T., Hentschel M., Fukushima T. and Harayama T.: 2007. *Phys. Rev. Lett.*, **98**, 033902
Tulek A. and Vardeny Z.V.: 2007. *Appl. Phys. Lett.*, **90**, 161106
Unterhinninghofen J., Wiersig J. and Hentschel M.: 2008. *Phys. Rev. E*, **78**, 016201
Vahala K.J. ed.: 2004. *Optical Microcavities*, World Scientific, Singapore
Wiersig J.: 2003. *J. Opt. A: Pure Appl. Opt.*, **5**, 53
Wiersig J. and Hentschel M.: 2006. *Phys. Rev. A*, **73**, 031802(R)
Wiersig J. and Hentschel M.: 2008. *Phys. Rev. Lett.*, **100**, 033901
Wiersig J., Kim S.W. and Hentschel M.: 2008. *Phys. Rev. A*, **78**, 053809
Yan C., Wang Q.J. et al: submitted. *Opt. Express*

# MESOSCOPIC INTERPLAY OF SUPERCONDUCTIVITY AND FERROMAGNETISM IN ULTRA-SMALL METALLIC GRAINS

S. Schmidt*
*Institute for Theoretical Physics, ETH Zurich, Zurich, Switzerland; schmidts@phys.ethz.ch*

Y. Alhassid
*Center for Theoretical Physics, Yale University, New Haven, Connecticut 06520, U.S.A.*

**Abstract.** We review the effects of electron–electron interactions on the ground-state spin and the transport properties of ultra-small chaotic metallic grains. Our studies are based on an effective Hamiltonian that combines a superconducting BCS-like term and a ferromagnetic Stoner-like term. Such terms originate in pairing and spin exchange correlations, respectively. This description is valid in the limit of a large dimensionless Thouless conductance. We present the ground-state phase diagram in the fluctuation-dominated regime where the single-particle mean level spacing is comparable to the bulk BCS pairing gap. This phase diagram contains a regime in which pairing and spin exchange correlations coexist in the ground-state wave function. We discuss the calculation of the tunneling conductance for an almost-isolated grain in the Coulomb-blockade regime, and present measurable signatures of the competition between superconductivity and ferromagnetism in the mesoscopic fluctuations of the conductance.

**Key words:** Metallic grains; Superconductivity; Ferromagnetism

## 1. Introduction

Superconductivity and ferromagnetism compete with each other. Pairing correlations lead to Cooper pairs of electrons with opposite spins and thus tend to minimize the total spin of the grain, while ferromagnetic correlations tend to maximize the total spin.

Nevertheless, it is well known that superconducting and ferromagnetic order can be present simultaneously in bulk systems when ferromagnetism is caused by localized paramagnetic impurities (Abrikosov and Gorkov 1960, 1961; Clogston 1962; Chandrasekhar 1962; Fulde and Ferrell 1964; Larkin and Ovchinnikov 1964, 1965). Recently, it was observed that both states of matter can coexist in high-Tc superconductors (Tallon et al. 1999; Bernhard et al. 1999) and in heavy fermion systems (Saxena et al. 2000; Pfleiderer et al. 2001; Aoki et al. 2001) even when the electrons that are

G. Casati and D. Matrasulov (eds.), *Complex Phenomena in Nanoscale Systems*,
NATO Science for Peace and Security Series B: Physics and Biophysics,

© Springer Science+Business Media B.V. 2009

responsible for superconductivity and ferromagnetism are the same. This surprising observation led to the search of new theoretical models that can describe this coexistence.

In ultra-small metallic grains, in which the bulk pairing gap $\Delta$ is comparable to the single-particle mean level spacing $\delta$, a coexistence regime of superconductivity and ferromagnetism was predicted (Falci et al. 2003; Ying et al. 2006; Schmidt et al. 2007). The ground state of the grain is described by a state where a few single-particle levels around the Fermi energy are singly occupied while all other electron are paired. This coexistence regime is characterized by spin jumps and its size can be tuned by an external Zeeman field.

However, it is difficult to measure the ground-state spin of a grain, and a more directly measurable quantity is the tunneling conductance through the grain (von Delft and Ralph 2001). In addition, one has to take into account the mesoscopic fluctuations that are typical for chaotic grains (Alhassid 2000). Effects of exchange correlations on the conductance statistics in quantum dots, in which pairing correlations are absent, were studied in Alhassid and Rupp (2003). In Schmidt and Alhassid (2008) we identified signatures of the coexistence of pairing and exchange correlations in the mesoscopic fluctuations of the conductance through a metallic grain that is weakly coupled to leads.

The fabrication and control of nano-size metallic devices is a challenging task. The first conductance measurements in ultra-small metallic grains were carried out in the mid-nineties (Ralph et al. 1995; Ralph et al. 1997; Black et al. 1996). The grains were produced by breaking nanowires and their size was difficult to control. Coulomb blockade, discrete levels and pairing effects were observed in a single grain by measuring the tunneling conductance (von Delft and Ralph 2001). During the last decade numerous technological advances led to an increase in control and tunability of ultra-small metallic grains. Break junction techniques (Park et al. 1999) and electromigration (Bolotin et al. 2004) were used for gating and establishing precise contact between leads and grain. A particularly important recent development has been the use of monolayers of organic molecules as tunnel barriers, enabling control of the size and shape of the grain (Kuemmeth et al. 2008). New materials have been tested as well. Cobalt nanoparticles were used to investigate the effect of ferromagnetism (Deshmukh et al. 2001; Kleff et al. 2001). Spin-orbit coupling and non-equilibrium excitations were studied in gold grains (Bolotin et al. 2004; Kuemmeth et al. 2008; Gueron et al. 1999). The recent discovery of superconductivity in doped silicon at atmospheric pressure and critical temperatures of a few hundred millikelvin (Bustarret et al. 2006) might further facilitate the development of mesoscopic superconducting devices. However, the competition between superconductivity and ferromagnetism has not been investigated so far.

Here we review the effects of electron–electron interactions on the ground-state spin and transport properties of ultra-small metallic grains. Our analysis is based on an effective Hamiltonian for chaotic or disordered systems that combines a BCS-like pairing term and a Stoner-like spin exchange term. This so-called universal Hamiltonian (Kurland et al. 2000; Aleiner et al. 2002) is valid in the limit of a large Thouless conductance. This universal Hamiltonian and its solution are described in Sect. 2. In Sect. 3 we present the phase diagram of the ground-state spin and discuss a regime in which superconductivity and ferromagnetism coexist. In Sect. 4 we review the mesoscopic fluctuations of the tunneling conductance through an almost-isolated metallic grain. In particular, we discuss signatures of the coexistence of pairing and exchange correlations in the conductance peak height and peak spacing statistics. We also propose specific materials for which such mesoscopic coexistence might be observed experimentally.

## 2. Model

We consider a chaotic metallic grain with a large dimensionless Thouless conductance. The low-energy excitations of such a grain are described by an effective universal Hamiltonian (Kurland et al. 2000; Aleiner et al. 2002)

$$\hat{H} = \sum_{k\sigma} \epsilon_k c^\dagger_{k\sigma} c_{k\sigma} - G\hat{P}^\dagger \hat{P} - J_s \hat{\mathbf{S}}^2, \tag{1}$$

where $c^\dagger_{k\sigma}$ is the creation operator for an electron in the spin-degenerate single-particle level $\epsilon_k$ with spin up ($\sigma = +$) or spin down ($\sigma = -$). The first term on the r.h.s. of (1) describes the single-particle Hamiltonian of an electron in the grain (i.e. kinetic energy plus confining single-particle potential). The second term is the pairing interaction with coupling constant $G$ and the pair creation operator $P^\dagger = \sum_i c^\dagger_{i+} c^\dagger_{i-}$. The third term is an exchange interaction where $\hat{\mathbf{S}} = \sum_{k\sigma\sigma'} c^\dagger_{k\sigma} \tau_{\sigma\sigma'} c_{k\sigma'}$ is the total spin operator ($\tau_i$ are Pauli matrices) and $J_s$ is the exchange coupling constant. Estimated values of $J_s$ for a variety of materials were tabulated in Gorokhov and Brouwer (2004). In (1) we have omitted the charging energy term $e^2\hat{N}^2/2C$ ($C$ is the capacitance of the grain and $\hat{N}$ is the number of electrons) since it is constant for a grain with a fixed number of electrons.

The universal Hamiltonian equation (1) conserves the total spin of the grain, i.e. $[\hat{H}, \hat{\mathbf{S}}] = 0$. Consequently, each eigenstate has a well-defined total spin $S$ and spin projection $M$. The pairing interaction scatters pairs of spin up/down electrons from doubly occupied to empty orbitals. Therefore the pairing interaction does not affect the singly occupied levels. This is known as the blocking effect and the singly occupied levels are also referred to as "blocked levels." On the other hand, these singly occupied levels are the only levels that contribute to the exchange interaction. Thus, each eigenstate of (1) factorizes into two parts. The first part $|\mathcal{U}\rangle$ is a zero spin eigenstate of the reduced BCS Hamiltonian $\sum_{k\sigma} \epsilon_k c^\dagger_{k\sigma} c_{k\sigma} - GP^\dagger P$, and is described as a superposition of Slater determinants that are constructed from the subset $\mathcal{U}$ of empty and doubly occupied levels. The second part of the eigenstate, $|\mathcal{B}, \gamma, S, M\rangle$, is obtained by coupling the set of singly occupied levels $\mathcal{B}$, each carrying spin $1/2$, to total spin $S$ and spin projection $M$ (Alhassid and Rupp 2003; Tureci and Alhassid 2006). Here, $\gamma$ denotes a set of quantum numbers distinguishing between eigenstates with the same spin and singly occupied levels. For a given set $\mathcal{B}$ of $b$ singly occupied levels, the allowed spin values vary between $S = 0$ ($S = 1/2$) for even (odd) number of electrons and $S = b/2$. Each of these spin values has a degeneracy of

$$d_b(S) = \binom{b}{S + b/2} - \binom{b}{S + 1 + b/2}. \tag{2}$$

The complete set of eigenstates is then given by

$$|i\rangle = |\mathcal{U}, \mathcal{B}, \gamma, S, M\rangle. \tag{3}$$

The reduced pairing Hamiltonian is characterized by a coupling constant $G$. However, the physical parameter that determines the low-energy spectrum of the grain (for $J_s = 0$) is $\Delta/\delta$, where $\Delta$ is the bulk pairing gap and $\delta$ the single-particle mean level spacing. We can truncate the total number of levels from $N_o$ to $N_r < N_o$, and renormalize $G$ such that the low-energy spectrum of the grain remains approximately the same. For a picketfence spectrum, the renormalized coupling constant is given by

$$\frac{G_r}{\delta} = \frac{1}{\operatorname{arcsinh}\left(\frac{N_r+1/2}{\Delta/\delta}\right)}. \tag{4}$$

The exchange interaction only affects the singly occupied levels, and we expect the renormalization (4) to hold as long as the number of singly occupied levels is small compared with $N_r$. In practice, we study the relevant observable as a function of truncated bandwidth $N_r$ and make sure that its value has converged for the largest bandwidth $N_r$ we can calculate.

## 3. Ground-State Phase Diagram

In this section we study the ground-state spin of the grain as a function of $J_s/\delta$ and $\Delta/\delta$. We find the lowest energy $E(S)$ for a given spin $S$ and then minimize $E(S)$ with respect to $S$. The ground-state spin of the grain is determined by

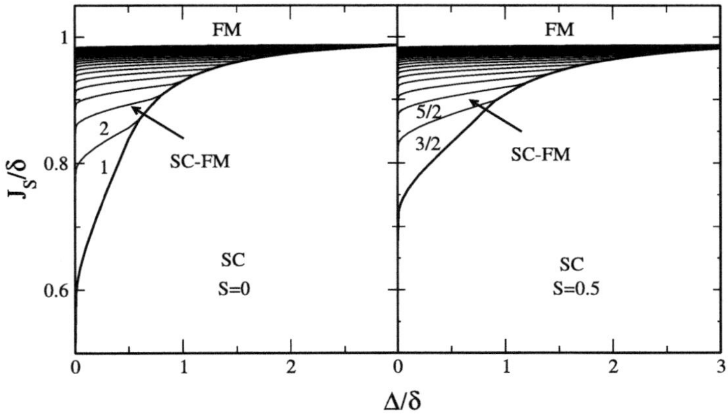


*Figure 1.* Ground-state phase diagram of a grain with picketfence single-particle spectrum in the $J_s/\delta$–$\Delta/\delta$ plane for an even (left panel) and odd (right panel) number of electrons. Numbers denote the spin values in the corresponding sectors. We observe an intermediate regime (SC-FM) in which the ground state is partly polarized and partly paired. Taken from Schmidt et al. (2007)

the competition between various terms in the Hamiltonian equation (1). The one-body part and pairing interaction favor minimal spin, while exchange interaction favors a maximally polarized state.

The ground-state phase diagram in the $\Delta/\delta$–$J_s/\delta$ plane of a grain with a picketfence single-particle spectrum is shown in Fig. 1. We find three different phases: a superconducting phase (SC) where the number of pairs is maximal and $S$ is minimal, a ferromagnetic phase (FM) where the system is fully polarized $S = N/2$, and an intermediate regime (SC-FM) where exchange and pairing correlations coexist. The ground-state wave function with spin $S$ in the coexistence regime is described by $b = 2S$ singly occupied levels closest to the Fermi energy while the remaining electrons are paired.

The coexistence regime is bounded by two critical values $J_s^{(1)}/\delta$ and $J_s^{(2)}/\delta$ of the exchange interaction that are function of $\Delta/\delta$. The lower value $J_s^{(1)}/\delta$ is a monotonically increasing function of $\Delta/\delta$ (stronger exchange is required to polarize a grain with stronger pairing correlations), while the higher value $J_s^{(2)}/\delta$ depends only weakly on $\Delta/\delta$.

It is interesting to follow the dependence of the ground-state spin as a function of the exchange coupling constant $J_s/\delta$ for a fixed value of $\Delta/\delta$. In the absence of pairing ($\Delta = 0$), this dependence follows a stepwise behavior known as the mesoscopic Stoner staircase (Kurland et al. 2000), where a transition from spin $S$ to spin $S + 1$ occurs for an exchange coupling of

$$\frac{J_s}{\delta} = \frac{2S + 1}{2S + 2} \qquad \text{at} \quad \Delta = 0. \tag{5}$$

The first step occurs at $J_s/\delta = 0.5$ (where the ground-state spin increases from $S = 0$ to 1) and continues up to $J_s = 0.75$ (where the $S = 1$ to 2 transition takes place). In the presence of pairing, the first step is shifted to higher values of the exchange and the Stoner staircase is compressed. For $\Delta/\delta < 0.6$, all steps have a height of $\Delta S = 1$. However, for $0.6 < \Delta/\delta < 0.8$, the first step has a height of $\Delta S = 2$, describing a spin jump from $S = 0$ to $S = 2$. This first step starts at $J_s/\delta \approx 0.87$ and ends at $J_s/\delta \approx 0.9$. The height of the first-step spin jump increases at larger values of $\Delta/\delta$. All subsequent steps are of height one.

Spin jumps also occur when superconductivity in metallic grains breaks down in the presence of a sufficiently large external Zeeman field (Braun et al. 1997). This "softened" first-order phase transition from a superconductor to a paramagnet was explained qualitatively using a finite-spin BCS approximation.

In the presence of exchange correlations, spin jumps are predicted to occur at $J_s/\delta > 0.87$. Such exchange coupling values are significantly larger than the exchange coupling values of most metals (see Fig. 9 in Gorokhov and Brouwer 2004). Moreover, the exchange strength is an intrinsic material property and is difficult to tune experimentally. In Schmidt et al. (2007) we have shown that the coexistence regime can be tuned to experimentally accessible values of the exchange interaction by applying an external Zeeman field.

## 4. Conductance

The determination of the ground-state spin of a grain is a difficult measurement. It is then desirable to identify signatures of coexistence of superconductivity and ferromagnetism in a quantity that is directly measurable, e.g. the conductance. Furthermore, the universal Hamiltonian equation (1) is only valid for chaotic (or disordered) grains, in which mesoscopic fluctuations are generic. Therefore, in order to make quantitative predictions for the experiment it is necessary to include the effect of mesocopic fluctuations. In this section we discuss the mesoscopic fluctuations of the tunneling conductance for an almost-isolated metallic grain in the Coulomb blockade regime. We find signatures of the coexistence of pairing and exchange correlations in the conductance statistics. Since the tunneling conductance can be measured in a single-electron transistor that uses the metallic grain as an island, our results are directly relevant for the experiment.

We consider grains that are weakly coupled to external leads. In the regime of sequential tunneling $\delta, T \gg \Gamma$ ($\Gamma$ is a typical tunneling width). Assuming the charging energy to be much larger than temperature ($e^2/2C \gg T$), the conductance displays a series of sharp peaks as a function of gate voltage. The $N$-th conductance peak describes a tunneling event in which the number of electrons in the dot changes from $N$ to $N+1$. and is determined by the many-body energies and transition rates between eigenstates of the $N$ and $N+1$ electrons. The conductance peak height and peak position are determined by solving a system of rate equations (Alhassid et al. 2004).

Here we present results for the conductance peak spacing and peak height statistics for an experimentally accessible temperature of $T = 0.1\,\delta$ (Ralph et al. 1997). In the absence of an external magnetic field, the single-particle Hamiltonian is described by the Gaussian orthogonal ensemble (GOE) of random matrices. For each random matrix realization of the one-body Hamiltonian, we use the Lanczos method to find the five lowest eigenstates of the many-body Hamiltonian equation (1). The calculations are carried out for a truncated bandwidth $N_r = 8$ and electron numbers $N = 16, 17, 18$ and 19. Using the many-body energies and wave functions, we calculate the tunneling matrix elements for the corresponding three tunneling events. We then solve the system of rate equations and determine the conductance as a function of gate voltage. The peak position and height are determined numerically. To ensure good statistics, the above procedure is repeated for $4,000$ realizations of the one-body Hamiltonian.

### 4.1. PEAK SPACING STATISTICS

The peak spacing distribution is shown in Fig. 2, where the spacing is measured relative the constant charging energy. For $\Delta = J_s = 0$ and at low temperatures, this distribution is bimodal because of the spin degeneracy of the single-particle levels (Alhassid 2000). The exchange interaction induces mesoscopic spin fluctuations and suppresses this bimodality (see top left panel of Fig. 2). This is known from the case of semiconductor quantum dots.

Pairing correlations can restore bimodality. For a moderate exchange value of $J_s = 0.3\,\delta$, bimodality is restored for a relatively weak pairing strength of $\Delta = 0.25\,\delta$. For $J_s = 0.6\,\delta$, this bimodality is suppressed but reappears at $\Delta/\delta = 0.5$ (see bottom left panel of Fig. 2). The left part of the peak spacing distribution describes even-odd-even (E-O-E) tunneling events (the parity refers to the number of electrons), and its right part describes odd-even-odd (O-E-O) transitions.

The bimodality of the peak spacing distribution in the presence of strong pairing correlations can be understood qualitatively in the $T = 0$ fixed-spin BCS approximation (Schmidt et al. 2007). For an E-O-E transition, the first conductance peak corresponds to the blocking of an additional single-particle level, while the second conductance peak corresponds to the removal of this blocked level by creating an additional Cooper pair. This leads to the estimate

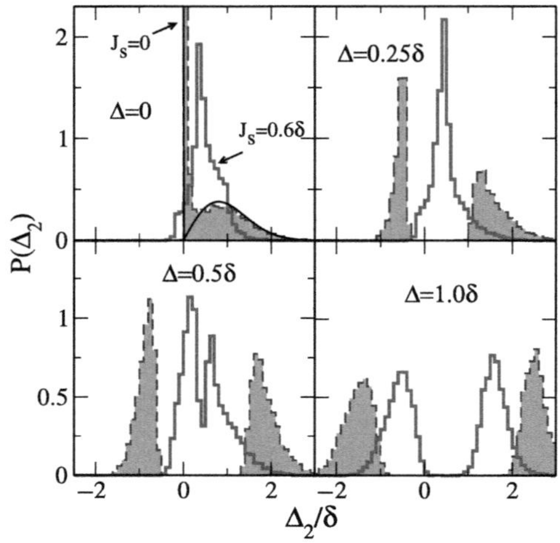


*Figure 2.* Peak spacing distributions at $T = 0.1\,\delta$ for several values of $\Delta/\delta$. Results are shown for both $J_s = 0$ (dashed, grey-filled histograms) and $J_s = 0.6\,\delta$ (solid histograms). For $\Delta = 0$ we also compare with the analytic result (Jalabert et al. 1992) at $T \ll \delta$ and $J_s = 0$ (solid line). The bimodality of the distribution at $\Delta = J_s = 0$ is suppressed at finite exchange ($J_s = 0.6\,\delta$) but is restored for $\Delta/\delta = 0.5$. Taken from Schmidt and Alhassid (2008)

$\Delta_2^{\text{EOE}} \approx -2\Delta + \frac{3}{2}J$. In a O-E-O tunneling sequence, these two events are reversed and we find $\Delta_2^{\text{OEO}} \approx 2\Delta - \frac{3}{2}J$. The contribution of the exchange interaction in these estimates is straightforward because, in the limit of strong pairing, the ground-state spin is always $S = 0$ ($S = 1/2$) for an even (odd) number electrons. The difference of these two peak spacing values is

$$\delta\Delta_2 = \Delta_2^{\text{OEO}} - \Delta_2^{\text{EOE}} \approx 4\Delta - 3J\,, \quad \Delta \gg \delta, \tag{6}$$

and bimodality becomes more pronounced when $\Delta/\delta$ increases. Since the exchange interaction strength for most metals is smaller or comparable to $J_s \sim 0.6\,\delta$, exchange correlations are insufficient to suppress the bimodality in the presence of strong pairing correlations.

## 4.2. PEAK HEIGHT STATISTICS

The peak height distribution is shown in Fig. 3. For $\Delta = J_s = 0$, this distribution is known analytically at $T \ll \delta$ (Jalabert et al. 1992), and, in the absence of an external magnetic field (GOE statistics), diverges at small values of the conductance. Finite temperature and exchange interaction have a similar effect on the peak height distribution; they both reduce the occurrence of small conductance values. While thermal fluctuations open an energy window in which states become available for tunneling and thus can contribute to the conductance, the exchange interaction increases the many-body

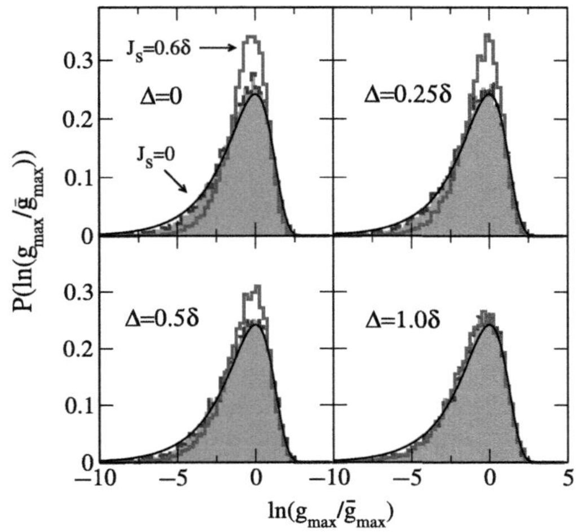


*Figure 3.* Peak height distributions at $T = 0.1\,\delta$ for several values of $\Delta/\delta$. We show results for $J_s = 0$ (dashed, grey-filled histograms) and $J_s = 0.6\,\delta$ (solid histograms). The solid line is the analytic distribution at $T \ll \delta$ and $J_s = 0$ (Jalabert et al. 1992). Since the conductance peak height $g_{\max}$ fluctuates over several order of magnitude, we show the distributions as a function of $\ln(g_{\max}/\bar{g}_{\max})$ where $\bar{g}_{\max}$ is the average conductance peak height. Taken from Schmidt and Alhassid (2008)

density of states around the Fermi energy and makes otherwise high-lying non-zero spin states available for tunneling. This effect is shown in Fig. 3. For $\Delta = 0$ we clearly observe that a finite exchange interaction suppresses the peak height distribution at small conductance values. The same behavior was observed for the Gaussian unitary ensemble (GUE) ensemble in semiconductor quantum dots, where the pairing interaction can be ignored and a closed solution for the conductance is available (Alhassid and Rupp 2003). There, the suppression of probability at small conductance values by the inclusion of exchange interaction leads to better agreement with the experimental results. Another signature of exchange correlations is the suppression of peak height fluctuations as described by $\sigma(g_{\max})/\bar{g}_{\max}$ (the ratio between is the standard deviation and average value of the peak height) (Schmidt and Alhassid 2008).

However, it is important to note that at very low temperatures, e.g. $T = 0.01\,\delta$, the small conductance values are no longer suppressed by an exchange interaction. For such very low temperatures, the increase in the density of states near the Fermi energy by the exchange interaction does not affect the conductance since only the ground state contributes significantly to transport.

The pairing interaction leads to an excitation gap that pushes states with large spin to higher energies. Thus, already for $\Delta = 0.5\,\delta$ we observe less suppression of the small conductance values by an exchange of $J = 0.6\,\delta$.

When the pairing interaction is strong enough to suppress all spin polarization of low-lying states (e.g. $\Delta \geq 1.0\,\delta$ for $J_s < 0.6\,\delta$), the peak height distribution becomes essentially independent of the exchange interaction and diverges again for small conductance values.

For $\Delta/\delta = 0.5$ and $J_s/\delta = 0.6$, we observe a signature of pairing correlations in the peak spacing distribution (bimodality) and a signature of exchange correlations in the peak height distribution (suppression of the probability of small conductance values). We can interpret these results to describe the mesoscopic coexistence of pairing and exchange correlations. A candidate for this mesoscopic coexistence is vanadium, which has $J_s/\delta \sim 0.57$–$0.63$ (Gorokhov and Brouwer 2004) and is superconducting in the bulk. Another candidate is platinum, which has $J_s/\delta \sim 0.59$–$0.72$ and is superconducting in granular form.

## 5. Conclusion

We reviewed the competition between superconductivity and ferromagnetism in ultra-small metallic grain. In particular, we presented the ground-state phase diagram in the $J_s/\delta - \Delta/\delta$ plane, and discussed a coexistence regime of superconductivity and ferromagnetism. This regime is characterized by spin jumps that are greater than unity. We also discussed the transport properties of the grain in its Coulomb blockade regime of weak coupling to leads and described the statistics of the conductance peaks in the presence of both pairing and exchange correlations. Of particular interest is a regime in which pairing and ferromagnetic correlations coexist. Such a regime is defined by the simultaneous occurrence of bimodality in the peak spacing distribution (caused by pairing correlations) and the suppression of the peak height distribution at small conductance values (caused by ferromagnetic correlations).

The exchange interaction strength is a material constant and might be difficult to tune experimentally. Alternatively, the coexistence regime can be controlled by an external Zeeman field while measurements are carried out at a fixed exchange interaction strength.

## Acknowledgements

This work was supported in part by the U.S. DOE grant No. DE-FG-0291-ER-40608. Computational cycles were provided by the NERSC high performance computing facility at LBL and the Yale Bulldog cluster.

## References

Abrikosov A.A. and Gorkov L.P.: 1960. *Zh. Eksp. Teor. Fiz.*, **39**, 1781
Abrikosov A.A. and Gorkov L.P.: 1961. *Sov. Phys. JETP*, **12**, 1243
Aleiner I.L., Brouwer P.W. and Glazman L.I.: 2002. *Phys. Rep.*, **358**, 309
Alhassid Y.: 2000. *Rev. Mod. Phys.*, **72**, 895
Alhassid Y. and Rupp T.: 2003. *Phys. Rev. Lett.*, **91**, 056801
Alhassid Y., Rupp T., Kaminski A. and Glazman L.I.: 2004. *Phys. Rev. B*, **69**, 115331
Aoki D. et al: 2001. *Nature*, **413**, 613
Bernhard C. et al: 1999. *Phys. Rev. B*, **59**, 14099
Black C.T., Ralph D.C. and Tinkham M.: 1996. *Phys. Rev. Lett.*, **76**, 688
Bolotin K.I., Kuemmeth F., Pasupathy A.N. and Ralph D.C.: 2004. *Appl. Phys. Lett.*, **84**, 3154
Braun F., von Delft J., Ralph D.C. and Tinkham M.: 1997. *Phys. Rev. Lett.*, **79**, 921
Bustarret E. et al: 2006. *Nature*, **444**, 465
Chandrasekhar B.S.: 1962. *Appl. Phys. Lett.*, **1**, 7
Clogston A.M.: 1962. *Phys. Rev. Lett.*, **9**, 266
von Delft J. and Ralph D.C.: 2001. *Phys. Rep.*, **345**, 661
Deshmukh M.M. et al: 2001. *Phys. Rev. Lett.*, **87**, 226801
Falci G., Fazio R. and Mastellone A.: 2003. *Phys. Rev. B*, **67**, 132501
Fulde P. and Ferrell R.A.: 1964. *Phys. Rev.*, **135**, A550
Gorokhov D.A. and Brouwer P.W.: 2004. *Phys. Rev. B*, **69**, 155417
Gueron S., Deshmukh M.M., Myers E.B. and Ralph D.C.: 1999. *Phys. Rev. Lett.*, **83**, 4148
Jalabert R.A., Stone A.D. and Alhassid Y.: 1992. *Phys. Rev. Lett.*, **68**, 3468
Kleff S., von Delft J., Deshmukh M. and Ralph D.C.: 2001. *Phys. Rev. B*, **64**, 220401
Kuemmeth F., Bolotin K.I., Shi S.-F. and Ralph D.C.: 2008. *Nano Lett.*, **8**, 4506
Kurland I.L., Aleiner I.L. and Altshuler B.L.: 2000. *Phys. Rev. B*, **62**, 14886
Larkin A.I. and Ovchinnikov Yu.N.: 1964. *Zh. Eksp. Teor. Fiz.*, **47**, 1136
Larkin A.I. and Ovchinnikov Yu.N.: 1965. *Sov. Phys. JETP*, **20**, 762
Park H., Lim A.K.L., Alivisatos A.P., Park J. and McEuen P.L.: 1999. *Appl. Phys. Lett.*, **75**, 301
Pfleiderer C. et al: 2001. *Nature*, **412**, 58
Ralph D.C., Black C.T. and Tinkham M.: 1995. *Phys. Rev. Lett.*, **74**, 3241
Ralph D.C., Black C.T. and Tinkham M.: 1997. *Phys. Rev. Lett.*, **78**, 4087
Saxena S.S. et al: 2000. *Nature*, **406**, 587
Schmidt S. and Alhassid Y.: 2008. *Phys. Rev. Lett.*, **101**, 207003
Schmidt S., Alhassid Y. and van Houcke K.: 2007. *Europhys. Lett.*, **80**, 47004
Tallon J. et al: 1999. *IEEE Trans. Appl. Superconduct.*, **9**, 1696
Tureci H.E. and Alhassid Y.: 2006. *Phys. Rev. B*, **74**, 165333
Ying Z., Couco M., Noce C. and Zhou H.: 2006. *Phys. Rev. B*, **74**, 012503

# TIME-REVERSAL MIRRORS IN CHAOTIC CAVITIES

H.L. Calvo and H.M. Pastawski
*Facultad de Matemática, Astronomía y Física and Instituto de Física (CONICET), Universidad Nacional de Córdoba, Ciudad Universitaria, 5000 Córdoba, Argentina*

R.A. Jalabert*
*Institut de Physique et Chimie des Matériaux de Strasbourg, Strasbourg, France; jalabert@ipcms.u-strasbg.fr*

**Abstract.** We study time-reversal mirrors in chaotic cavities using a semiclassical approach. We present calculations of the reconstructed signal under various conditions. The refocused signal is shown to build up in the region of the source, by multiple interference, in the form of a reversed wave. We demonstrate that the optimal reconstruction is achieved at the refocusing time and at the point of the original emission. This refocused signal scales linearly with the recording interval and inversely with the area of the cavity. We stress the importance of the underlying dynamics, and its role in the signal reconstruction when the time-reversal mirror operates under an external perturbation.

**Key words:** Quantum chaos; Time-reversal mirror; Loschmidt echo

## 1. Introduction

The time-reversal mirror is an experimental technique allowing to focus various kinds of waves in a desired region of space and time (Derode et al. 1995; Fink 2001; Draeger and Fink 1997; Catheline et al. 2008; Negreira and Fint 2007). The protocol starts with an initially localized excitation that propagates and is registered by a set of transducers surrounding the emission region. The recorded signals are then played-back in the inverse temporal sequence and the pulse is reconstructed around the initial point. Albeit not perfect (Pastawski et al. 2007), the pulse reconstruction can be done quite faithfully, and with an arbitrary level of amplification. Various applications of this technique in the fields of medical physics, oceanography and telecommunications are underway and envisioned (Fink 1997; Ribay et al. 2005; Heinemann et al. 2002).

Surprisingly, the focalization signal improves when the wave propagation occurs in a disordered media or inside a chaotic cavity, as compared with

G. Casati and D. Matrasulov (eds.), *Complex Phenomena in Nanoscale Systems*,
NATO Science for Peace and Security Series B: Physics and Biophysics,

© Springer Science+Business Media B.V. 2009

the homogeneous or integrable cases. In closed chaotic cavities, the focalization was successfully performed even with a single transducer (Draeger and Fink 1997). The underlying classical dynamics of the wave-problem appears as a crucial factor for the quality of the refocusing and its stability against external perturbations or the use of partial information. This observation renders the problem of time-reversal focusing particularly interesting for quantum-chaology studies (Casati et al. 2000).

In this work we present a detailed calculation of the focalization amplitude for times close to refocalization and positions near the source point where the recovering takes place. We use the semiclassical approach and follow the lines of Calvo et al. (2008) to evaluate the scaling of the focalization amplitude with the temporal re-emission interval and the parameters of the cavity in order to compare with the experimental and numerical results of Draeger and Fink (1997) for time reversal of elastic waves in a chaotic cavity by using a single transducer. We stress the similarities with the related case of the Loschmidt echo, where the reconstruction of an initially localized quantum state is obtained by inversion of the dynamics after some propagating time, and where the underlying classical dynamics was shown to play an important role (Jalabert and Pastawski 2001; Cucchietti et al. 2004).

## 2. Semiclassical Theory

A time-reversal mirror experiment starts with a high frequency excitation emitted by a source located at a point $\mathbf{r}_0$. Such an excitation can be viewed as an initial wave packet $\psi_{\mathbf{p}_0}(\mathbf{r}')$, centered around $\mathbf{r}_0$ and with momentum $\mathbf{p}_0$, that propagates through the cavity reaching the transducer at position $\mathbf{r}_i$. We consider only one transducer, but the generalization to an array is straightforward. The recording begins at time $t_1$ and the signal detected by the transducer for later times $t > t_1$ is

$$\psi(\mathbf{r}_i, t) = \int d\mathbf{r}' G(\mathbf{r}_i, \mathbf{r}', t)\, \psi_{\mathbf{p}_0}(\mathbf{r}'), \tag{1}$$

where $G(\mathbf{r}_i, \mathbf{r}', t)$ is the propagator describing the wave propagation from $\mathbf{r}'$ to $\mathbf{r}_i$ in the time $t$. The recording process takes place until a time $t_2 = t_1 + \Delta T$. After a waiting time $t_W > t_2$, the registered signal is time-reversed and emitted between the subsequent times $t_2' = 2t_W - t_2$ and $t_1' = 2t_W - t_1$. These waves propagate from the transducer into the system and interfere in the cavity, eventually giving rise to the focusing. Once the emission has finished ($t > t_1'$), the signal detected in the source region can be computed as

$$\mathcal{F}_{\mathbf{p}_0}(\mathbf{r}, t) = \int_{t_2'}^{t_1'} d\tau\, G(\mathbf{r}, \mathbf{r}_i, t - \tau) \int d\mathbf{r}'\, G^*(\mathbf{r}_i, \mathbf{r}', 2t_W - \tau)\, \psi^*_{\mathbf{p}_0}(\mathbf{r}'). \tag{2}$$

Performing the change of variables $\tau' = 2t_W - \tau$, redefining the focalization time $2t_W$ as the new time origin ($t' = t - 2t_W$), and dropping the primes in order to simplify the notation, we have (Tanner and Søndergaard 2007)

$$\mathcal{F}_{\mathbf{p}_0}(\mathbf{r}, t) = \int_{t_1}^{t_2} \mathrm{d}\tau\, G(\mathbf{r}, \mathbf{r}_i, t + \tau) \int \mathrm{d}\mathbf{r}'\, G^*(\mathbf{r}_i, \mathbf{r}', \tau)\, \psi^*_{\mathbf{p}_0}(\mathbf{r}'). \tag{3}$$

Working with two-dimensional systems, the initial excitation can be faithfully represented by a Gaussian wave packet

$$\psi_{\mathbf{p}_0}(\mathbf{r}') = \frac{1}{\sqrt{\pi}\sigma} \exp\left[\frac{\mathrm{i}}{\hbar}\mathbf{p}_0 \cdot (\mathbf{r}' - \mathbf{r}_0) - \frac{(\mathbf{r}' - \mathbf{r}_0)^2}{2\sigma^2}\right], \tag{4}$$

centered around $\mathbf{r}_0$ and with dispersion $\sigma$. The momentum $\mathbf{p}_0$ gives the main energy and direction of the excitation. The choice of a quantum formalism to represent the ray picture is motivated by convenience, as we are leaving aside the delicate issue concerning a quantal recording-emission process. In the semiclassical approximation, the propagator can be expanded as a sum over classical trajectories $s(\mathbf{r}', \mathbf{r}, \tau)$ linking the points $\mathbf{r}'$ and $\mathbf{r}$ in a time $\tau$ (Brack 1997),

$$G(\mathbf{r}, \mathbf{r}', \tau) = \sum_{s(\mathbf{r}', \mathbf{r}, \tau)} G_s(\mathbf{r}, \mathbf{r}', \tau), \tag{5}$$

according to the Van Vleck's formula

$$G_s(\mathbf{r}, \mathbf{r}', \tau) = \frac{1}{2\pi\mathrm{i}\hbar}\, C_s^{1/2} \exp\left[\frac{\mathrm{i}}{\hbar} S_s(\mathbf{r}, \mathbf{r}', \tau) - \frac{\mathrm{i}}{2}\pi\mu_s\right]. \tag{6}$$

Here, $S_s(\mathbf{r}, \mathbf{r}', \tau)$ denotes the action integral over the classical path $s$, while $\mu_s$ is the Maslov index accounting for the number of conjugates points along $s$, and the stability factor $C_s = \left|\det(-\partial S_s^2/\partial\mathbf{r}'\partial\mathbf{r})\right|$ sets the weight of the trajectory. Since the initial wave packet is focused on $\mathbf{r}_0$ we can expand the action integral in terms of those trajectories $\tilde{s}$ departing from the center of the wave packet and reaching $\mathbf{r}_i$ in a time $\tau$ as

$$S_s(\mathbf{r}_i, \mathbf{r}', \tau) \simeq S_{\tilde{s}}(\mathbf{r}_i, \mathbf{r}_0, \tau) - \mathbf{p}_{\tilde{s}} \cdot (\mathbf{r}' - \mathbf{r}_0)\,, \tag{7}$$

with $\mathbf{p}_{\tilde{s}}$ the initial momentum of the trajectory $\tilde{s}$. Therefore, the detected signal at the transducer point takes the form (Jalabert and Pastawski 2001; Goussev et al. 2008)

$$\psi(\mathbf{r}_i, \tau) = \left(4\pi\sigma^2\right)^{1/2} \sum_{s(\mathbf{r}_0, \mathbf{r}_i, \tau)} G_s(\mathbf{r}_i, \mathbf{r}_0, \tau) \exp\left[-\frac{\sigma^2}{2\hbar^2}(\mathbf{p}_s - \mathbf{p}_0)^2\right]. \tag{8}$$

We evaluate the focalization signal for times $t$ close to the refocusing one and positions $\mathbf{r}$ near $\mathbf{r}_0$. Thus, the action integral in the propagator of the returning waves can be expanded as

$$S_s(\mathbf{r}, \mathbf{r}_i, t+\tau) \simeq S_{\tilde{s}}(\mathbf{r}_0, \mathbf{r}_i, \tau) + E_{\tilde{s}}\, t + \mathbf{p}_{\tilde{s}} \cdot (\mathbf{r} - \mathbf{r}_0). \tag{9}$$

Notice that the sign of the momentum becomes opposite as compared to (7) since we are dealing with trajectories arriving to $\mathbf{r}_0$. Furthermore, the expansion around the focalization time gives an additional term that accounts for the energy $E_{\tilde{s}}$ of the trajectory. Hence, the focalization signal writes as a double sum involving trajectories going from $\mathbf{r}_0$ to $\mathbf{r}_i$ and back

$$\begin{aligned}\mathcal{F}_{\mathbf{p}_0}(\mathbf{r}, t) &= \left(4\pi\sigma^2\right)^{1/2} \int_{t_1}^{t_2} \mathrm{d}\tau \sum_{s'(\mathbf{r}_i,\mathbf{r}_0,\tau)} \sum_{s(\mathbf{r}_0,\mathbf{r}_i,\tau)} G_{s'}(\mathbf{r}_0, \mathbf{r}_i, \tau)\, G_s^*(\mathbf{r}_i, \mathbf{r}_0, \tau) \\ &\quad \times \exp\left[-\frac{\sigma^2}{2\hbar^2}(\mathbf{p}_s - \mathbf{p}_0)^2 + \frac{\mathrm{i}}{\hbar} E_{s'}\, t + \frac{\mathrm{i}}{\hbar} \mathbf{p}_{s'} \cdot (\mathbf{r} - \mathbf{r}_0)\right].\end{aligned} \tag{10}$$

The main contribution comes from the diagonal term in which only the same trajectories are kept ($s = s'$). Reciprocity in the propagators leads to a focalization signal given by

$$\mathcal{F}_{\mathbf{p}_0}(\mathbf{r}, t) = \frac{\sigma}{2\pi^{3/2}\hbar^2} \int_{t_1}^{t_2} \mathrm{d}\tau \sum_{s(\mathbf{r}_i,\mathbf{r}_0,\tau)} C_s \exp\left[-\frac{\sigma^2}{2\hbar^2}(\mathbf{p}_s - \mathbf{p}_0)^2\right. \tag{11}$$

$$\left. + \frac{\mathrm{i}}{\hbar} E_s\, t + \frac{\mathrm{i}}{\hbar} \mathbf{p}_s \cdot (\mathbf{r} - \mathbf{r}_0)\right]. \tag{12}$$

In billiards the same path $\hat{s}(\mathbf{r}_0, \mathbf{r}_i)$ can correspond to different traveling times, depending on the magnitude of the momentum. The energy and momentum can be expressed in terms of the length $L_{\hat{s}}$ and the traveling time $\tau$ as $E_s = \mathrm{p}_s^2/2m$ and $\mathrm{p}_s = mL_{\hat{s}}/\tau$ respectively. The stability factor $C_s$ decreases for long times as $\exp(-\lambda_s\tau)$, with $\lambda_s$ the largest Lyapunov exponent, whereas for the short time limit, it presents a ballistic behavior. Assuming a uniformly hyperbolic dynamics (Goussev et al. 2008), and using $\lambda_s\tau = \hat{\lambda}L_{\hat{s}}$ (with $\hat{\lambda}$ an inverse length), we can write

$$C_s = \frac{2m^2\hat{\lambda}L_{\hat{s}}}{\tau^2} \exp(-\hat{\lambda}L_{\hat{s}}). \tag{13}$$

The sum over trajectories can be converted into an integral over lengths by introducing the density (Sieber 1999)

$$\frac{\mathrm{d}N(L)}{\mathrm{d}L} = \frac{\pi}{\hat{\lambda}\mathcal{A}} \exp(hL), \tag{14}$$

with $\mathcal{A}$ the area of the cavity. For closed systems $h = \hat{\lambda}$, and the long paths linking $\mathbf{r}_0$ to the transducer can be considered uniformly distributed along all directions. The sum over trajectories can then be replaced by a double integral over lengths and initial angles leading to

$$\begin{aligned}\mathcal{F}_{\mathbf{p}_0}(\mathbf{r},t) &= \frac{m^2\sigma}{\sqrt{\pi}\hbar^2\mathcal{A}}\int_{t_1}^{t_2}\frac{\mathrm{d}\tau}{\tau^2}\int_0^{2\pi}\frac{\mathrm{d}\varphi}{2\pi}\int_{L_\mathrm{d}}^{\infty}\mathrm{d}L\,L\,\exp\left[-\frac{\sigma^2}{2\hbar^2}\left(\frac{mL}{\tau}\mathbf{e}(\varphi)-\mathbf{p}_0\right)^2\right.\\ &\quad\left.+\frac{\mathrm{i}}{\hbar}\frac{mL^2}{2\tau^2}t+\frac{\mathrm{i}}{\hbar}\frac{mL}{\tau}\mathbf{e}(\varphi)\cdot(\mathbf{r}-\mathbf{r}_0)\right], \qquad (15)\end{aligned}$$

where $\mathbf{e}(\varphi)$ is an unitary vector that gives the direction of the initial momentum and $L_\mathrm{d}$ is the length of the shortest trajectory linking $\mathbf{r}_0$ and $\mathbf{r}_i$.

## 3. Focalization in the Optimal Condition

As a first step in the evaluation of the focalization, we start with the optimal condition given by $t = 0$ and $\mathbf{r} = \mathbf{r}_0$. Defining $\varphi$ as the angle of $\mathbf{e}(\varphi)$ with $\mathrm{p}_0$, we have $\mathbf{e}(\varphi)\cdot\mathbf{p}_0 = \mathrm{p}_0\cos\varphi$, and the argument in the exponential of (15) can be written as

$$-\frac{\sigma^2}{2\hbar^2}\left(\mathrm{p}_0^2\sin^2\varphi+\left(\frac{mL}{\tau}-\mathrm{p}_0\cos\varphi\right)^2\right). \qquad (16)$$

The corresponding $\tau$-integral of (15) is then obtained as

$$\begin{aligned}\mathcal{I}_1(L,\varphi) &= \int_{t_1}^{t_2}\frac{\mathrm{d}\tau}{\tau^2}\exp\left[-\frac{\sigma^2}{2\hbar^2}\left(\frac{mL}{\tau}-\mathrm{p}_0\cos\varphi\right)^2\right]\\ &= \frac{\sqrt{\pi}\hbar}{\sqrt{2}m\sigma L}\left(\mathrm{erf}(\eta_1)-\mathrm{erf}(\eta_2)\right), \qquad (17)\end{aligned}$$

where $\mathrm{erf}(x)$ stands for the error function, $\eta_i = \eta(t_i)$ for $i = \{1, 2\}$, and

$$\eta(\tau) = \frac{\sigma}{\sqrt{2}\hbar}\left(\frac{mL}{\tau}-\mathrm{p}_0\cos\varphi\right). \qquad (18)$$

The next step is to calculate the integral over length given by

$$\mathcal{I}_2(\varphi) = \int_{L_\mathrm{d}}^{\infty}\mathrm{d}L\,L\,\mathcal{I}_1(L,\varphi). \qquad (19)$$

Defining $L_i = vt_i$, $l_i = L/L_i$, and $l_\varphi = \sigma\mathrm{p}_0\cos\varphi/\sqrt{2}\hbar$ with mean spreading velocity of the wave packet

$$v = \frac{\sqrt{2}\hbar}{m\sigma}, \qquad (20)$$

the integral over length writes

$$\begin{aligned}\mathcal{I}_2(\varphi) &= \frac{\sqrt{\pi}\hbar^2}{(m\sigma)^2}\Big[t_2\int_{L_\mathrm{d}/L_2}^{\infty}\mathrm{d}l_2\left[1-\mathrm{erf}(l_2-l_\varphi)\right] \\ &\quad - t_1\int_{L_\mathrm{d}/L_1}^{\infty}\mathrm{d}l_1\left[1-\mathrm{erf}(l_1-l_\varphi)\right]\Big] \\ &= \frac{\sqrt{\pi}\hbar^2}{(m\sigma)^2}\Big[(t_2-t_1)\int_{L_\mathrm{d}/L_2}^{\infty}\mathrm{d}l\left[1-\mathrm{erf}(l-l_\varphi)\right] \\ &\quad - t_1\int_{L_\mathrm{d}/L_2}^{L_\mathrm{d}/L_1}\mathrm{d}l\left[1-\mathrm{erf}(l-l_\varphi)\right]\Big]. \end{aligned} \tag{21}$$

Regardless of the value of $\mathrm{p}_0$, the recording process is supposed to start once the wave packed have been spread out on the whole cavity. This means that the involved lengths satisfy $L_\mathrm{d} \ll L_1 < L_2$ and $\mathcal{I}_2$ can be approximated as

$$\begin{aligned}\mathcal{I}_2(\varphi) &\simeq \frac{\sqrt{\pi}\hbar^2\Delta T}{(m\sigma)^2}\int_0^{\infty}\mathrm{d}l\left[1-\mathrm{erf}(l-l_\varphi)\right] \\ &\simeq \frac{\sqrt{\pi}\hbar^2\Delta T}{(m\sigma)^2}\left[\frac{\exp(-l_\varphi^2)}{\sqrt{\pi}} + l_\varphi\left[1+\mathrm{erf}(l_\varphi)\right]\right].\end{aligned} \tag{22}$$

The optimal focalization then writes

$$\begin{aligned}\mathcal{F}_{\mathbf{p}_0}(\mathbf{r}_0,0) &= \frac{\Delta T}{\sigma\mathcal{A}}e^{-\sigma^2\mathrm{p}_0^2/2\hbar^2}\Big[\frac{1}{\sqrt{\pi}} \\ &\quad + \int_0^{2\pi}\frac{\mathrm{d}\varphi}{2\pi}K\cos\varphi\, e^{K^2\cos^2\varphi}\left[1+\mathrm{erf}\left(K\cos\varphi\right)\right]\Big],\end{aligned} \tag{23}$$

where $K = \sigma\mathrm{p}_0/\sqrt{2}\hbar$. Because of the periodic behavior in the arguments, the first term of the angular integral vanishes. In the Appendix we show that

$$\mathcal{I}_3 = \int_0^{2\pi}\frac{\mathrm{d}\varphi}{2\pi}K\cos\varphi\,\exp\left[K^2\cos^2\varphi\right]\mathrm{erf}\left(K\cos\varphi\right) = \frac{e^{K^2}-1}{\sqrt{\pi}}, \tag{24}$$

allowing us to write

$$\mathcal{F}_{\mathbf{p}_0}(\mathbf{r}_0,0) = \mathcal{F}_{\max} = \frac{\Delta T}{\sqrt{\pi}\sigma\mathcal{A}}. \tag{25}$$

This result has been obtained for the case $\mathbf{p}_0 = 0$ in Calvo et al. (2008). The linear scaling with the temporal window of re-emission is a natural result observed in Draeger and Fink (1997), while the scaling with $\mathcal{A}$ has not been systematically tested so far.

## 4. Focalization as a Function of Time and Space

Having evaluated the focalization signal at the optimal condition, we now analyze the more general situation in which the time and position are not fixed. In this case the argument of the exponential of (15) can be written as

$$-\frac{\sigma^2}{2\hbar^2}\left(\left(\mathrm{p}_0^2-\frac{\left(\mathrm{p}_0'\cos\varphi\right)^2}{\gamma}\right)+\gamma\left(\frac{mL}{\tau}-\frac{\mathrm{p}_0'\cos\varphi}{\gamma}\right)^2\right), \tag{26}$$

where we have defined

$$\gamma = 1-\mathrm{i}\frac{\hbar t}{m\sigma^2},\quad \mathbf{p}_0' = \mathbf{p}_0+\mathrm{i}\frac{\hbar}{\sigma^2}(\mathbf{r}-\mathbf{r}_0), \tag{27}$$

and $\mathbf{e}(\varphi)\cdot\mathbf{p}_0' = \mathrm{p}_0'\cos\varphi$. The focalization amplitude takes the form

$$\mathcal{F}_{\mathbf{p}_0}(\mathbf{r},t) = \frac{m^2\sigma e^{-\sigma^2\mathrm{p}_0^2/2\hbar^2}}{\sqrt{\pi}\hbar^2\mathcal{A}}\int\frac{\mathrm{d}\varphi}{2\pi}\exp\left[\frac{\sigma^2}{2\hbar^2\gamma}\left(\mathrm{p}_0'\cos\varphi\right)^2\right]\int_{L_\mathrm{d}}^{\infty}\mathrm{d}L\, L\,\mathcal{I}_1(L,\varphi), \tag{28}$$

with

$$\mathcal{I}_1(L,\varphi) = \int_{t_1}^{t_2}\frac{\mathrm{d}\tau}{\tau^2}\exp\left[-\frac{\sigma^2}{2\hbar^2}\gamma\left(\frac{mL}{\tau}-\frac{\mathrm{p}_0'\cos\varphi}{\gamma}\right)^2\right].$$

The only difference with the previous case (17) is that the argument in the exponential has been extended into the complex domain. We take $\gamma = e^{\mathrm{i}\alpha}\,|\gamma|$ and

$$\eta(\tau) = \frac{\sigma e^{\mathrm{i}\alpha/2}}{\sqrt{2}\hbar}\,|\gamma|^{1/2}\left(\frac{mL}{\tau}-\frac{\mathrm{p}_0'\cos\varphi}{\gamma}\right), \tag{29}$$

leading to the resulting integral

$$\mathcal{I}_1(L,\varphi) = \frac{\sqrt{2\pi}\hbar}{2m\sigma L}\frac{e^{-\mathrm{i}\alpha/2}}{|\gamma|^{1/2}}\left(\mathrm{erf}(\eta_1)-\mathrm{erf}(\eta_2)\right). \tag{30}$$

In analogy with (19) we write

$$\begin{aligned}\mathcal{I}_2(\varphi) &= \int_{L_\mathrm{d}}^{\infty}\mathrm{d}L\, L\,\mathcal{I}_1(L,\varphi)\\ &= \frac{\sqrt{2\pi}\hbar}{2m\sigma}\frac{e^{-\mathrm{i}\alpha/2}}{|\gamma|^{1/2}}\int_{L_\mathrm{d}}^{\infty}\mathrm{d}L\,\left(\mathrm{erf}(\eta_1)-\mathrm{erf}(\eta_2)\right).\end{aligned} \tag{31}$$

Defining $L_i = v't_i$, $l_i = L/L_i$, and $l_\varphi = \sigma\,|\gamma|^{1/2}\,\mathrm{p}_0'\cos\varphi/\sqrt{2}\hbar\gamma$ with

$$v' = \frac{\sqrt{2}\hbar}{m\sigma\,|\gamma|^{1/2}} \tag{32}$$

we have

$$\begin{aligned}\mathcal{I}_2(\varphi) &= \left(\frac{\hbar}{m\sigma}\right)^2 \frac{\sqrt{\pi}e^{-i\alpha/2}}{|\gamma|}\left[(t_2-t_1)\int_{L_d/L_2}^{\infty} dl\left(1-\operatorname{erf}\left[e^{i\alpha/2}(l-l_\varphi)\right]\right)\right.\\ &\quad \left. + t_1\int_{L_d/L_2}^{L_d/L_1} dl\left(1-\operatorname{erf}\left[e^{i\alpha/2}(l-l_\varphi)\right]\right)\right]. \qquad (33)\end{aligned}$$

Assuming again $L_d \ll L_1 < L_2$, the first term, proportional to $\Delta T$, dominates and can be calculated as

$$\mathcal{I}_2(\varphi) = \left(\frac{\hbar}{m\sigma}\right)^2 \frac{\sqrt{\pi}e^{-i\alpha/2}}{|\gamma|}\Delta T\left[\frac{e^{-i\alpha/2}}{\sqrt{\pi}}\exp(-e^{i\alpha}l_\varphi^2) + l_\varphi\left[1+\operatorname{erf}\left(e^{i\alpha/2}l_\varphi\right)\right]\right]. \qquad (34)$$

Therefore, we obtain for the focalization

$$\begin{aligned}\mathcal{F}_{\mathbf{p}_0}(\mathbf{r},t) &= \frac{\Delta T e^{-\sigma^2 p_0^2/2\hbar^2}e^{-i\alpha}}{|\gamma|\,\sigma\mathcal{A}}\left[\frac{1}{\sqrt{\pi}}\right.\\ &\quad \left. + \int\frac{d\varphi}{2\pi}\, e^{i\alpha/2}l_\varphi\,\exp\left(e^{i\alpha}l_\varphi^2\right)\left[1+\operatorname{erf}\left(e^{i\alpha/2}l_\varphi\right)\right]\right]. \qquad (35)\end{aligned}$$

Defining

$$\mathcal{I}_4 = \int_0^{2\pi}\frac{d\varphi}{2\pi}e^{i\alpha/2}l_\varphi\exp\left(e^{i\alpha}l_\varphi^2\right)\left[1+\operatorname{erf}\left(e^{i\alpha/2}l_\varphi\right)\right], \qquad (36)$$

noticing that

$$e^{i\alpha/2}l_\varphi = \frac{e^{-i\alpha/2}}{|\gamma|^{1/2}}\frac{\sigma}{\sqrt{2}\hbar}p_0'\cos\varphi = K\cos\varphi, \qquad (37)$$

and that the angular integration over the first term in the integrand of (36) vanishes, we have $\mathcal{I}_4 = \mathcal{I}_3$. Therefore

$$\mathcal{F}_{\mathbf{p}_0}(\mathbf{r},t) = \frac{\Delta T e^{-\sigma^2 p_0^2/2\hbar^2}e^{-i\alpha}}{\sqrt{\pi}\,|\gamma|\,\sigma\mathcal{A}}\exp\left(\frac{\sigma^2}{2\hbar^2}\frac{p_0'^2}{\gamma}\right), \qquad (38)$$

where

$$p_0'^2 = p_0^2 + 2i\frac{\hbar}{\sigma^2}\mathbf{p}_0\cdot(\mathbf{r}-\mathbf{r}_0) - \frac{\hbar^2}{\sigma^4}(\mathbf{r}-\mathbf{r}_0)^2. \qquad (39)$$

The focalization then becomes

$$\mathcal{F}_{\mathbf{p}_0}(\mathbf{r},t) = \frac{\mathcal{F}_{\mathbf{p}_0}(\mathbf{r}_0,0)e^{i\alpha}}{\sqrt{1+(\hbar t/m\sigma^2)^2}}\exp\left[-\frac{(\mathbf{r}-\mathbf{r}_0)^2}{2\sigma^2\gamma} + \frac{i}{\hbar}\frac{\mathbf{p}_0}{\gamma}\cdot\left[\frac{\mathbf{p}_0 t}{2m} + (\mathbf{r}-\mathbf{r}_0)\right]\right], \qquad (40)$$

which has the form of a reversed wave respect to the original one. The magnitude of the focalization amplitude is thus given by

$$\left|\mathcal{F}_{\mathbf{p}_0}(\mathbf{r},t)\right| = \frac{\mathcal{F}_{\max}}{\sqrt{1+(\hbar t/m\sigma^2)^2}} \exp\left[-\frac{(\mathbf{r}-\mathbf{r}_0+\mathbf{p}_0 t/m)^2}{2\sigma^2}\frac{1}{1+(\hbar t/m\sigma^2)^2}\right]. \tag{41}$$

This result was obtained in Calvo et al. (2008) using the ergodic approach (Argaman 1996) of trading the actual dynamics by phase-space averages. The procedure presented in this work is more laborious than the ergodic approach, but allows to establish the necessary conditions for the refocalization and it has the advantage of being generalizable to the case where we introduce perturbations between the recording and injection phases.

## 5. Focalization Under External Perturbations

The time-reversal mirror technique has been shown to be robust against external perturbations. These perturbations may be due to uncontrollable evolution of the environment, unavoidable errors in the reversal protocol, but also intentional changes in the setup between the recording and injection phases (Derode et al. 1995; Fink 2001). In the spirit of the Loschmidt echo studies (Jalabert and Pastawski 2001), we can model the effect of a generic perturbation by writing the reconstructed signal as

$$\mathcal{F}_{\mathbf{p}_0}(\mathbf{r},t) = \int_{t_1}^{t_2} \mathrm{d}\tau\, \tilde{G}(\mathbf{r},\mathbf{r}_i,t+\tau) \int \mathrm{d}\mathbf{r}'\, G^*(\mathbf{r}_i,\mathbf{r}',\tau)\, \psi^*_{\mathbf{p}_0}(\mathbf{r}'). \tag{42}$$

$G^*(\mathbf{r}_i,\mathbf{r}',\tau)$ describes, like in (3), the propagation of the recording process, which is governed by a Hamiltonian $H$. We assume that the emission process is governed by a slightly different Hamiltonian $\tilde{H}$, leading to the modified propagator $\tilde{G}(\mathbf{r},\mathbf{r}_i,t+\tau)$. Within a semiclassical approach the effect of the perturbation $\tilde{H}-H$ can be accounted by affecting the contribution of each trajectory $s$ with an additional factor $\langle \exp[\mathrm{i}\Delta S_s/\hbar]\rangle$, where $\Delta S_s$ represents the action difference for traveling along $s$ under each of the Hamiltonians. The angular brackets represent an average that depends on the kind of perturbation chosen. A particularly useful choice is that of fictitious static impurities characterized by an effective elastic mean-free-path $\tilde{l} = \hbar^2 \mathrm{v}_0^2 \left(\int \mathrm{d}q C(q)\right)^{-1}$, where $\mathrm{v}_0 = p_0/m$ and $C(q)$ is the correlation function describing the disorder (Jalabert and Pastawski 2001). Taking the accumulated action as a Gaussian random variable, the previous average can be expressed as $\exp[-L_{\hat{s}}/2\tilde{l}] = \exp[-\gamma\tau^2/L_{\hat{s}}]$, with $\gamma = (1/2\hbar^2)\left(\int \mathrm{d}q C(q)\right)^{-1}$ giving the strength of the perturbation (and independent of the parameters of the trajectories).

Under the optimal focalization condition we now have

$$\begin{aligned} \mathcal{F}_{\max}(\gamma) &= \frac{m^2\sigma}{\sqrt{\pi}\hbar^2\mathcal{A}} \int_{t_1}^{t_2} \frac{\mathrm{d}\tau}{\tau^2} \int_{L_\mathrm{d}}^{\infty} \mathrm{d}L\, L \exp\left[-\left(\frac{L}{v\tau}\right)^2 - \frac{\gamma\tau^2}{L}\right] \\ &= \frac{2v}{\sqrt{\pi}\sigma\mathcal{A}\gamma} \left\{ \left( \int_{L_\mathrm{d}/L_2}^{L_\mathrm{d}/L_1} \mathrm{d}\xi\, \xi^2 e^{-\xi^2} \left( \exp\left[-\frac{\gamma L_\mathrm{d}}{\xi^2 v^2}\right] - \exp\left[-\frac{\gamma L_1}{\xi^2 v^2}\right]\right)\right.\right. \\ &\quad \left.\left. + \int_{L_\mathrm{d}/L_2}^{\infty} \mathrm{d}\xi\, \xi^2\, e^{-\xi^2} \left( \exp\left[-\frac{\gamma L_1}{\xi v^2}\right] - \exp\left[-\frac{\gamma L_2}{\xi v^2}\right]\right)\right)\right\}. \end{aligned} \tag{43}$$

For the second equality we have made the change of variables $\xi = L/v\tau$ and then exchanged the order of the $L$ and $\xi$ integrals. Since we work in the case $L_\mathrm{d} \ll L_1 < L_2$ only the last term of (43) is relevant, and furthermore its lower integration limit can be taken to be 0.

In the limit of $\gamma \to \infty$ we trivially recover $\mathcal{F}_{\max}$ as in (25). The characteristic time defined by the perturbation is $\tilde{\tau} = v/(2\sqrt{\pi}\gamma)$. For very weak perturbations such a time is larger than the maximum recording time, and we simply have

$$\mathcal{F}_{\max}(\gamma) = \mathcal{F}_{\max}\left(1 - \frac{t_2 + t_1}{4\tilde{\tau}}\right) \qquad \text{for} \quad t_2 \ll \tilde{\tau}. \tag{44}$$

For larger perturbations the previous approximation is not valid, but if we stay in the regime where the recording interval is small compared with $\tilde{\tau}$ we see that the contributions of all trajectories are affected by almost the same reduction factor. We thus have

$$\mathcal{F}_{\max}(\gamma) = \mathcal{F}_{\max}\, \exp\left[-\frac{t_2 + t_1}{4\tilde{\tau}}\right] \qquad \text{for} \quad t_2 - t_1 \ll \tilde{\tau}. \tag{45}$$

When the perturbation is strong enough to differentiate trajectories in the recording interval, the focalization will be dominated by the smallest time $t_1$, and we lose the proportionality between the refocused signal and the recording interval. Writing $\exp\left[-\gamma L_1/\xi v^2\right]$ as a Gaussian integral over the variable $\rho$, we can cast the $L_1$-dependent contribution in the last term of (43) as

$$\mathcal{F}_{\max}(\gamma) = \frac{2v}{\sqrt{\pi}\sigma\mathcal{A}\gamma} \int_0^{\infty} \mathrm{d}\xi\, \xi^2 \int_{-\infty}^{\infty} \mathrm{d}\rho\, \exp\left[-\xi^2 - \xi\rho^2 + \frac{2\mathrm{i}\sqrt{\gamma L_1}}{v}\rho\right]. \tag{46}$$

Exchanging the order of the integrals, and assuming a strong perturbation, we have

$$\begin{aligned} \mathcal{F}_{\max}(\gamma) &= \frac{2v}{\sqrt{\pi}\sigma\mathcal{A}\gamma} \frac{[\Gamma(7/4)]^{3/2}}{2[\Gamma(9/4)]^{1/2}} \exp\left[-\frac{\Gamma(7/4)}{\Gamma(9/4)}\frac{\gamma t_1}{v^2}\right] \\ &= \frac{c\tilde{\tau}}{\sqrt{\pi}\sigma\mathcal{A}} \exp\left[-\frac{c' t_1}{2\tilde{\tau}}\right] \qquad \text{for} \quad t_2 > \tilde{\tau}. \end{aligned} \tag{47}$$

In the last expression we have defined the numerical constants $c = 2.94$ and $c' = 0.46$. Leaving aside the trivial case of (44), we will go from the regime of (45) to that of (47) while increasing the strength of the perturbation. The most important change is the breakdown of the linearity of $\mathcal{F}_{\max}(\gamma)$ with $\Delta T$ when going from one regime to the other. For current experiments the regime of (47) is the most relevant one. In this case the exponential suppression of the signal with the typical recording times appears as a stringent limitation. Depending on the model chosen for the perturbation, the dependence of the typical time $\tilde{\tau}$ on the parameters of the system might be quite different. For instance, choosing as a perturbation the mass distortion in a system represented by a Lorentz gas would lead to a different scaling of $\tilde{\tau}$ than the above presented (Jalabert and Pastawski 2001) such that it increases with the chaoticity of the system, which would result in an improved focusing. The role of the perturbation in limiting the refocalization is an important aspect that deserves further study.

## 6. Conclusion

In conclusion, we have presented detailed semiclassical calculations of the refocalization signal for the time-reversal mirror procedure. The chaotic nature of the underlying classical dynamics appears as a key ingredient of the reconstruction. Our calculations are valid for the chaotic regime and agree quite accurately with the corresponding numerical simulations (Calvo et al. 2008). Our semiclassical calculations cannot be directly applied to the case of integrable geometries, where the experimental and numerical results show a poorer signal reconstruction with lower signal compression, the appearance of focalization satellites, and an important sensitivity with respect to the position of the transducers.

## Appendix

In this appendix we calculate the integral $\mathcal{I}_3$ of (24). The expansion of the exponentials in the corresponding integrand leads to

$$\mathcal{I}_3 = \frac{1}{\pi^{3/2}} \sum_{n=0}^{\infty} \sum_{m=0}^{\infty} \frac{(-1)^m}{n!m!(2m+1)} \int_0^{2\pi} \mathrm{d}\varphi \, [K \cos\varphi]^{2(n+m+1)} .$$

Performing the replacement $l = n + m$, we have

$$\mathcal{I}_3 = \frac{1}{\pi^{3/2}} \sum_{l=0}^{\infty} \frac{1}{l!} \sum_{m=0}^{l} \binom{l}{m} \frac{(-1)^m}{2m+1} \int_0^{2\pi} \mathrm{d}\varphi \, [K \cos\varphi]^{2(l+1)} .$$

The sum over $m$ can be obtained from

$$\sum_{m=0}^{l} \binom{l}{m} \frac{(-1)^m}{2m+1} = \int_0^1 (1-x^2)^l \mathrm{d}x = \int_0^{\pi/2} \mathrm{d}\phi \cos^{2l+1}\phi = \frac{\sqrt{\pi}}{2} \frac{\Gamma(l+1)}{\Gamma(l+3/2)} .$$

On the other hand, the remaining integral takes the value

$$\int_0^{2\pi} \mathrm{d}\varphi \cos^{2(l+1)}\varphi = 2\sqrt{\pi}\, \frac{\Gamma(l+3/2)}{\Gamma(l+2)},$$

yielding

$$\mathcal{I}_3 = \frac{1}{\sqrt{\pi}} \sum_{l=0}^{\infty} \frac{K^{2(l+1)}}{l!} \frac{\Gamma(l+1)}{\Gamma(l+2)} = \frac{1}{\sqrt{\pi}} \sum_{l=0}^{\infty} \frac{K^{2(l+1)}}{(l+1)!} = \frac{1}{\sqrt{\pi}} \left(e^{K^2} - 1\right) ,$$

in agreement with (24) of the text.

## Acknowledgements

We thank A. Goussev, T. Kramer and K. Richter for valuable discussions and D. Weinmann for a careful reading of the manuscript. We acknowledge financial support of SeCyT-UNC, CONICET, and the French-Argentinian program ECOS-Sud. H.M.P. acknowledges the hospitality of the MPI-PKS of Dresden.

## References

Argaman N.: 1996. *Phys. Rev. B*, **53**, 7035
Brack M.: 1997. Semiclassical Physics. *Frontiers in Physics*, vol. 96, Westview Press, Boulder
Calvo H.L., Jalabert R.A. and Pastawski H.M.: 2008. *Phys. Rev. Lett.*, **101**, 240403
Casati G., Guarneri I. and Smilansky U. eds.: 2000. New directions in quantum chaos. In: *Proceedings of the International School of Physics Enrico Fermi*, Course CXLIII, IOS Press Amsterdam; see also Jalabert R.A., references within Casati et al. (2000)
Catheline S., Benech N., Brum J. and Negreira C.: 2008. *Phys. Rev. Lett.*, **100**, 064301
Cucchietti F.M., Pastawski H.M. and Jalabert R.A.: 2004. *Phys. Rev. B*, **70**, 035311
Derode A., Roux P. and Fink M.: 1995. *Phys. Rev. Lett.*, **75**, 4206
Draeger C. and Fink M.: 1997. *Phys. Rev. Lett.*, **79**, 407
Fink M.: 1997. *Phys. Today*, **20**, 34
Fink M.: 2001. *Phys. Scr.* **T90**, 268
Goussev A., Waltner D., Richter K. and Jalabert R.A.: 2008. *New J. Phys.*, **10**, 093010
Heinemann M.G., Larraza A. and Smith K.B.: 2002. *Appl. Phys. Lett.*, **80**, 694
Jalabert R.A. and Pastawski H.M.: 2001. *Phys. Rev. Lett.*, **86**, 2490
Negreira C. and Fink M.: 2007. *Wave Random Complex*, **17**, 67

Pastawski H.M., Danieli E.P., Calvo H.L. and Foa Torres L.E.F.: 2007. *Europhys. Lett.*, **77**, 40001
Ribay G, de Rosny J. and Fink M.: 2005. *J. Acoust. Soc. A.* **117**, 2866
Sieber M.: 1999. *J. Phys. A*, **32**, 7679
Tanner G. and Søndergaard N.: 2007. *J. Phys. A*, **40**, R443

# ENTANGLEMENT AND LOCALIZATION OF WAVEFUNCTIONS

O. Giraud, B. Georgeot*, and J. Martin
*Université de Toulouse; UPS; Laboratoire de Physique Théorique (IRSAMC); CNRS; LPT (IRSAMC), Toulouse, France; georgeot@irsamc.ups-tlse.fr*

**Abstract.** We review recent works that relate entanglement of random vectors to their localization properties. In particular, the linear entropy is related by a simple expression to the inverse participation ratio, while next orders of the entropy of entanglement contain information about e.g. the multifractal exponents. Numerical simulations show that these results can account for the entanglement present in wavefunctions of physical systems.

**Key words:** Quantum information; Entanglement; Random vectors; Localization; Multifractals

## 1. Introduction

Quantum mechanics has always seemed puzzling since its first construction in the first half of the twentieth century. Many properties are different from the world of classical physics in which our intuition is built. The development of quantum information science in the last decades has exemplified this aspect. Indeed, it was realized that it is in principle possible to exploit the features of quantum mechanics to treat information in a different way from what a classical computer would do. In this context, the specific properties of quantum mechanics are put forward as new resources which enable to treat information in completely new ways.

One of the most peculiar properties of quantum mechanics is entanglement, that is the possibility to construct quantum states of several subsystems that cannot be factorized into a product of individual states of each subsystem. Such entangled states are the most common in quantum mechanics, and they display correlations which cannot be seen in a classical world, exemplified by e.g. the Einstein–Podolsky–Rosen "paradox." Entanglement is also a resource for quantum information (see Nielsen and Chuang 2000 and references therein), and has been widely studied as such in the past few years.

Despite intensive work, entanglement remains a somewhat mysterious property of physical systems. The structure of entanglement of systems even

G. Casati and D. Matrasulov (eds.), *Complex Phenomena in Nanoscale Systems*,
NATO Science for Peace and Security Series B: Physics and Biophysics,

© Springer Science+Business Media B.V. 2009

with small numbers of particles is hard to characterize. Even properly measuring the entanglement present in a system is difficult for mixed states. This is all the more important since recent results have shown that (at least for pure states) if a process creates a sufficiently low level of entanglement, it can be simulated efficiently by a classical computer (Jozsa and Linden 2003; Vidal 2003). This gives a limit on the speedup over classical computation a quantum computer can achieve, and also gives rise to interesting proposals for building classical algorithms simulating weakly entangled quantum systems (Verstraete et al. 2004).

In this paper, we review recent results we obtained (details can be found in Giraud et al. 2007, 2009), which concern the relationship of entanglement to localization properties of a quantum state. Our strategy is to consider $n$-qubit systems, and to study entanglement of quantum states relative to their localization properties in the $2^n$-dimensional Hilbert space in the computational basis. We obtain analytical results for *random states*, that is ensemble of quantum states sharing some properties. Such random states have been recently studied in the literature. They are interesting in themselves, since it has been shown for example in quantum information that they are useful in various quantum protocols (Harrow et al. 2004; Hayden et al. 2004; Bennett et al. 2005; Cappellaro et al. 2005). This motivated a recent activity in the quantum information community to try and produce efficiently such random vectors or random operators through quantum algorithms (Emerson et al. 2003; Weinstein and Hellberg 2005), and to characterize their entanglement properties (Scott 2004; Sommers and Zyczkowski 2004; Giraud 2007a, b; Zindaric 2007; Zindaric et al. 2007; Facchi et al. 2008). In addition to their intrinsic usefulness, random states are important since they can describe typical states of a "complex" system. For example, it has been known for some times now that random vectors built from Random Matrix Theory (RMT) can describe faithfully the properties of quantum Hamiltonian systems whose classical limit is chaotic, and more generally of many complex quantum systems (Giannoni et al. 1991). Such random vectors are ergodic, and the entanglement they contain has been calculated some time ago (Lubkin 1978; Page 1993). However, in many quantum systems, the wavefunctions are not ergodic but localized. This can correspond to electrons in a disordered potential, which are exponentially localized due to Anderson localization. It can also be seen in many-body interacting systems, where the presence of a moderate interaction can lead to states partially localized in energy. Some systems are in a well-defined sense neither ergodic neither localized: they correspond to e.g. states at the Anderson transition between localized and delocalized states, and can show multifractal properties (Mirlin 2000; Evers and Mirlin 2007).

In this paper, we calculate the amount of entanglement present in ensembles of random vectors displaying these various degrees of localization. Besides generalizing the result for RMT-type random vectors, this gives the entanglement present in a "typical state" of such localized or partially localized systems. This enables to estimate the complexity of simulating such systems on classical computers, but also sheds light on the entanglement itself, since in these cases it is related through simple formulas to quantities characterizing the degree of localization of the system.

Our results show that for random vectors which are localized on the computational basis, the linear entropy which approximates the amount of entanglement in the vector is simply related to the Inverse Participation Ratio (IPR), a popular measure of localization. The next term in the approximation is related to higher moments, and in particular to the multifractal exponents for multifractal systems. In order to assess the usefulness of these results to physical systems, we compare them to the entanglement numerically computed for several models. After a general discussion on entanglement of random vectors (Sect. 2), we consider the entanglement of one qubit with the others (Sect. 3), and give explicitly the first and second order of the expansion of the entropy of entanglement around its maximum. Section 4 generalizes these results to other bipartitions, and Sect. 5 compares the formula obtained with the numerical results for two physical systems. Section 6 considers the physically important case of vectors localized not on a random subset of the basis vectors, but on a subset composed of adjacent basis vectors (that is the states are localized on computational basis states which are adjacent when the basis vectors are ordered according to the number which labels them). showing that the results become profoundly different. Section 7 presents the conclusions.

## 2. Entanglement of Random Vectors

Random vectors are ensembles of vectors whose components are distributed according to some probability distribution. If for example the system considered is composed of $n$ qubits, the Hilbert space is of dimension $N = 2^n$, and random vectors distributed according to the uniform measure on the $N$-dimensional sphere describe typical quantum states of the $n$ qubits. Such states are ergodically distributed in the computational basis, and their entanglement has been already studied in Lubkin (1978) and Page (1993). In this paper, we are interested in random vectors which are not ergodically distributed. Ensembles of such states will be characterized by localization properties. The simplest example of such localized random vectors can be constructed by taking $M$ components ($M < N$) with equal amplitudes and uniformly distributed random phases, and setting all the others to zero. The random vectors will all be exactly localized on $M$ basis states. A more physically relevant example consists in still choosing $M < N$ nonzero components, and giving them the distribution of column vectors of $M \times M$ random unitary matrices drawn from the Circular Unitary Ensemble of random matrices (CUE vectors). In general our result will be averaged both over the distribution of the nonzero components and the position of these nonzero components in the computational basis. This corresponds to classes of random vectors sharing the same localization length. Our results will in fact generalize to any such distribution of random vector whose localization properties are fixed. In addition, we shall see that if we impose that the distribution of the position of nonzero components is such that there are always adjacent in the computational basis, the results change drastically.

The localization properties of the random vectors can be probed using the moments of the distribution

$$p_q = \sum_{i=1}^{N} |\psi_i|^{2q} \tag{1}$$

The second moment is $p_2 = 1/\xi$ where $\xi$ is the Inverse Participation Ratio (IPR) which is often used in the mesoscopic physics literature to measure the localization length. Indeed, for a state uniformly spread on exactly $M$ basis vectors, one has $\xi = M$. The scaling of $p_2$ and higher moments with the size also probes the multifractal properties of the wavefunction.

The random states we consider are built on the $N$-dimensional Hilbert space of a $n$-qubit system with $N = 2^n$. We are interested in bipartite entanglement between subsystems defined by different partitions of the $n$ qubits into two sets. In general, bipartite entanglement of a pure state belonging to a Hilbert space $\mathcal{H}_A \otimes \mathcal{H}_B$ is measured through the entropy of entanglement, which has been shown to be a unique entanglement measure (Popescu and Rohrlich 1997). We consider pure states belonging to $\mathcal{H}_A \otimes \mathcal{H}_B$ where $\mathcal{H}_A$ is a set of $\nu$ qubits and $\mathcal{H}_B$ a set of $n - \nu$ qubits. If $\rho_A$ is the density matrix obtained by tracing out subsystem $B$, then the entropy of entanglement of the state $\rho$ with respect to the bipartition $(A, B)$ is the von Neumann entropy of $\rho_A$, that is $S = -\mathrm{tr}(\rho_A \log_2 \rho_A)$.

## 3. Entanglement of One Qubit with All the Others

To obtain an approximation for the entropy, one can expand $S$ around its maximal value. In the case of the partition of the $n$ qubits into 1 and $n - 1$ qubits, the entropy can be written as a function of $\tau$, with

$$\tau = 4 \det \rho_A \tag{2}$$

(in the case of 2 qubits this quantity is called the *tangle* and corresponds to the square of the generalized concurrence, Rungta and Caves 2003). One has

$$S(\tau) = h\left(\frac{1+\sqrt{1-\tau}}{2}\right), \tag{3}$$

where $h(x) = -x\log_2 x - (1-x)\log_2(1-x)$. The series expansion of $S(\tau)$ up to order $m$ in $(1-\tau)$ reads

$$S_m(\tau) = 1 - \frac{1}{\ln 2}\sum_{n=1}^{m}\frac{(1-\tau)^n}{2n(2n-1)}. \tag{4}$$

The first order corresponds to $\tau$ itself up to constants and its average over the choice of the $(1, n-1)$ partition is known as the linear entropy or Meyer–Wallach entropy $Q$ (Meyer and Wallach 2002; Brennen 2003). Our results show that for our class of random vectors, the average linear entropy is given by

$$\langle\tau\rangle = \frac{N-2}{N-1}(1-\langle p_2\rangle) = \frac{N-2}{N-1}(1-\langle 1/\xi\rangle). \tag{5}$$

This formula was obtained first by considering a random vector which is nonzero only on $M$ basis vectors among $N$, and summing explicitly the combinatorial terms. It can also be obtained in a more general setting by taking $M = N$ and summing up all the localization properties of the vector in the IPR $\xi$ alone. For any $(1, n-1)$ partition of the $n$ qubits, the components of the vector can be divided in two sets according to the value of the first qubit. Assuming no correlation among these sets enables to get (5) (details on the calculations can be found in Giraud et al. 2007).

It is interesting to compare this formula with a similar one obtained in Viola and Brown (2007) and Brown et al. (2008) using different assumptions, in particular without average over random phases. The formula obtained relates entanglement to the mean inverse participation ratio calculated in three different bases, a quantity that is often delicate to evaluate. In our case, the additional assumption of random phases enables to obtain a formula which involves only the IPR in one basis, a quantity that can be easily evaluated in many cases. For example, it enables to compute readily the entanglement for localized CUE vectors. However there are instances of systems (e.g. spin systems) where these different formulas give the same results.

In particular, our formula (5) allows to compute $\langle\tau\rangle$ e.g. for a CUE vector localized on $M$ basis vectors; in this case $\xi = (M+1)/2$, and we get

$$\langle\tau\rangle = \frac{M-1}{M+1}\frac{N-2}{N-1}. \tag{6}$$

In Lubkin (1978), $\langle \tau \rangle$ was calculated for non-localized CUE vectors of length $N$, giving $\langle \tau \rangle = (N-2)/(N+1)$. Consistently, our formula yields the same result if we take $M = N$. For a vector with constant amplitudes and random phases on $M$ basis vectors, $\xi = M$ and

$$\langle \tau \rangle = \frac{M-1}{M}\frac{N-2}{N-1}. \tag{7}$$

The next order in the expansion (4) can be obtained by similar methods that we do not detail here (see Giraud et al. 2009 for details); summing up all terms involved in $\tau^2$ we get

$$\langle \tau^2 \rangle = N(N-2)(N^2-6N+16)c_{1111}+4N(N-2)(N-4)c_{211}+4N(N-2)c_{22}. \tag{8}$$

with

$$c_{22} = \frac{\langle p_2^2 \rangle - \langle p_4 \rangle}{N(N-1)}, \quad c_{211} = \frac{\langle p_2 \rangle - \langle p_2^2 \rangle - 2\langle p_3 \rangle + 2\langle p_4 \rangle}{N(N-1)(N-2)},$$
$$c_{1111} = \frac{1 - 6\langle p_2 \rangle + 8\langle p_3 \rangle + 3\langle p_2^2 \rangle - 6\langle p_4 \rangle}{N(N-1)(N-2)(N-3)}. \tag{9}$$

This gives the next order of the entropy of entanglement in terms of the moments up to order 4 of the vector. What this means is that at this order, the average entanglement of random vectors with fixed moments will be related to them through (8). Although more complicated than (5), the formula indicates that e.g. for states having multifractal properties, since moments scale with system size according to quantities called multifractal exponents, the behavior of the entanglement at this order will be also controlled by these multifractal exponents.

The $n$th order of the expansion (4) can similarly be obtained and has been derived in Giraud et al. (2009). It is interesting to note that in the case of a CUE random vector of size $N$, resummation of the whole series for $S(\tau)$ yields, after some algebra,

$$\langle S(\tau) \rangle = \frac{1}{\ln 2} \sum_{k=N/2+1}^{N-1} \frac{1}{k}, \tag{10}$$

which has been obtained earlier by a different method (Page 1993).

A general conclusion obtained from these formulas is that the entanglement associated to such bipartition goes to the maximal value for large $N$ and large $\xi$, even if $\xi$ remains smaller than $N$. For fixed $\xi$, it tends for large $N$ to a constant nonzero value which depends on $\xi$. We will see in Sect. 6 that this result can change drastically if we impose a localisation on fixed locations in Hilbert space.

## 4. Entanglement of Random Vectors: Other Partitions

Up to now we have considered the entanglement of one qubit with all the others, i.e. the $(1, n-1)$ partition of $n$ qubits. What about bipartite entanglement relative to other bipartitions $(\nu, n-\nu)$, where $\nu$ is any number between 1 and $n-1$? In this case, it is convenient to define the linear entropy as $S_L = \frac{d}{d-1}(1 - \mathrm{tr}\rho_A^2)$, where $d = \dim \mathcal{H}_A \leq \dim \mathcal{H}_B$. The scaling factor is such that $S_L$ varies in $[0, 1]$.

A similar calculation as above enables then to obtain the first order of the mean von Neumann entropy, given by

$$\langle S \rangle \approx \nu - \frac{2^\nu - 1}{2\ln 2}\left(1 - \frac{N - 2^\nu}{N-1}\left\langle \frac{1}{\xi} \right\rangle\right), \tag{11}$$

with $p_2 = 1/\xi$, which generalizes (5).

Higher-order terms can be obtained as well, although the calculations become tedious. To this end, the entropy $S = -\mathrm{tr}(\rho_A \log_2 \rho_A)$ is expanded around the maximally mixed state $\rho_0 = \mathbf{1}/2^\nu$, as

$$S = \nu + \frac{1}{\ln 2}\sum_{n=1}^{\infty} \frac{(-2^\nu)^n}{n(n+1)} \mathrm{tr}((\rho_A - \rho_0)^{n+1}). \tag{12}$$

We remark that again the linear entropy 11 tends to the maximal possible value when $N$ and $\xi$ become large, as for the $(1, n-1)$ partition.

## 5. Entanglement of Random Vectors: Application to Physical Systems

In order to test these results on physical systems, we compared them to numerical results obtained from different models.

The first one corresponds to a diagonal Hamiltonian matrix to which a two-body interaction is added.

$$H = \sum_i \Gamma_i \sigma_i^z + \sum_{i<j} J_{ij}\sigma_i^x \sigma_j^x \tag{13}$$

This system can describe a quantum computer in presence of static disorder (Georgeot and Shepelyansky 2000a, b). Here the $\sigma_i$ are the Pauli matrices for the qubit $i$, energy spacing between the two states of qubit $i$ is given by $\Gamma_i$ are randomly and uniformly distributed in the interval $[\Delta_0 - \delta/2, \Delta_0 + \delta/2]$, and $J_{ij}$ uniformly distributed in the interval $[-J, J]$ represent a random static interaction. Entanglement of eigenvectors of this Hamiltonian was already considered in a different context in Mejia-Monasterio et al. (2005). It is known (Georgeot and Shepelyansky 2000a) that in this model a transition to quantum chaos takes place for sufficiently large coupling strength $J$. In this regime, eigenvectors of (13) are spread among all noninteracting eigenstates, which correspond to the computational basis, but in a certain window of energy, and are distributed according to the Breit–Wigner (Lorentzian) distribution. Thus these wavefunctions are distributed among a certain subset of the computational basis, although they are not strictly zero outside it, and the distribution is not uniform, but rather Lorentzian. Nevertheless, our data show (see Figs. 1 and 2) that the behavior of the bipartite entanglement of eigenvectors of this model is well described by the results derived for random vectors. The agreement becomes very accurate if the eigenvector components are randomly shuffled to lower correlations.

We also considered another model, based on $N \times N$ matrices of the form

$$U_{kl} = \frac{e^{i\phi_k}}{N} \frac{1 - e^{2i\pi N\gamma}}{1 - e^{2i\pi(k-l+N\gamma)/N}}, \tag{14}$$

where $\phi_k$ are random variables independent and uniformly distributed in $[0, 2\pi[$. This model introduced in Bogomolny and Schmit (2004) is the randomized version of a simple quantum map introduced in Giraud et al. (2004). The eigenvectors of (14) have multifractal properties in the momentum representation (Martin et al. 2008) for rational $\gamma$, although again the components are nonzero everywhere. The results of Figs. 2 and 3 shows that again the results for random vectors describes very well the entanglement for this system for randomly shuffled components, and that even the first order is already a good approximation.

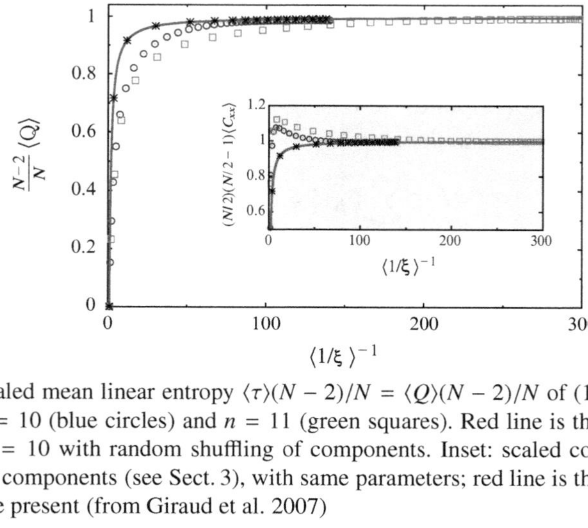


*Figure 1.* Scaled mean linear entropy $\langle\tau\rangle(N-2)/N = \langle Q\rangle(N-2)/N$ of (13) vs. mean IPR for $\delta = \Delta_0$, $n = 10$ (blue circles) and $n = 11$ (green squares). Red line is the theory, crosses the data for $n = 10$ with random shuffling of components. Inset: scaled correlator between the two sets of components (see Sect. 3), with same parameters; red line is the result when no correlations are present (from Giraud et al. 2007)

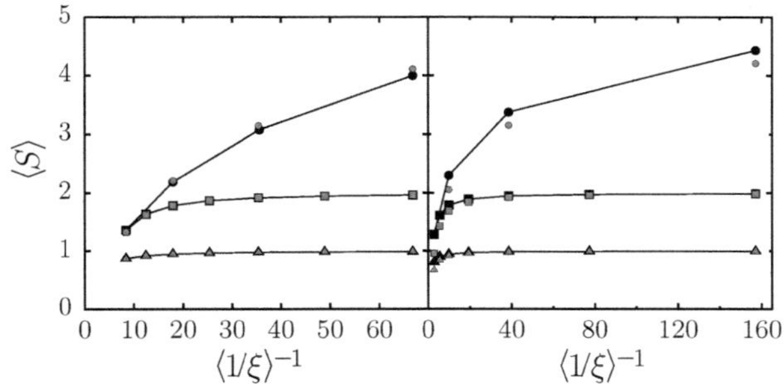


*Figure 2.* Mean entropy of entanglement $S$ for different bipartitions $(\nu, n-\nu)$ as a function of the mean IPR. Left: eigenvectors of (14) with $\gamma = 1/3$; the average is taken over $10^6$ eigenvectors. Right: eigenvectors of (13) with $\delta = \Delta_0$ and $J/\delta = 1.5$; average over $\approx 3\times10^5$ vectors. Triangles correspond to $\nu = 1$, squares to $\nu = 2$ and circles to $\nu = n/2$, with $n = 4-10$. Black symbols are the theoretical predictions for the mean value of $S$ (obtained from (11) and green (gray) symbols are the computed mean values of the von Neumann entropy (from Giraud et al. 2009)

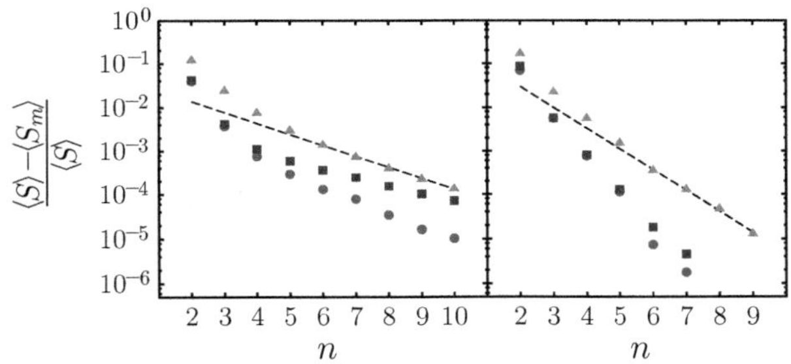


*Figure 3.* Relative difference of the entropy of entanglement (3) and its successive approximations $S_m$ $(m = 1, 2)$ with respect to the number of qubits for eigenvectors of (14) for (left) $\gamma = 1/3$ and (right) $\gamma = 1/7$. The average is taken over $10^7$ eigenvectors, yielding an accuracy $\lesssim 10^{-6}$ on the computed mean values. Green triangles correspond to the first order expansion $S_1$, blue squares and red circles to the second order expansion $S_2$. The difference between the latter two is that for blue squares $\langle p_2^2\rangle$ appearing in (9) has been replaced by $\langle p_2\rangle^2$ yielding a less accurate approximation. Dashed line is a linear fit yielding $1 - \langle S_1\rangle/\langle S\rangle \sim N^{-0.84}$ for $\gamma = 1/3$ and $N^{-1.58}$ for $\gamma = 1/7$ (from Giraud et al. 2009)

## 6. Entanglement of Adjacent Random Vectors

In the preceding sections we discussed formulas for entanglement of ensembles of random vectors where the components over each basis vector are independent. If we relax this assumption, the result may change. A particular important case corresponds e.g. to random vectors localized on $M$ computational basis states which are adjacent when the basis vectors are ordered according to the number which labels them (if the two states of a qubit are denoted $|0\rangle$ and $|1\rangle$, each state in the computational basis corresponds to a sequence of 0 and 1 and thus can be labelled naturally by a number between 0 and $2^n - 1$). In this case, we had to use combinatorial methods; summing all contributions together we get for the linear entropy of $(1, n-1)$ partitions

$$\langle \tau \rangle = \Bigg[ \bigg( \frac{M-2}{M-1} r_0 + \frac{2(2^{r_0}-1)}{M(M-1)} + \frac{4}{3} \frac{(M+1)(2^n - 2^{r_0})}{2^{n+r_0}} - \frac{1}{M(M-1)} \sum_{r=0}^{r_0-1} \chi_r(m_r) \bigg) \bigg( 1 - \langle \frac{1}{\xi} \rangle \bigg) \Bigg] \frac{1}{n}, \quad (15)$$

where $r_0$ is such that $2^{r_0-1} < M \leq 2^{r_0}$ and $\chi_r(x) = \chi_r(2^{r+1} - x) = x^2 - \frac{2}{3}x(x^2 - 1)/2^r$ for $0 \leq x \leq 2^r$. Equation (15) is an exact formula for $M \leq N/2$. For fixed $M$ and $n \to \infty$, $n\langle Q \rangle$ converges to a constant $C$ which is a function of $M$ and $\xi$. For $M = 2^{r_0}$, $r_0 < n$, (15) simplifies to

$$\langle \tau \rangle = \Bigg[ \bigg( \frac{(r_0 + \frac{4}{3})M^2 - 2(r_0 - 1)M - \frac{10}{3}}{M(M-1)} - \frac{4(M+1)}{3N} \bigg) \bigg( 1 - \langle \frac{1}{\xi} \rangle \bigg) \Bigg] \frac{1}{n}. \quad (16)$$

Numerically, this expression with $r_0 = \log_2 M$ gives a very good approximation to (15) for all $M$.

Equation (15) is exact for e.g. uniform and CUE vectors, and can be applied even if the vector is not strictly zero outside a $M$-dimensional subspace. Indeed, for $N$-dimensional CUE vectors with exponential envelope $\exp(-x/l)$, $\langle Q \rangle$ is in excellent agreement with (15) with $\xi = l$ and $M = 2\xi$ (stars in inset of Fig. 4).

In order to compare these findings to those of a physical system with such property of localization on adjacent basis vectors, Fig. 4 shows the theory (15) together with the entropy for the one-dimensional Anderson model. This model corresponds to a one-dimensional chain of vertices with nearest-neighbour coupling and randomly distributed on-site disorder, described by the Hamiltonian $H_0 + V$. Here $H_0$ is a diagonal operator whose elements $\epsilon_i$ are Gaussian random variables with variance $w^2$, and $V$ is a tridiagonal matrix with non-zero elements only on the first diagonals, equal to the coupling strength, set to 1. It is known that eigenstates of this system, which modelizes electrons in a disordered potential, have envelopes of the form $\exp(-|x - x_0|/l)$, where $l$ is the localization length. It was shown in Pomeransky and Shepelyansky (2004) and Giraud et al. (2005) that this model can be simulated efficiently on a quantum computer, and the wavefunction of the computer during the algorithm will be localized on adjacent basis vectors, which correspond to the position of vertices. Figure 4 shows that the asymptotic behavior of the linear entropy of the eigenstates (with all correlations left between components, i.e. no random shuffling) is well captured by (15).

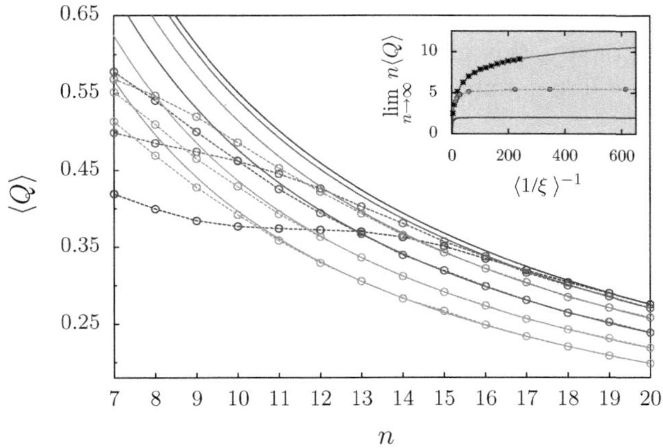


*Figure 4.* Mean linear entropy $\langle \tau \rangle = \langle Q \rangle$ of partitions $(1, n-1)$ vs. number of qubits for the one-dimensional Anderson model with disorder from top to bottom $w = 0.2$ (blue), 0.5 (red), 1.0 (green), 1.5 (magenta), 2.0 (cyan), and 2.5 (orange). Average is over 10000 eigenstates. Solid lines are the $C/n$ fits of the tails. Inset: Value of $C = \lim_{n\to\infty} n\langle Q \rangle$ as a function of IPR $\xi$ (green dots) for the values of $w$ above and $w = 0.4$, together with analytical result of (15) (red line, top) and by $\frac{26}{9} - \frac{4}{M} - \frac{8(3r_0+1)}{9M^2}$ for $M = 2\xi$ (blue line, bottom). Stars are the $C$ values resulting from a $C/n$ fit of the numerical data for CUE vectors of size $N$ with exponential envelope $\exp(-x/l)$ (from Giraud et al. 2007)

Thus random vectors localized on adjacent basis vectors correspond to a drastically different behavior compared to the vectors of Sect. 3: indeed, for fixed $\xi$ the entanglement (at least the linear entropy) always tends to zero for large $N$, even if it does it rather slowly (as $\sim 1/\ln N$).

## 7. Conclusion

The results above indicate that the entanglement of random vectors can be directly related to the fact that they are localized, multifractal or extended. The numerical simulations for different physical systems show that these results obtained for random vectors describe qualitatively the entanglement present in several physical systems, and reproduce it accurately if correlations are averaged out.

Thus the results are interesting to predict the amount of entanglement present in random vectors, and also can be applied to physical systems for which such random vectors describe typical states. This gives insight on the difficulty to simulate classically such systems, since systems with low amounts of entanglement can be simulated classically efficiently. This also can be applied to estimate the changes in entanglement at a quantum phase transition (Amico et al. 2008), in particular for the Anderson transition between localized and extended states (see Giraud et al. 2009 for more details). Additionally, this gives also insight on the nature of entanglement itself by relating it to simple physical properties of the system.

## Acknowledgements

We thank CalMiP for access to their supercomputers. This work was supported by the Agence Nationale de la Recherche (project ANR-05-JCJC-0072 INFOSYSQQ) and the European program EC IST FP6-015708 EuroSQIP.

## References

Amico L., Fazio R., Osterloh A. and Vedral V.: 2008. *Rev. Mod. Phys.*, **80**, 517
Bennett C.H., Hayden P., Leung D., Shor P. and Winter A.: 2005. *IEEE Trans. Inf. Theory*, **51**, 56
Bogomolny E. and Schmit C.: 2004. *Phys. Rev. Lett.*, **93**, 254102
Brennen G.K.: 2003. *Quant. Inf. Comp.*, **3**, 619
Brown W.G., Santos L.F., Starling D.J. and Viola L.: 2008. *Phys. Rev. E*, **77**, 021106
Cappellaro P., Emerson J., Boulant N., Ramanathan C. and Cory D.G.: 2005. *Phys. Rev. Lett.*, **94**, 020502
Emerson J., Weinstein Y.S., Saraceno M., Lloyd S. and Cory D.S.: 2003. *Science*, **302**, 2098
Evers F. and Mirlin A.D.:2007. arXiv:0707.4378
Facchi P., Marzolino U., Parisi G., Pascazio S. and Scardicchio A.: 2008. *Phys. Rev. Lett.*, **101**, 050502
Georgeot B. and Shepelyansky D.L.: 2000a. *Phys. Rev. E*, **62**, 3504
Georgeot B. and Shepelyansky D.L.: 2000b *Phys. Rev. E* **62**, 6366
Giannoni M.-J., Voros A. and Zinn-Justin J. (Eds.): 1991. *Proceedings of the 52th Les Houches Summer School*, North-Holland, Amsterdam
Giraud O.: 2007a. *J. Phys. A*, **40**, 2793
Giraud O.: 2007b. *J. Phys. A*, **40**, F1043
Giraud O., Marklof J. and O'Keefe S.: 2004. *J. Phys. A.*, **37**, L303
Giraud O., Georgeot B. and Shepelyansky D.L.: 2005. *Phys. Rev. E*, **72**, 036203
Giraud O., Martin J. and Georgeot B.: 2007. *Phys. Rev. A*, **76**, 042333
Giraud O., Martin J. and Georgeot B.: 2009. *Phys. Rev. A*, **79**, 032308
Harrow A., Hayden P. and Leung D.: 2004. *Phys. Rev. Lett.*, **92**, 187901
Hayden P., Leung D., Shor P. and Winter A.: 2004. *Commun. Math. Phys.*, **250**, 317
Jozsa R. and Linden N.: 2003. *Proc. R. Soc. London Ser. A*, **459**, 2011
Lubkin E.: 1978. *J. Math. Phys.*, **19**, 1028
Martin J., Giraud O. and Georgeot B.: 2008. *Phys. Rev. E*, **77**, R035201
Mejia-Monasterio C., Benenti G., Carlo G.G. and Casati G.: 2005. *Phys. Rev. A*, **71**, 062324
Meyer A.D. and Wallach N.R.: 2002. *J. Math. Phys.*, **43**, 4273
Mirlin A.D.: 2000. *Phys. Rep.*, **326**, 259
Nielsen M.A. and Chuang I.L.: 2000, Quantum computation and quantum information. Cambridge University Press, Cambridge
Page D.N.: 1993. *Phys. Rev. Lett.*, **71**, 1291
Pomeransky A.A. and Shepelyansky D.L.: 2004. *Phys. Rev. A*, **69**, 014302

Popescu S. and Rohrlich D.: 1997. *Phys. Rev. A*, **56**, R3319
Rungta P. and Caves C.M.: 2003. *Phys. Rev. A*, **67**, 012307
Scott A.J.: 2004. *Phys. Rev. A*, **69**, 052330
Sommers H.-J. and Zyczkowski K.: 2004. *J. Phys. A*, **37**, 8457
Verstraete F., Porras D. and Cirac J.I.: 2004. *Phys. Rev. Lett.*, **93**, 227205
Vidal G.: 2003. *Phys. Rev. Lett.*, **91**, 147902
Viola L. and Brown W.G.: 2007. *J. Math. Phys.*, **43**, 8109
Weinstein Y.S. and Hellberg C.S.: 2005. *Phys. Rev. Lett.*, **95**, 030501
Znidaric M.: 2007. *J. Phys. A*, **40**, F105
Znidaric M., Prosen T., Benenti G. and Casati G.: 2007. *J. Phys. A*, **40**, 13787

# EXACT ANALYSIS OF ADIABATIC INVARIANTS IN TIME DEPENDENT HARMONIC OSCILLATOR

M. Robnik* and V.G. Romanovski
*CAMTP – Center for Applied Mathematics and Theoretical Physics, University of Maribor, Krekova 2, SI-2000 Maribor, Slovenia; robnik@uni-mb.si*

**Abstract.** The theory of adiabatic invariants has a long history, and very important implications and applications in many different branches of physics, classically and quantally, but is rarely founded on rigorous results. Here we treat the general time-dependent one-dimensional linear (harmonic) oscillator, whose Newton equation $\ddot{q} + \omega^2(t)q = 0$ cannot be solved in general. We follow the time-evolution of an initial ensemble of phase points with sharply defined energy $E_0$ at time $t = 0$ (microcanonical ensemble) and calculate rigorously the distribution of energy $E_1$ after time $t = T$, which is fully (all moments, including the variance $\mu^2$) determined by the first moment $\bar{E}_1$. For example, $\mu^2 = E_0^2[(\bar{E}_1/E_0)^2 - (\omega(T)/\omega(0))^2]/2$, and all higher even moments are powers of $\mu^2$, whilst the odd ones vanish identically. This distribution function does not depend on any further details of the function $\omega(t)$ and is in this sense universal, it is a normalized distribution function given by $P(x) = \pi^{-1}(2\mu^2 - x^2)^{-\frac{1}{2}}$, where $x = E_1 - \bar{E}_1$. $\bar{E}_1$ and $\mu^2$ can be calculated exactly in some cases. In ideal adiabaticity $\bar{E}_1 = \omega(T)E_0/\omega(0)$, and the variance $\mu^2$ is zero, whilst for finite $T$ we calculate $\bar{E}_1$, and $\mu^2$ for the general case using exact WKB-theory to all orders. We prove that if $\omega(t)$ is of class $C^m$ (all derivatives up to and including the order $m$ are continuous) $\mu^2 \propto T^{-2(m+1)}$, whilst for class $C^\infty$ it is known to be exponential $\mu^2 \propto \exp(-\alpha T)$. Due to the positivity of $\mu^2$ we also see that *the adiabatic invariant $I = \bar{E}_1/\omega(T)$ at the average energy $\bar{E}_1$ never decreases.*

**Key words:** Nonlinear dynamics; Nonautonomous Hamiltonian systems; Adiabatic invariants; Energy evolution; Statistical mechanics; Microcanonical ensemble

## 1. Introduction

This paper considers the adiabatic invariants and the energy evolution as well as its statistical properties in the time-dependent linear oscillator, with the oscillation frequency being an *arbitrary* function of time, and also allowing for arbitrary external forcing. It comprises a series of four papers so far (Robnik and Romanovski 2006a, b; Robnik et al. 2006; Kuzmin and Robnik 2007), although not much will be presented about external forcing analyzed in Kuzmin and Robnik (2007). We omit many details due to lack of space, but they can be found in our review paper (Robnik and Romanovski 2008).

G. Casati and D. Matrasulov (eds.), *Complex Phenomena in Nanoscale Systems*,
NATO Science for Peace and Security Series B: Physics and Biophysics,

© Springer Science+Business Media B.V. 2009

In time-independent (autonomous) Hamiltonian systems the total energy of the system is conserved by construction, i.e. due to the Hamilton equations of motion, and the Liouville theorem applies, because the phase flow velocity vector field has vanishing divergence. In time-dependent (nonautonomous) Hamilton systems the total energy is not conserved, whilst the Liouville theorem still applies (the phase space volume is preserved). If (some parameter of) the Hamilton function varies in time, the energy of the system generally also changes. But, if the changing of the parameter is very slow, on the typical time scale $T$, there might be a quantity $I$, a function of the said parameter, of the energy $E$ and of other dynamical quantities, which is approximately conserved. It might be even exactly conserved if $T \to \infty$, i.e. if the variation is infinitely slow, to which case we refer as the ideal adiabatic variation. Such a conserved quantity is called *adiabatic invariant*, and it plays an important role in the dynamical analysis of a long-time evolution of nonautonomous Hamilton systems.

The theory of adiabatic invariants is aimed at finding the adiabatic invariants $I$ and analyzing the error of its preservation at finite $T$. Namely, the statement of exactness in preservation of $I$ is asymptotic in the sense that the conservation is exact in the limit $T \to \infty$, whilst for finite $T$ we see the deviation $\Delta I = I_1 - I_0$ of final value of $I_1$ from its initial value $I_0$ and would like to calculate $\Delta I$. Thus for finite $T$ the final values of $I_1$, like the final values of the energy $E_1$, will have some distribution with nonvanishing variance. Indeed, for one-dimensional harmonic oscillator it is known since Lorentz and Einstein (Einstein 1911) that the adiabatic invariant for $T = \infty$ is $I = E/\omega$, which is the ratio of the total energy $E = E(t)$ and the frequency of the oscillator $\omega(t)$, both being a function of time. Of course, $2\pi I$ is exactly the area in the phase plane $(q, p)$ enclosed by the energy contour of constant $E$, i.e. $I$ is the classical action (variable). A general introductory account of the theory of adiabatic invariants can be found in Robnik (2005) and references therein, especially Landau and Lifshitz (1996) and Reinhardt (1994).

However, in the literature this $I$ and $\Delta I$ are not even precisely defined. It turns out that $I$ must be generally considered as a function of the initial conditions, and then it turns out that it is conserved for some initial conditions but not for some others. As a consequence of that there is a considerable confusion about its meaning. Let us just mention the case of periodic parametric resonance (see the relevant section below), with otherwise arbitrarily slowly changing $\omega(t)$, in one-dimensional harmonic oscillator, where the total energy of the system can grow indefinitely for certain (almost all) initial conditions, and since $\omega(t)$ is bounded, $I = E(t)/\omega(t)$ simply cannot be conserved for the said initial conditions for all times, but only for sufficiently small times. In this paper we consider $I$ as a function of the initial conditions, and give a precise meaning to these and similar statements.

Therefore to be on rigorous side we must carefully define what we mean by $I$ and $\Delta I$. This can be done by considering an ensemble of initial conditions at time $t = 0$ just before the adiabatic process starts. Of course, there is a vast freedom in choosing such ensembles. Let us consider the one degree of freedom systems, which is the topic of this paper. If the initial conditions are on a closed contour $\mathcal{K}_0$ in the phase space at time $t = 0$, then at the end of the adiabatic process at time $t = T$ they are also on a closed contour $\mathcal{K}_T$ which in general is different from $\mathcal{K}_0$, but due to the Liouville theorem the area inside the contour is constant for all $T$. If $\mathcal{K}_0$ is a contour of constant energy at time $t = 0$, then $\mathcal{K}_T$ generally is not a contour of constant energy of the Hamiltonian at time $T$. The final energies of the system, depending on the initial conditions, are spread and thus distributed between some minimal and maximal energy, $E_{min}$ and $E_{max}$, respectively. In the said case of periodic $\omega(t)$ and parametric resonance in a harmonic oscillator the contour $\mathcal{K}_T$ is squeezed in one direction and expanded in the other one (the transformation is a linear map), and this contraction and expansion is exponential in time. Therefore, $I = E(T)/\omega(T)$ can not be conserved.

Thus we must always study $I$ as a function of initial conditions, by looking at the ensembles of initial conditions. There is a considerable multitude of possibilities in choosing such ensembles, but usually we do not know very much about the system except e.g. just the energy.

Therefore in an integrable conservative Hamiltonian system the most natural and the most important case (choice) is taking as the initial ensemble all phase points uniformly distributed on the initial $N$-torus, i.e. uniform w.r.t. the angle variables. We call it *uniform canonical ensemble of initial conditions*. Such an ensemble has a sharply defined initial energy $E_0$. Then we let the system evolve in time, not necessarily slowly, and calculate the probability distribution $P(E_1)$ of the final energy $E_1$, or of other dynamical quantities. Typically $E_1$ is distributed on an interval $(E_{min}, E_{max})$, and $P(E_1)$ can be universal there, as it does not depend on any further properties of $\omega(t)$ except for $\bar{E}_1$, like in one-dimensional linear oscillator. More general ensembles of initial energies $w(E_0)$ can be described in terms of the uniform canonical ensembles, as explained in the last section.

To describe $P(E_1)$ is in general a difficult problem, but in this work we confine ourselves to the one-dimensional general time-dependent harmonic oscillator, so $N = 1$. We do not consider the external forcing any further, but the details can be found in Kuzmin and Robnik (2007). The system is described by the Newton equation

$$\ddot{q} + \omega^2(t)q = 0 \tag{1}$$

and we work out rigorously $P(E_1)$. Given the general dependence of the oscillator's frequency $\omega(t)$ on time $t$ the calculation of $q(t)$ is already a very difficult, in fact unsolvable, problem.[1] Nevertheless, we can calculate $P(E_1)$ and it turns out to be surprisingly simple and universal (independent of $\omega(t)$), namely we find the so-called arc sine density (or distribution) (Feller 1971)

$$P(x) = \frac{1}{\pi\sqrt{2\mu^2 - x^2}}, \tag{2}$$

where $x = E_1 - \bar{E}_1$.

In performing our analysis, we shall answer the questions as to when is $I = E(T)/\omega(T)$ conserved, and if it is not conserved, what is the spread or variance $\mu^2$ of the energy, and the higher moments etc. Then (the not sharply defined) $\Delta I$ in the literature is proportional to $\mu$, namely $\Delta I \approx \mu/\omega$. After performing the exact analysis, we provide a powerful technique based on the WKB method (Robnik and Romanovski 2000) to calculate $\mu^2$, and show that it gives exact leading asymptotic terms when $T \to \infty$, and moreover, generally we can do the expansion to all orders, exactly. We treat several exactly solvable cases, and compare them with the WKB results, and finally prove the theorem as for how $\mu^2$ behaves when $\omega(t)$ is of class $C^m$, which means having $m$ continuous derivatives.

We give a brief historical review of contributions to this field. After Einstein (1911), Kulsrud (1957) was the first to show, using a WKB-type method, that for a finite $T$, $I$ is preserved to all orders, for harmonic oscillator, if all derivatives of $\omega$ vanish at the beginning and at the end of the time interval, whilst in case of a discontinuity in one of the derivatives he estimated $\Delta I$ but did not give our explicit general expressions (42) and (43). Hertweck and Schluter (1957) did the same thing independently for a charged particle in slowly varying magnetic field for infinite time domain. Kruskal, as reported in Gardner (1959) and Lenard (1959) studied more general systems, whilst Gardner (1959) used the classical Hamiltonian perturbation theory. Courant and Snyder (1958) have studied the stability of synchrotron and analyzed $I$ employing the transition matrix. The interest then shifted to the infinite time domain. Littlewood (1963) showed for the harmonic oscillator that if $\omega(t)$ is an analytic function, $I$ is preserved to all orders of the *adiabatic parameter* $\epsilon = 1/T$. Kruskal (1962) developed the asymptotic theory of Hamiltonian and other systems with all solutions nearly periodic. Lewis (1968), using the

[1] In the sense of mathematical physics (1) is exactly equivalent to the one-dimensional stationary Schrödinger equation: the coordinate $q$ appears instead of the probability amplitude $\psi$, time $t$ appears instead of the coordinate $x$ and $\omega^2(t)$ plays the role of $\mathcal{E} - V(x)$ = energy − potential. If $\mathcal{E}$ is greater than any local maximum of $V(x)$ then the scattering problem is equivalent to our 1D harmonic oscillator problem.

Kruskal's method, discovered a connection between $I$ of the 1-dim harmonic oscillator and another nonlinear differential equation. Later on Symon (Lewis 1970) used the Lewis'es results to calculate the (canonical) ensemble average of the $I$ and its variance, which is the analogue of our $\bar{E}_1$ and $\mu^2$. Finally, Knorr and Pfirsch (1966) proved $\Delta I \propto \exp(-const/\epsilon)$. Meyer (1973a) relaxed some conditions and calculated the constant $const$. Exponential preservation of $I$ for an analytic $\omega$ on $(-\infty, +\infty)$ with constant limits at $t \to \pm\infty$, is thus well established (Landau and Lifshitz 1996).

Due to the limited space we cannot present all the details of our derivations and calculations, some of which are quite extensive, but most of them can be found in Robnik and Romanovski (2008).

## 2. Transition Map and Exact Results

We begin our discussion by defining the system by giving its Hamilton function $H = H(q, p, t)$, whose numerical value $E(t)$ at time $t$ is precisely the total energy of the system at time $t$, and for the general time-dependent one-dimensional harmonic oscillator this is

$$H = \frac{p^2}{2M} + \frac{1}{2}M\omega^2(t)q^2, \tag{3}$$

where $q, p, M, \omega$ are the coordinate, the momentum, the mass and the frequency of the linear oscillator, respectively. The dynamics is linear in $q, p$, as described by (1), but nonlinear as a function of $\omega(t)$ and therefore is subject to the nonlinear dynamical analysis. By using the index 0 and 1 we denote the initial ($t = t_0$) and final ($t = t_1$) value of the variables, and by $T = t_1 - t_0$ we denote the length of the time interval of changing the parameters of the system.

We consider the phase flow map (we shall call it transition map)

$$\Phi : \begin{pmatrix} q_0 \\ p_0 \end{pmatrix} \mapsto \begin{pmatrix} q_1 \\ p_1 \end{pmatrix}. \tag{4}$$

Because equations of motion are linear in $q$ and $p$, and since the system is Hamiltonian, $\Phi$ is a linear area preserving map, that is,

$$\Phi = \begin{pmatrix} a & b \\ c & d \end{pmatrix}, \tag{5}$$

with $\det(\Phi) = ad - bc = 1$. Let $E_0 = H(q_0, p_0, t = t_0)$ be the initial energy and $E_1 = H(q_1, p_1, t = t_1)$ be the final energy, that is,

$$E_1 = \frac{1}{2}\left(\frac{(cq_0 + dp_0)^2}{M} + M\omega_1^2(aq_0 + bp_0)^2\right). \tag{6}$$

Introducing the new coordinates, namely the action $I = E/\omega$ and the angle $\phi$,

$$q_0 = \sqrt{\frac{2E_0}{M\omega_0^2}} \cos\phi, \; p_0 = \sqrt{2ME_0} \sin\phi \tag{7}$$

from (6) we obtain

$$E_1 = E_0(\alpha \cos^2\phi + \beta \sin^2\phi + \gamma \sin 2\phi), \tag{8}$$

where

$$\alpha = \frac{c^2}{M^2\omega_0^2} + a^2\frac{\omega_1^2}{\omega_0^2}, \quad \beta = d^2 + \omega_1^2 M^2 b^2, \quad \gamma = \frac{cd}{M\omega_0} + abM\frac{\omega_1^2}{\omega_0}. \tag{9}$$

Given the uniform probability distribution of initial angles $\phi$ equal to $1/(2\pi)$, which defines our initial *uniform canonical ensemble* (*microcanonical ensemble*) at time $t = t_0$, we can now calculate the averages. Thus

$$\bar{E}_1 = \frac{1}{2\pi}\oint E_1 d\phi = \frac{E_0}{2}(\alpha + \beta). \tag{10}$$

That yields $E_1 - \bar{E}_1 = E_0(\delta \cos 2\phi + \gamma \sin 2\phi)$ and

$$\mu^2 = \overline{(E_1 - \bar{E}_1)^2} = \frac{E_0^2}{2}\left(\delta^2 + \gamma^2\right), \tag{11}$$

where we have denoted $\delta = (\alpha - \beta)/2$. It follows from (9), (10) that we can write (11) also in the form

$$\mu^2 = \overline{(E_1 - \bar{E}_1)^2} = \frac{E_0^2}{2}\left[\left(\frac{\bar{E}_1}{E_0}\right)^2 - \left(\frac{\omega_1}{\omega_0}\right)^2\right]. \tag{12}$$

As we shall see, in an ideal adiabatic process $\mu = 0$, and therefore $E_1 = \bar{E}_1 = \omega_1 E_0/\omega_0$, and consequently $P(E_1)$ is a delta function,

$$P(E_1) = \delta(E_1 - \omega_1 E_0/\omega_0). \tag{13}$$

Now we calculate higher moments of $P(E_1)$. We can show that in general for arbitrary positive integer $m$

$$\overline{(E_1 - \bar{E}_1)^{2m-1}} = 0 \tag{14}$$

and

$$\overline{(E_1 - \bar{E}_1)^{2m}} = \frac{(2m-1)!!}{m!}\left(\overline{(E_1 - \bar{E}_1)^2}\right)^m. \tag{15}$$

Knowing all the moments of our distribution function (2) we can derive it by calculating first its characteristic function (which is nothing but the Fourier transform of $P(E_1)$, and the coefficients in its Taylor expansion are just the moments times powers of the imaginary unit), and then take its inverse Fourier transform (Gradshteyn and Ryzhik 1994), resulting in (2).

Another way is the direct algebraic derivation of the energy distribution function (2). By definition we have (see the expression before (11))

$$P(E_1) = \frac{1}{2\pi}\sum_{i=1}^{4}\left|\frac{d\phi}{dE_1}\right|_{\phi=\phi_j(E_1)}, \tag{16}$$

where we have to sum up contributions from all four branches of the function $\phi(E_1)$. Let us denote $x = E_1 - \bar{E}_1$, so that we have

$$x = E_0\,(\delta\cos(2\phi) + \gamma\sin(2\phi)) = \mu\sqrt{2}\sin(2\phi + \psi), \tag{17}$$

where $\delta = (\alpha - \beta)/2$, and $\alpha$, $\beta$ and $\gamma$ are expressed in terms of $a, b, c, d$ as shown in (9), the variance is as in (11), namely $\mu^2 = \frac{E_0^2}{2}(\delta^2 + \gamma^2)$, and $\tan\psi = \delta/\gamma$, so that $\phi = \frac{1}{2}\arcsin\frac{x}{\mu\sqrt{2}} - \frac{\psi}{2}$. Therefore $|\frac{d\phi_i}{dx}| = \frac{1}{2\sqrt{2\mu^2 - x^2}}$ for all four solutions $i = 1, 2, 3, 4$ and from (16) we get (2) at once.

## 3. Some Exactly Solvable Cases

Our general study now focuses on the calculation of the transition map (5), namely its matrix elements $a, b, c, d$. Starting from the Hamilton function (3) and its Newton equation (1) we consider two linearly independent solutions $\psi_1(t)$ and $\psi_2(t)$ and introduce the matrix

$$\Psi(t) = \begin{pmatrix} \psi_1(t) & \psi_2(t) \\ M\dot{\psi}_1(t) & M\dot{\psi}_2(t) \end{pmatrix}. \tag{18}$$

Consider a solution $\hat{q}(t)$ of (1) such that

$$\hat{q}(t_0) = q_0, \quad \dot{\hat{q}}(t_0) = p_0/M. \tag{19}$$

Because $\psi_1$ and $\psi_2$ are linearly independent, we can look for $\hat{q}(t)$ in the form

$$\hat{q}(t) = A\psi_1(t) + B\psi_2(t). \tag{20}$$

Then $A$ and $B$ are determined by

$$\begin{pmatrix} A \\ B \end{pmatrix} = \Psi^{-1}(t_0)\begin{pmatrix} q_0 \\ p_0 \end{pmatrix}. \tag{21}$$

Let $q_1 = \hat{q}(t_1)$, $p_1 = M\dot{\hat{q}}(t_1)$. Then from (19)–(21) we see that

$$\begin{pmatrix} q_1 \\ p_1 \end{pmatrix} = \Psi(t_1)\Psi^{-1}(t_0)\begin{pmatrix} q_0 \\ p_0 \end{pmatrix}. \tag{22}$$

We recognize the matrix on the right-hand side of (22) as the transition map $\Phi$, that is,

$$\Phi = \begin{pmatrix} a & b \\ c & d \end{pmatrix} = \Psi(t_1)\Psi^{-1}(t_0). \tag{23}$$

The theory of adiabatic invariants is rarely founded on rigorous results, therefore the study of exactly solvable cases is of fundamental importance, namely we can use them to test various analytic approximations and also the accuracy of numerical calculations. Because the ideal adiabatic processes are infinitely slow and refer to the limit $T \to \infty$ we must deal necessarily with the asymptotic behaviour of dynamical systems, which is difficult to approximate, because all simple minded perturbational and other approximation techniques fail (break down) after some finite time, and so typically give wrong predictions for the asymptotic behaviour in the limit $T \to \infty$. The same difficulty occurs in numerical calculations. We shall see that the WKB methods (Robnik and Romanovski 2000) can be successfully applied. Here we deal only with the linear model, in which $\omega^2(t)$ is a linear function of time,

$$\omega^2(t) = \begin{cases} \omega_0^2 & \text{if } t \le 0 \\ \omega_0^2 + \frac{(\omega_1^2 - \omega_0^2)}{T}t & \text{if } 0 < t < T \\ \omega_1^2 & \text{if } t \ge T \end{cases} \cdot \tag{24}$$

Thus $\omega(t)$ has discontinuous first derivative at $t = 0$ and $t = T$, and belongs to the class $C^0$. Introducing the notation $\tilde{a} = \omega_0^2$, $\tilde{b} = \omega_1^2 - \omega_0^2$ we obtain that on the interval $(0, T)$ (1) has the form

$$\ddot{q} + \left(\tilde{a} + \frac{\tilde{b}t}{T}\right)q = 0. \tag{25}$$

Two linear independent solutions of (25) are given by the Airy functions:

$$\psi_1(t) = Ai\left(\frac{\tilde{b}t + \tilde{a}T}{\tilde{b}^{2/3}T^{1/3}}\right) \tag{26}$$

and

$$\psi_2(t) = Bi\left(\frac{\tilde{b}t + \tilde{a}T}{\tilde{b}^{2/3}T^{1/3}}\right). \tag{27}$$

The elements $a, b, c, d$ of the matrix $\Phi_T$ are defined by (22) and (23).

The exact analytic expression for

$$\mu^2 = \overline{(E_1 - \bar{E}_1)^2} = \frac{E_0}{2}\left(\delta^2 + \gamma^2\right) \tag{28}$$

is very complex, and we do not show it here. However, for $\omega_0^2 = 1, \omega_1^2 = 2$, $E_0 = 1$, using the asymptotic expansion 10.4.60,62,64,66 of Abramowitz and Stegun (1972) (pp. 448–449), we obtain the following approximation

$$\overline{(E_1 - \bar{E}_1)^2} \approx \frac{\epsilon^2}{128}\left(9 - 4\sqrt{2}\cos\left(\frac{4 - 8\sqrt{2}}{3\epsilon}\right)\right), \tag{29}$$

where we introduce *the adiabatic parameter* $\epsilon$,

$$\epsilon = \frac{1}{T}. \tag{30}$$

It can be seen that the exact expression (28) and its leading asymptotic approximation (29) practically coincide for $\epsilon \leq 0.05$, which demonstrates the power of the asymptotic expansion of the relevant expressions containing the Airy functions. Observe that the decay of $\mu^2$ to zero as $\epsilon \to 0$ is oscillatory but quadratic on the average, namely as $y = \frac{9}{128}\epsilon^2$. This is always the case when $\omega(t)$ is of class $C^0$. As we will see, in general, if $\omega(t)$ is of class $C^m$, then $\mu^2$ goes to zero oscillatory but in the mean as a power $\epsilon^{2(m+1)}$. This theorem will be proven in the next section using the exact formulation of the WKB method, applied to the relevant (but arbitrarily high) order. We shall see that the leading WKB term precisely reproduces the exact leading term in (29).

## 4. The Application of the WKB Theory

We proceed with the calculation of the transition map $\Phi$ in the general case, and because (1) is generally not solvable, we have ultimately to resort to some approximations. Since the adiabatic limit $\epsilon \to 0$ is the asymptotic regime that we would like to understand, the application of the rigorous WKB theory (up to all orders) is most convenient, and usually it turns out that the leading asymptotic terms are well described by just the leading WKB terms if $\epsilon$ is sufficiently small. In using the WKB method we refer to our work (Robnik and Romanovski 2000), where we have derived the explicit analytic expressions for all WKB orders in closed form, except for the exact rational coefficients, which can be easily obtained from a recurrence formula.[2] The

[2] There is substantial literature on WKB method, which due to limited space cannot be reviewed here. But we should mention the classic works by N. Fröman and P.O. Fröman, who have found a number of interesting relationships, e.g. a relation between the even and odd order terms (Fröman 1966), although we do not use it here, so that our exposition is selfcontained.

WKB method, when worked out to all orders, and when the series is convergent, has the potential to be exact and rigorous provided that the underlying series converges.

We introduce re-scaled and dimensionless time $\lambda$

$$\lambda = \epsilon t, \qquad \epsilon = 1/T, \tag{31}$$

so that (1) is transformed to the equation

$$\epsilon^2 q''(\lambda) + \omega^2(\lambda) q(\lambda) = 0. \tag{32}$$

By prime we denote the differentiation w.r.t. $\lambda$. Let $q_+(\lambda)$ and $q_-(\lambda)$ be two linearly independent solutions of (32). Then the matrix (18) takes the form

$$\Psi_\lambda = \begin{pmatrix} q_+(\lambda) & q_-(\lambda) \\ \epsilon M q'_+(\lambda) & \epsilon M q'_-(\lambda) \end{pmatrix} \tag{33}$$

and taking into account that $\lambda_0 = \epsilon t_0$, and $\lambda_1 = \epsilon t_1$, we obtain for the matrix (5) the expression

$$\Phi = \begin{pmatrix} a & b \\ c & d \end{pmatrix} = \Psi_\lambda(\lambda_1)\Psi_\lambda^{-1}(\lambda_0). \tag{34}$$

We now use the WKB method in order to obtain the coefficients $a, b, c, d$ of the matrix $\Phi$. To do so, we look for solution of (32) in the form

$$q(\lambda) = w \exp\left\{\frac{1}{\epsilon}\sigma(\lambda)\right\} \tag{35}$$

where $\sigma(\lambda)$ is a complex function that satisfies the differential equation

$$(\sigma'(\lambda))^2 + \epsilon\sigma''(\lambda) = -\omega^2(\lambda) \tag{36}$$

and $w$ is some constant with dimension of length. The WKB expansion for the phase is

$$\sigma(\lambda) = \sum_{k=0}^{\infty} \epsilon^k \sigma_k(\lambda). \tag{37}$$

Substituting (37) into (36) and comparing like powers of $\epsilon$ gives the recursion relation

$$\sigma_0'^2 = -\omega^2(\lambda), \qquad \sigma'_n = -\frac{1}{2\sigma'_0}\left(\sum_{k=1}^{n-1} \sigma'_k \sigma'_{n-k} + \sigma''_{n-1}\right). \tag{38}$$

Here we apply our WKB notation and formalism from our work (Robnik and Romanovski 2000) and we can choose $\sigma'_{0,+}(\lambda) = \mathrm{i}\omega(\lambda)$ or $\sigma'_{0,-}(\lambda) =$

$-\mathrm{i}\omega(\lambda)$. That results in two linearly independent solutions of (32) given by the WKB expansions with the coefficients

$$\sigma_{0,\pm}(\lambda) = \pm\mathrm{i}\int_{\lambda_0}^{\lambda} \omega(x)dx, \quad \sigma_{1,\pm}(\lambda) = -\frac{1}{2}\log\frac{\omega(\lambda)}{\omega(\lambda_0)}, \tag{39}$$

$$\sigma_{2,\pm} = \pm\frac{\mathrm{i}}{8}\int_{\lambda_0}^{\lambda} \frac{3\,\omega'(x)^2 - 2\,\omega(x)\,\omega''(x)}{\omega(x)^3}\,dx, \;\ldots \tag{40}$$

and so on, reminding that all coefficents are known as explicit formulae in terms of $\omega(t)$, a fundamental result obtained in Robnik and Romanovski (2000). Since $\omega(\lambda)$ is a real function we deduce from (38) that all functions $\sigma'_{2k+1}$ are real and all functions $\sigma'_{2k}$ are pure imaginary and $\sigma'_{2k,+} = -\sigma'_{2k,-}$ and $\sigma'_{2k+1,+} = \sigma'_{2k+1,-}$ where $k = 0, 1, 2, \ldots$, and thus we have $\sigma'_+ = A(\lambda) + \mathrm{i}B(\lambda)$, $\sigma'_- = A(\lambda) - \mathrm{i}B(\lambda)$ where $A(\lambda) = \sum_{k=0}^{\infty}\epsilon^{2k+1}\sigma'_{2k+1}(\lambda)$, $B(\lambda) = -\mathrm{i}\sum_{k=0}^{\infty}\epsilon^{2k}\sigma'_{2k,+}(\lambda)$ are both real quantities. Integration of the above equations yields

$$\sigma_+ = r(\lambda) + \mathrm{i}s(\lambda), \quad \sigma_- = r(\lambda) - \mathrm{i}s(\lambda), \tag{41}$$

where $r(\lambda) = \int_{\lambda_0}^{\lambda} A(x)\,dx$, $s(\lambda) = \int_{\lambda_0}^{\lambda} B(x)\,dx$. Below we shall denote $s_1 = s(\lambda_1)$. To simplify the expressions let us denote $A_0 = A(\lambda_0)$, $A_1 = A(\lambda_1)$, $B_0 = B(\lambda_0)$ and $B_1 = B(\lambda_1)$.

Using this notation we find (after a very long calculation) that the elements of the matrix $\Phi_\lambda$ are

$$\begin{aligned}
a &= -\tfrac{1}{\sqrt{B_0B_1}}\left[A_0\sin\left(\tfrac{s_1}{\epsilon}\right) - B_0\cos\left(\tfrac{s_1}{\epsilon}\right)\right], \\
b &= \tfrac{1}{M\sqrt{B_0B_1}}\sin\left(\tfrac{s_1}{\epsilon}\right), \\
c &= -\tfrac{M}{\sqrt{B_0B_1}}\left[(A_0A_1 + B_0B_1)\sin\left(\tfrac{s_1}{\epsilon}\right) + (A_0B_1 - A_1B_0)\cos\left(\tfrac{s_1}{\epsilon}\right)\right], \\
d &= \tfrac{1}{\sqrt{B_0B_1}}\left[A_1\sin\left(\tfrac{s_1}{\epsilon}\right) + B_1\cos\left(\tfrac{s_1}{\epsilon}\right)\right].
\end{aligned} \tag{42}$$

Thus we can obtain the final result for expression (10), exact to all orders,

$$\begin{aligned}
\alpha + \beta = \frac{1}{B_0B_1}\Bigg[&\sin^2\left(\frac{s_1}{\epsilon}\right)\left(\frac{B_0^2B_1^2}{\omega_0^2} + \omega_1^2\right) + \cos^2\left(\frac{s_1}{\epsilon}\right)\left(B_0^2\frac{\omega_1^2}{\omega_0^2} + B_1^2\right) + \\
&\sin^2\left(\frac{s_1}{\epsilon}\right)\left(A_0^2\frac{\omega_1^2}{\omega_0^2} + \frac{A_0^2A_1^2}{\omega_0^2} + \frac{2A_0A_1B_0B_1}{\omega_0^2} + A_1^2\right) + \\
&\cos^2\left(\frac{s_1}{\epsilon}\right)\left(\frac{A_0^2B_1^2}{\omega_0^2} + \frac{A_1^2B_0^2}{\omega_0^2} - \frac{2A_0A_1B_0B_1}{\omega_0^2}\right) + \\
&\sin\left(\frac{s_1}{\epsilon}\right)\cos\left(\frac{s_1}{\epsilon}\right)\times
\end{aligned} \tag{43}$$

$$\left(-2A_0B_0\frac{\omega_1^2}{\omega_0^2}+2A_1B_1+\frac{2}{\omega_0^2}(A_0A_1+B_0B_1)(A_0B_1-A_1B_0)\right)\Bigg],$$

reminding that $\bar{E}_1 = (E_0/2)(\alpha+\beta)$.

So far the result is exact. Let us consider the first order WKB approximation, which is the generic case, that is

$$A(\lambda) \approx \epsilon\sigma'_{1,+}(\lambda), \quad B(\lambda) \approx \frac{\sigma'_{0,+}(\lambda)}{\mathrm{i}} = \omega(\lambda). \tag{44}$$

Substituting these values of $A(\lambda)$ and $B(\lambda)$ into (43) we find

$$\alpha+\beta = 2\frac{\omega_1}{\omega_0} + \epsilon^2\left(\frac{\omega_1\omega'(\lambda_0)^2}{4\omega_0^5} - \frac{\cos\left(\frac{2\int_{\lambda_0}^{\lambda_1}\omega(x)\,dx}{\epsilon}\right)\omega'(\lambda_0)\omega'(\lambda_1)}{2\omega_0^3\omega_1} + \frac{\omega'(\lambda_1)^2}{4\omega_0\omega_1^3}\right) + O(\epsilon^3). \tag{45}$$

As shown in (10)–(12), $\bar{E}_1 = \frac{\alpha+\beta}{2}E_0$ and $\frac{\overline{(\Delta E_1)^2}}{E_0^2} = \frac{1}{2}\left[\left(\frac{\bar{E}_1}{E_0}\right)^2 - \left(\frac{\omega_1}{\omega_0}\right)^2\right]$. Therefore

$$\frac{\overline{(\Delta E_1)^2}}{E_0^2} = \epsilon^2\left(\frac{\omega_1^2\omega'(\lambda_0)^2}{8\omega_0^6} - \frac{\cos\left(\frac{2\int_{\lambda_0}^{\lambda_1}\omega(x)\,dx}{\epsilon}\right)\omega'(\lambda_0)\omega'(\lambda_1)}{4\omega_0^4} + \frac{\omega'(\lambda_1)^2}{8\omega_0^2\omega_1^2}\right) + O(\epsilon^3). \tag{46}$$

Substituting into the last formula $\omega(\lambda) = \sqrt{1+\lambda}$ we obtain exactly the approximation (29). Thus the WKB approach yields exactly the leading asymptotic term for $\bar{E}_1$ and $\mu^2$ for general $\omega(t)$, and not only that, but also all the higher power terms of $\epsilon$ if desired. In Robnik and Romanovski (2008) we explain how all the higher order WKB terms can be calculated using our closed form formula (Robnik and Romanovski 2000). Please note that the expression in the big brackets of (46) is positive definite (its minimal value is a complete square of a real quantity), as it must be.

This result can be generalized to the case that all derivatives of $\omega(t)$ are zero at $t_0$ and $t_1$ up to and including order $(n-1)$. In such case $\mu^2$ goes to zero with $\epsilon$ oscillating like in (46), but in the mean as a power of $\epsilon^{2n}$. From this we conclude (*theorem*) that if the function $\omega(t)$ is of class $C^m$ (all derivatives up to and including order $m$ are continuous), then $\mu^2$ goes to zero with $\epsilon$ oscillating but in the mean as a power of $\mu^2 \propto \epsilon^{2(m+1)}$. The details can be found in Robnik and Romanovski (2008).

## 5. Some Other Results and Considerations

If $\omega(t)$ is a periodic function of time with period $\tau$ some further general conclusions can be made. First, after an integer multiple of $\tau$ the mean energy $\bar{E}_1$ never decreases. The transition map $\Phi_n$ over an integer number $n$ of periods $\tau$ is simply a power of $\Phi_1$, namely $\Phi_n = (\Phi_1)^n$. Then, since the determinant of $\Phi_n$ is unity for any $n$ (Liouville theorem; area preserving map) everything depends on the trace $S = Tr(\Phi_1)$. If $|S|$ is greater than 2, the eigenvalues of $\Phi_1$ are real reciprocals and the energy of the system grows exponentialy with $n$ for almost all initial conditions, except for those which are exactly on the exponentially contracting manifold. Since $\omega(t)$ is bounded, the adiabatic invariant $I = E(t)/\omega(t)$ is *not* conserved, but we can describe the system, which in fact in such case exhibits parametric resonance, and both $\bar{E}_1$ and $\mu^2$ increase exponentially with time. If $|S|$ is less than 2 then the eigenvalues are complex conjugates on the unit circle and we have just the oscillation of $\bar{E}_1$ and $\mu^2$.

One can consider also other than microcanonical ensembles of initial conditions. For example, if we have a shell of microcanonical ensembles with the distribution $w(E_0)$ of initial energies, then the final energy can be written as

$$P(E_1) = \int G(E_1; E_0)w(E_0)dE_0. \tag{47}$$

where $G(E_1; E_0) = P(E_1)$ is a kind of "Green function" for the energy evolution of the system with the intitial sharp energy $E_0$ of the underlying microcanonical ensemble. Even more general aspects of composition and of generalization of ensembles of initial conditions can be obtained. More details are given in Robnik and Romanovski (2008).

Finally, we mention that one can generalize the results by including the arbitrary external forcing, derived in Kuzmin and Robnik (2007), which cannot be presented here due to lack of space.

## 6. Discussion and Conclusions

The purpose of this paper is to review the ideas, methods and results of the series of our papers (Robnik and Romanovski 2006a, b; Robnik et al. 2006) on the dynamics of a time-dependent linear oscillator, where the oscillation frequency $\omega(t)$ is an arbitrary function of time. External forcing can be analyzed in a similar way, but could not be covered due to the lack of space, but the interested reader should see the paper Kuzmin and Robnik (2007). Most of the details can be found in Robnik and Romanovski (2008).

We have studied the evolution of the energy of such a system over a time interval of length $T$, and also its statistical properties, in particular the distribution function $P(E_1)$ of the final energies $E_1$ under the assumption that the initial ensemble of initial conditions is *a uniform canonical ensemble of initial conditions* (i.e. uniform w.r.t. the angle variables)[3] at the sharply defined initial energy $E_0$. The crucial point is that for the linear oscillator we know the phase flow or the so-called transition map explicitly in the form of a 2 x 2 matrix. Using this, we can explicitly calculate the distribution function $P(x) = \pi^{-1}(2\mu^2 - x^2)^{-\frac{1}{2}}$, where $x = E_1 - \bar{E}_1$, and $\bar{E}_1$ and $\mu^2$ are the final average energy and variance, respectively. Thus the distribution function is universal, and this result is exact and rigorous. In nonlinear time-dependent oscillators the main and major difficulty is that we do not know the phase flow globally, and consequently we can not perform averages of the final quantities in terms of the initial conditions. Practically all perturbation schemes break down at finite time, so asymptotic results $T \to \infty$ cannot be obtained in an easy way.

We are able to calculate rigorously all the moments of $P(E_1)$. Odd moments are exactly zero, the even moments are powers of the second moment $\mu^2$. The latter one is a function of the first moment. Therefore everything is determined by the first moment $\bar{E}_1$. In order to calculate $\bar{E}_1$ and $\mu^2$ etc. we can use the WKB method as developed in our paper (Robnik and Romanovski 2000), and explained in detail in Robnik and Romanovski (2008).

In our analysis we clearly see when the adiabatic invariant $I(t) = E(t)/\omega(t)$ is conserved or not. In the (ideal) adiabatic limit $T \to \infty$ it is conserved, the variance $\mu^2$ is zero and $E_1 = \bar{E}_1 = \omega_1 E_0/\omega_0$, and $P(E_1)$ is a delta function $P(E_1) = \delta(E_1 - \bar{E}_1)$. If it is not conserved exactly, when $T$ is finite, we find $\mu^2 = E_0^2[(\bar{E}_1/E_0)^2 - (\omega(T)/\omega(0))^2]/2 > 0$, and it can be calculated using a WKB method analytically in a closed form, which is a major achievement of our work. From this follows at once the conclusion that *the adiabatic invariant $I = \bar{E}_1/\omega_1$ at the average energy $\bar{E}_1$ never decreases*, which is a kind of irreversibility statement.

We have also studied in Robnik and Romanovski (2008) three specific solvable models and have demonstrated the power of the WKB expansion, where already the leading WKB term usually very well describes the asymptotic behaviour of $\mu^2$ when $\epsilon = 1/T$ goes to zero. In this paper we have

[3] In case of $N = 1$ this is also *microcanonical ensemble*, in terminology of statistical mechanics, unlike the case of higher dimensional tori $N \geq 2$, which we do not analyze here.

discussed only the linear model. We also discuss what happens if $\omega(t)$ is smooth and of class $C^m$, having $m$ continuous derivatives, calculating and showing that $\mu^2$ oscillates as $\epsilon$ goes to zero, but in the mean vanishes as $\propto \epsilon^{2(m+1)}$. If $\omega(t)$ is analytic, thus it also is of class $C^\infty$, it is known from the literature that $\mu^2$ must decay exponentially $\propto \exp(-const/\epsilon)$. If $\omega(t)$ is periodic, $\bar{E}_1$ can grow exponentially, and so does the variance $\mu^2$, in which case $I(t) = E(t)/\omega(t)$ is not conserved, but we can describe the system.

We have introduced the so-called $G$-function, which is a kind of a Green function for the evolution of the energy and derived a composition formula for it when the interval of evolution is decomposed into a finite number of subintervals, for which the corresponding $G_j$-function is known for all subintervals $j$ and is uniform canonical there. This formula applies also to nonlinear systems and might be a good starting basis to describe them. The theory for nonlinear systems remains open and is a subject of the current research (Robnik and Romanovski 2009). The case of the separatrix crossing has been investigated numerically in Robnik and Wood (2006).

**Acknowledgements**

This work was supported by the Slovenian Research Agency, by the Nova Kreditna Banka Maribor and Telekom Slovenije. We thank Dr. Andrei V. Kuzmin (Minsk) for fruitful collaboration in paper (Kuzmin and Robnik 2007) where we treat the time dependent linear oscillator with arbitrary forcing.

## References

Abramowitz M. and Stegun I.A.: 1972. *Handbook of Mathematical Functions*, New York, Dover

Courant E.D. and Snyder H.S.: 1958. *Ann. Phys.*, **3**, 1

Einstein A.: 1911. *Inst. Intern. Phys. Solway, Rapports et discussions*, **1**, 450

Feller W.: 1971. *An Introduction to Probability Theory and its Applications*, Vol. II, New York, Wiley

Fröman N.: 1966. *Arkiv för Physik*, **32**, 541

Gardner C.S.: 1959. *Phys. Rev.*, **115**, 791

Gradshteyn I.S. and Ryzhik I.M.: 1994. *Table of Integrals, Series and Products*, A. Jeffrey, ed., San Diego, Academic

Hertweck H. and Schluter A.: 1957. *Zeitschrift fur Naturforschung*, **12A**, 844

Knorr G. and Pfirsch D.: 1966. *Zeitschrift fur Naturforschung*, **21**, 688

Kruskal M.: 1962. *J. Math. Phys.*, **3**, 806

Kulsrud R.M.: 1957. *Phys. Rev.*, **106**, 205

Kuzmin A.V. and Robnik M.: 2007. *Rep. Math. Phys.*, **60**, 69

Landau L.D. and Lifshitz E.M.: 1996. *Mechanics: Course of Theoretical Physics*, Oxford, Butterworth-Heineman

Lenard A.: 1959. *Ann. Phys.*, **6**, 261
Lewis H.R.: 1968. *J. Math. Phys.*, **9**, 1976
Lewis H.R.: 1970. *J. Math. Phys.*, **11**, 1320
Littlewood J.E.: 1963. *Ann. Phys.*, **21**, 223
Meyer R.E.: 1973a. *Z. Angew. Math. Phys.*, **24**, 293
Meyer R.E.: 1973b. *Z. Angew. Math. Phys.*, **24**, 517
Reinhardt W.P.: 1994. *Prog. Theor. Phys. Suppl.*, **116**, 179
Robnik M.: 2005. *Encyclopedia of Nonlinear Science*, A. Scott, ed., New York, Routledge, pp. 2–5
Robnik M. and Romanovski V.G.: 2000. *J. Phys. A: Math. Gen.*, **33**, 5093
Robnik M. and Romanovski V.G.: 2006a. *J. Phys. A: Math. Gen.*, **39**, L35
Robnik M. and Romanovski V.G.: 2006b. *Open Syst. Inform. Dyn.*, **13**, 197
Robnik M. and Wood T.A.: 2006. *Nonlin. Phenom. Complex Syst.*, **9**, 141
Robnik M., Romanovski V.G. and Stöckmann H.J.: 2006. *J. Phys. A: Math. Gen.*, **39**, L551
Robnik M. and Romanovski G.V.: 2008. Let's Face Chaos through Nonlinear Dynamics. Melville, New York *AIP Conf.Proc.*, Vol. 1076, 185–212
Robnik M. and Romanovski V.G.: 2009. *in preparation*

# TIME-DEPENDENT QUANTUM BILLIARDS

P. Schmelcher and F. Lenz
*Physikalisches Institut, Universität Heidelberg, Philosophenweg 12, 69120 Heidelberg, Germany; lenz@physi.uni-heidelberg.de*

D. Matrasulov*, Z.A. Sobirov, and S.K. Avazbaev
*Heat Physics Department of the Uzbek Academy of Sciences, 28 Katartal Street, 100135 Tashkent, Uzbekistan; dmatrasu@phys.ualberta.ca, sobirovzar@yahoo.com, sanat_avazbaev@yahoo.com*

**Abstract.** Quantum dynamics of a particle in time-dependent rectangular and elliptical billiards is studied. Exact analytical solution of the Schrödinger equation for monotonically expanding (contracting) rectangular billiard is obtained. It is shown that for rectangular billiard with two harmonically oscillating walls the quantum average energy can be time-periodic for certain values of the oscillation amplitudes. For harmonically oscillating elliptical quantum billiard the time-dependence of the average energy is calculated. It is found that the average energy can be time-periodic in the adiabatic regime. Possibility for application of the developed approach for arbitrary time-dependent (shape-keeping) billiard geometry is shown.

**Key words:** Time-dependent billiards; Fermi acceleration

## 1. Introduction

Nonlinear dynamics of time-dependent billiards has been the subject of extensive research during the past two decades (Koiller et al. 1995; Lenz et al. 2008). On of the main motivations for the studying particle dynamics in billiard geometries with time-dependent boundaries is the fact that Fermi-acceleration is possible in such systems. The dependence of the dynamics on the geometry of the billiard makes the acceleration process in two-dimensions more attractive compared to the one in one dimension. Classical dynamics of time-dependent billiards has been considered for different geometries, both integrable and non-integrable ones. First Koiller et al. (1995) studied billiards with periodically moving boundaries within the Hamiltonian formalism. One of the most simplest time-dependent billiards, the breathing circle was explored from the viewpoint of nonlinear dynamics and Fermi acceleration in Kamphorst and de Carvalho (1999). It was shown that the energy gain in such a system is bounded and no infinite acceleration is possible. A more systematic treatment of time-dependent billiards was done by Loskutov et al. (2000) and Loskutov and Ryabov (2002). In particular, they analyzed chaotic

G. Casati and D. Matrasulov (eds.), *Complex Phenomena in Nanoscale Systems*,
NATO Science for Peace and Security Series B: Physics and Biophysics,

© Springer Science+Business Media B.V. 2009

(stadium-like) billiards with time-dependent boundaries and conjectured that a sufficient condition for the possibility of Fermi acceleration is the nonintegrability of the unperturbed billiard. Another type of nonintegrable billiards, time-dependent annular billiards was studied recently in de Carvalho et al. (2006). By considering concentric and eccentric annular billiards the authors showed that Fermi acceleration is possible in the eccentric case (de Carvalho et al. 2006). However, most recently it was shown that unbounded energy gain is even possible in a time-dependent billiard with integrable static counterpart. Namely, it was found that in time-dependent elliptical billiard Fermi acceleration can be observed and this acceleration is tunable (Lenz et al. 2007, 2008). Despite a certain progress made on the study of classical dynamics of time-dependent billiards, the corresponding quantum problem has been rarely explored. In quantum case the problem of time-dependent billiards corresponds to solving the two-dimensional Schrödinger equation with time-dependent boundary conditions. We note that the one-dimensional Schrödinger equations with time-dependent boundary conditions have been extensively studied during the past four decades.

In particular, an early study of the time-dependent infinite square well dates back to Doescher and Rice (1969). A mathematical more detailed treatment of the one-dimensional Schrödinger equation with time-varying boundary conditions is given by Munier et al. (1981). Later, Pinder has studied the case of a contracting square quantum well (Pinder 1990). Scheininger and Kleber treated the case of periodically time-dependent boundaries by solving the problem in terms of the full-cycle propagator (Scheininger and Kleber 1991). Seba studied the time-periodic boundary conditions in terms of Floquet operators (Seba 1990). Systematic studies of the one-dimensional Schrödinger equation were done by Makowski and Dembinski (1991), Makowski and Peplowski (1992), and Makowski (1992), they classified the cases when the problem can be solved exactly. We mention also Badrinarayanan et al. (1995); Cohen and Wisniacki (2003); Jose and Gordery (1986); Kamphorst et al. (2007); Karner (1989); Koiller et al. (1996); Moralez et al. (1994); Pereshogin and Pronin (1991); Razavy (1991); Willemsen (1994); and Yuce (2004), where different aspects of the one-dimensional Schrödinger equation with time-dependent boundary conditions are treated. Most recently, Glasser et al. (2009) studied the quantum dynamics of a particle confined in a 1D-box with an oscillating wall. We note that in most cases the time-dependent boundary conditions are reduced to static ones by transforming the original Schrödinger equation to a time-dependent harmonic oscillator.

In this work we address the two-dimensional extension of the above mentioned time-dependent 1D box. We consider boundary conditions given on:

1. Time-dependent rectangular billiards
2. Time-dependent elliptical billiards
3. Some more general time-dependent geometries

This work is organized as follows: in the next section we briefly recall the solution of the Schrödinger equation for time-dependent 1D box. Section 3 treats the time-dependent rectangular billiard. In Sect. 4 we present the solution for the time-dependent elliptical billiard. Section 5 presents a short discussion of the extension of the used methods to other geometries. Finally, Sect. 6 presents some concluding remarks.

## 2. Time-Dependent 1D Box

In the case of time-dependent one-dimensional box we have the Schrödinger equation ($\hbar = m = 1$)

$$i\frac{\partial \Psi}{\partial t} = -\frac{1}{2}\frac{\partial^2 \Psi}{\partial x^2}, \tag{1}$$

where the wave function $\Psi = \Psi(x,t)$ satisfies the boundary and initial conditions

$$\Psi|_{x=0} = \Psi|_{x=L(t)} = 0, \quad \Psi|_{t=0} = \Psi_0(x), \tag{2}$$

with $L(t)$ is being the width of the box. Using the following time-dependent coordinate transformation

$$y = \frac{x}{L(t)},$$

and the substitution

$$\Psi(y,t) = \frac{1}{L(t)} \exp\left(\frac{i}{2}\dot{L}(t)L(t)y^2\right)\phi(y,t),$$

Equation (1) can be reduced to the Schrödinger equation for time-dependent harmonic oscillator (Makowski and Dembinski 1991):

$$iL^2\frac{\partial \phi}{\partial t} = -\frac{1}{2}\frac{\partial^2 \phi}{\partial y^2} + \frac{1}{2}L^3\ddot{L}y^2\phi. \tag{3}$$

The boundary and initial conditions for (3) are given by

$$\phi|_{y=0} = \phi|_{y=1} = 0, \quad t \geq 0, \quad \phi|_{t=0} = \phi_0(y),$$

where

$$\phi_0(y) = L(0)\exp\left[-\frac{i}{2}\dot{L}(0)L(0)y^2\right]\Psi_0(yL(0)).$$

Time and coordinate variables in (3) can be separated if $L(t)$ satisfies the following equation:

$$L^3\ddot{L} = const. \tag{4}$$

The solutions of (4) provide us with the motions of the wall for which the original problem can be solved analytically. The general solution of (4) can be written as

$$L^3\ddot{L} = -\frac{D}{4}, \tag{5}$$

$$L(t) = \sqrt{At^2 + Bt + C} \text{ for } D = B^2 - 4AC. \tag{6}$$

Special solutions are:

$$L(t) = At + B \text{ for } D = 0,$$

and

$$L(t) = \sqrt{At + B} \text{ for } D = B^2.$$

With the condition of (5), (3) can be rewritten as

$$i\frac{\partial\phi}{\partial\tau} = -\frac{1}{2}\frac{\partial^2\phi}{\partial y^2} - \frac{D}{8}\phi, \tag{7}$$

where

$$\tau = \int_0^t \frac{ds}{L^2(s)}.$$

For $D > 0$, time and coordinate variables can be separated using the substitution $\phi(\tau, y) = f(\tau)\Phi(y)$ and for $\Phi(y)$ we get the Kummer equation (Abramowitz and Stegun 1964)

$$z\frac{d^2U}{dz^2} + \left(\frac{1}{2} - z\right)\frac{dU}{dz} + \frac{1}{4}(k^2 - 1)U = 0, \tag{8}$$

where $U(z) = M(\alpha,\beta,z)$, $M(\alpha,\beta,z)$ is the Kummer function (Abramowitz and Stegun 1964), with

$$z = \frac{i\sqrt{D}}{2}y^2, \quad k^2 = \frac{4K}{i\sqrt{D}}, \quad U(z) = \exp\left(\frac{z}{2}\right)\Phi(z),$$

here $K$ is the separation constant. Thus the complete set of solutions of (1) can be written as

$$\Psi_n = N_n \cdot L^{-3/2} x \exp\left(-iK_n\tau(t)\right) \exp\left(\frac{i}{2}x^2\left(\frac{\dot{L}}{L} - \frac{\sqrt{D}}{2L^2}\right)\right) M\left(\frac{3}{4} + i\frac{K_n}{\sqrt{D}}, \frac{3}{2}, \frac{i\sqrt{D}}{2}\frac{x^2}{L^2}\right), \tag{9}$$

where the $N_n$ are the normalization constants and the $K_n$ are defined from the following equation:

$$M\left(\frac{3}{4}+i\frac{K_n}{\sqrt{D}},\frac{3}{2},\frac{i\sqrt{D}}{2}\right)=0.$$

For $D = 0$, i.e. for $L(t) = At + B$, the complete set of solutions of (1) can be written as

$$\Psi_n = \sqrt{\frac{2}{L}}\exp\left(\frac{i}{2}\frac{\dot{L}}{L}x^2 - \frac{in^2\pi^2}{2}\tau(t)\right)\sin\left(\frac{n\pi x}{L}\right). \tag{10}$$

For an arbitrary time-dependence of the boundary, an exact analytical solution is not possible. However, the problem can be solved e.g. by expanding the wave function in terms of the eigenfunctions of the corresponding static box (Makowski and Dembinski 1991).

## 3. Time-Dependent Rectangular Billiard

The static rectangular billiard is a simple integrable system which allows exact analytical solution both in the classical and in the quantum case. The motion of a quantum particle in a static rectangular billiard is described by the following time-independent Schrödinger equation:

$$-\frac{1}{2}\left(\frac{\partial^2}{\partial x^2}+\frac{\partial^2}{\partial y^2}\right)\psi_{nl} = \lambda_{nl}^2\psi_n, \tag{11}$$

where $\lambda_{nl} = \sqrt{E_{nl}}$, $E_{nl}$ are the energy eigenvalues. The corresponding boundary conditions are given by:

$$\psi(a,y) = \psi(x,b) = \psi(0,y) = \psi(x,0) = 0, \tag{12}$$

where $a$ and $b$ are the side lengths of the rectangle. The eigenfunctions $\psi_{nl}(x,y)$ and eigenvalues $\lambda_{nl}$ are given by

$$\psi_{nl}(x,y) = \frac{1}{2}\sin\left(\frac{n\pi x}{a}\right)\sin\left(\frac{l\pi y}{b}\right), \tag{13}$$

$$\lambda_{nl} = \frac{\pi}{\sqrt{2}}\sqrt{\frac{n^2}{a^2}+\frac{l^2}{b^2}}. \tag{14}$$

In the case of the time-dependent rectangular billiard we have the following Schrödinger equation

$$i\frac{\partial\Psi(x,y,t)}{\partial t} = -\frac{1}{2}\left(\frac{\partial^2}{\partial x^2}+\frac{\partial^2}{\partial y^2}\right)\Psi. \tag{15}$$

We consider rectangular billiards with two (perpendicular) moving walls, for which the boundary conditions are given by

$$\Psi|_{x=0} = \Psi|_{x=a(t)} = \Psi|_{y=0} = \Psi|_{y=b(t)} = 0. \tag{16}$$

Using the coordinate transformations $\xi = x/a(t)$ and $\eta = y/b(t)$, we can remove the time-dependence of the boundary conditions and get

$$i\frac{\partial\Psi(\xi,\eta,t)}{\partial t} = \left[-\frac{1}{2a^2}\frac{\partial^2\Psi}{\partial\xi^2} + i\frac{\dot{a}}{a}\xi\frac{\partial\Psi}{\partial\xi}\right] + \left[-\frac{1}{2b^2}\frac{\partial^2\Psi}{\partial\eta^2} + i\frac{\dot{b}}{b}\eta\frac{\partial\Psi}{\partial\eta}\right]. \tag{17}$$

The substitution

$$\Psi(\xi,\eta,t) = \frac{1}{\sqrt{ab}}\exp\left[\frac{i}{2}\left(\dot{a}a\xi^2 + \dot{b}b\eta^2\right)\right]\phi(\xi,\eta,t), \tag{18}$$

reduces (17) into the form

$$i\frac{\partial\phi}{\partial t} = \left[-\frac{1}{2a^2}\frac{\partial^2\phi}{\partial\xi^2} + \frac{1}{2}\xi^2\ddot{a}a\phi\right] + \left[-\frac{1}{2b^2}\frac{\partial^2\phi}{\partial\eta^2} + \frac{1}{2}\eta^2\ddot{b}b\phi\right], \tag{19}$$

with boundary conditions that vanish on a unit square:

$$\phi|_{\xi=0} = \phi|_{\xi=1} = 0, \quad \phi|_{\eta=0} = \phi|_{\eta=1} = 0.$$

Time and coordinate variables of (19) can be separated if the following equations are obeyed:

$$a^3\ddot{a} = const = D_1$$

and

$$b^3\ddot{b} = const = D_2.$$

Let $g_1(\xi,t)$ and $g_2(\eta,t)$ be the solutions of the equations

$$ia^2\frac{\partial g_1}{\partial t} = -\frac{1}{2}\frac{\partial^2 g_1}{\partial\xi^2} - \frac{D_1}{8}\xi^2 g_1, \tag{20}$$

and

$$ib^2\frac{\partial g_2}{\partial t} = -\frac{1}{2}\frac{\partial^2 g_2}{\partial\eta^2} - \frac{D_2}{8}\eta^2 g_2. \tag{21}$$

Then the solution of (19) can be represented as $\phi(\xi,\eta,t) = g_1(\xi,t)g_2(\eta,t)$. The functions $g_1$ and $g_2$ can be written explicitly as

$$g_{jn}(\xi,t) = \xi\exp\left(\frac{i}{4}\sqrt{D_j}\xi^2\right)M\left(\frac{3}{4} + \frac{iK_n^2}{\sqrt{D_j}}, \frac{3}{2}, \frac{i\sqrt{D_j}}{2}\xi^2\right)T_n,\ j = 1, 2,\ n = 1, 2, 3, ...,$$

where

$$T_n = \exp\left(-i\int\limits_0^t \frac{K_n^2}{a^2}d\tau\right).$$

For $D_j = 0$ we have

$$g_{jn}(\xi,t) = \sin\left(\sqrt{2}K_n\xi\right)T_n, \quad j = 1,2, \quad n = 1,2,3,....$$

Now we consider the case of oscillating walls, i.e. when $a(t) = a_1 + a_2\cos\omega t$ and $b(t) = b_1 + b_2\cos\omega t$. In this case time and coordinate variables of (19) cannot be separated, rather we solve it numerically by expanding the wave function in terms of the eigenfunctions of the static square billiard:

$$\phi(\xi,\eta,t) = \sum_{n,l} c_{nl}(t)\psi_{nl}(\xi,\eta),$$

where

$$\psi_{nl}(\xi,\eta) = 2\sin(\pi n\xi)\sin(\pi l\eta), \quad n,l = 1,2,3,...$$

Inserting this expansion into (19) one can obtain a system of first-order differential equations for the coefficients $c_{nl}(t)$, whose solution allows us to construct the solution of (17).

The time-dependence of the average energy can be calculated as

$$\langle E(t)\rangle = \int_0^{a(t)}\int_0^{b(t)} \Psi^*(x,y,t)H\Psi(x,y,t)dxdy. \tag{22}$$

In Fig. 1, the time-dependence of the average energy is plotted for different oscillation amplitudes of the billiard walls. Specifically, we use Fig. 1a $a(t) = 20+\cos(2.5t)$, $b(t) = 20+\cos(2.5t)$ and Fig. 1b $a(t) = 20+2\cos(2.5t)$, $b(t) = 20 + 2\cos(2.5t)$. The expansion coefficients for the initial state are chosen as $c_{nl}(0) = \frac{1}{\sqrt{2^{n+l}}}$. As can be seen from Fig. 1, for the amplitude $a_2 = b_2 = 2$ the time-dependence of the average energy is periodic, while for $a_2 = b_2 = 1$ it is not and the growth of $\langle E(t)\rangle$ is suppressed. The regime of higher amplitudes corresponds to higher perturbation strength. In this context we mention the results of Glasser et al. (2009), where it was shown that in the oscillating 1D-box, in the regime of high oscillation frequencies, the average energy becomes time-periodic. In our opinion, a similar situation appears in the regime of higher amplitudes in time-dependent rectangular billiard. In other words, for smaller oscillation amplitudes the velocity of the wall (for fixed oscillation frequency) is small and the system is kept in the adiabatic regime, where a quantum particle follows the wall movement, while for high enough oscillation amplitudes (i.e. for higher velocities of the wall), the particle does not feel the confinement of the boundary to the region with minimal

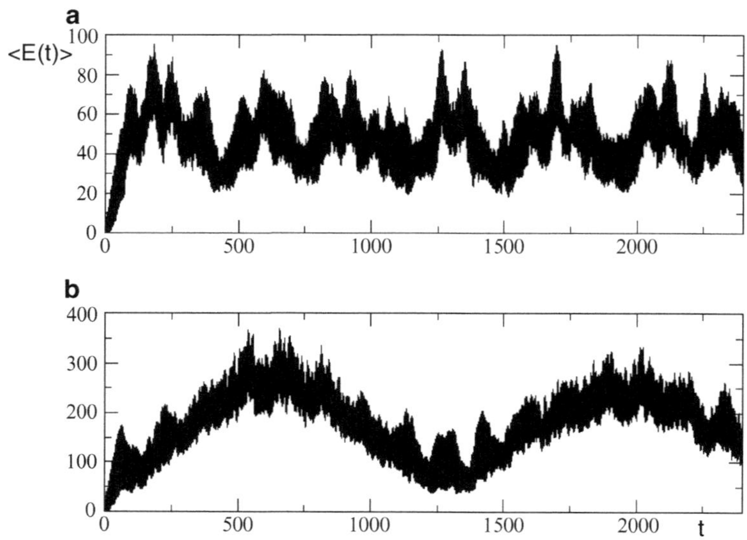


*Figure 1.* The time-dependence of the quantum average energy for harmonically oscillating rectangular billiard, (**a**) $a(t) = 20 + \cos(2.5t)$, $b(t) = 20 + \cos(2.5t)$, (**b**) $a(t) = 20 + 2\cos(2.5t)$, $b(t) = 20 + 2\cos(2.5t)$

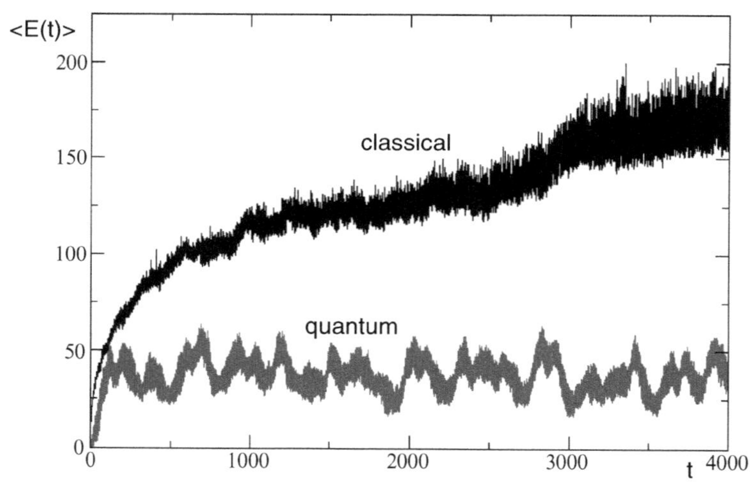


*Figure 2.* Classical and quantum average energies for time-dependent rectangular billiard, $a(t) = 20 + 0.5\cos(2.5t)$, $b(t) = 20 + 0.5\cos(\sqrt{3} \times 2.5t)$

area. Figure 2 compares the average energy for classical and the quantum cases. The growth of the energy is strongly suppressed in the quantum case compared to the corresponding classical system.

## 4. Elliptical Billiard with Moving Boundaries

The classical and quantum dynamics of static elliptical billiards has been extensively studied during the last two decades (Waalkens et al. 1997; Traiber et al. 1989). The classical time-dependent elliptical billiard was recently studied in detail in Lenz et al. (2008), where the possibility of tunable Fermi acceleration was shown.

The Schrödinger equation for the static elliptical billiard is separable in elliptic coordinates which are defined as

$$x = f \cos\theta \cosh\rho, \qquad y = f \sin\theta \sinh\rho, \tag{23}$$

with

$$0 < \rho < \rho_0, \qquad 0 < \theta < 2\pi. \tag{24}$$

Writing the wave function as

$$\psi = G(\rho)F(\theta), \tag{25}$$

and substituting it into the Schrödinger equation

$$-\frac{1}{2}\left(\frac{\partial^2}{\partial x^2} + \frac{\partial^2}{\partial y^2}\right)\psi = E\psi, \tag{26}$$

one can find that $G$ and $F$ obey the Mathieu and associated Mathieu equations, respectively (Waalkens et al. 1997; Traiber et al. 1989):

$$\frac{d^2G}{d\rho^2} + (k^2 f^2 \cosh^2\rho - E)G = 0, \tag{27}$$

$$\frac{d^2F}{d\theta^2} + (E - k^2 f^2 \cos^2\theta)F = 0. \tag{28}$$

The boundary conditions for the angular $F$ and radial $G$ functions are given by

$$F(\theta + 2\pi) = F(\theta), \qquad G(\rho_0) = 0.$$

Here, we consider the case when the ellipse is given by

$$\frac{x^2}{a^2\cosh^2\rho_0} + \frac{y^2}{a^2\sinh^2\rho_0} = 1, \tag{29}$$

where $\rho_0 = const$, $a = a(t) > 0$ is the half of the distance between the foci. It is clear that for this kind time dependence the eccentricity of the ellipse remains constant. In this case the Schrödinger equation can be written as

$$i\frac{\partial\Psi}{\partial t} = -\frac{1}{2}\left(\frac{\partial^2}{\partial x^2} + \frac{\partial^2}{\partial y^2}\right)\Psi, \tag{30}$$

with the initial condition

$$\Psi|_{t=0} = \Psi_0(x, y). \tag{31}$$

Using the coordinate transformations

$$\xi = \frac{x}{a(t)}, \qquad \eta = \frac{y}{a(t)}, \tag{32}$$

and the substitution

$$\Psi(\xi, \eta, t) = \frac{2}{a} \exp\left(\frac{i}{2} a\dot{a}\left(\xi^2 + \eta^2\right)\right) \phi(\xi, \eta, t), \tag{33}$$

we can rewrite the Schrödinger equation in the form

$$ia^2 \frac{\partial \phi}{\partial t} = -\frac{1}{2}\left(\frac{\partial^2}{\partial \xi^2} + \frac{\partial^2}{\partial \eta^2}\right)\phi + \frac{1}{2} a^3 \ddot{a}(\xi^2 + \eta^2)\phi \tag{34}$$

and the wave function has to vanish on the time-independent ellipse:

$$\frac{\xi^2}{\cosh^2 \rho_0} + \frac{\eta^2}{\sinh^2 \rho_0} = 1. \tag{35}$$

In elliptic coordinates, (34) can be rewritten as

$$ia^2 \frac{\partial \phi}{\partial t} = -\frac{1}{2(\cosh 2\rho - \cos 2\theta)}\left(\frac{\partial^2}{\partial \rho^2} + \frac{\partial^2}{\partial \theta^2}\right)\phi - a^3 \ddot{a}\,(\cosh 2\rho - \cos 2\theta)\,\phi. \tag{36}$$

In these coordinates the boundaries of the elliptical domain are given by

$$\{(\rho, \theta) : 0 \le \rho < \rho_0, 0 \le \theta \le 2\pi\}$$

and for the boundary conditions we have

$$\phi|_{\rho=\rho_0} = 0, \quad \phi|_{\theta=0} = \phi|_{\theta=2\pi}, \quad \frac{\partial}{\partial \theta}\phi\Big|_{\theta=0} = \frac{\partial}{\partial \theta}\phi\Big|_{\theta=2\pi}.$$

Time and coordinate variables in (36) can be separated if $a^3\ddot{a} = const = -\frac{D}{4}$. With the separation ansatz

$$\phi(\rho, \theta, t) = R(\rho)F(\theta)T(t),$$

we get

$$ia^2 \dot{T}(t) = kT(t), \tag{37}$$

$$\frac{d^2 R}{d\rho^2} + \left(\frac{1}{8} D \cosh 4\rho + 2k \cosh 2\rho + \lambda\right) R = 0, \tag{38}$$

$$\frac{d^2 F}{d\theta^2} - \left(\frac{1}{8} D \cos 4\theta + 2k \cos 2\theta + \lambda\right) F = 0, \tag{39}$$

where $k$ and $\lambda$ are the separation constants. We note that for $D = 0$ (38) and (39) coincide with the Mathieu equations.

In the case of oscillating boundaries, i.e. $a^3\ddot{a} \neq const$, the problem can be solved numerically by expanding the wave function in terms of the eigenfunctions of the corresponding static elliptical billiard.

$$\phi(\rho, \theta, t) = \sum_{n,l} c_{nl}(t)\psi_{nl}(\rho, \theta), \tag{40}$$

where $\psi_{nl}(\xi, \eta)$ is the solution of

$$-\frac{1}{2(\cosh 2\rho - \cos 2\theta)}\left(\frac{\partial^2}{\partial\rho^2} + \frac{\partial^2}{\partial\theta^2}\right)\phi = E_{nl}\phi, \tag{41}$$

with the boundary conditions given by (35).

To obtain the eigenfunctions of the static elliptical billiard we use the plane wave decomposition method described in Heller (1984) and Cohen et al. (2004).

Substituting (40) into (36) we obtain a system of coupled first order differential equations for the time-dependent coefficients $c_{nl}(t)$. With the $c_{nl}(t)$, we can calculate the average energy using (22). In Fig. 3, the quantum average energy is shown as a function of time for a periodic driving law given by

$$a(t) = 0.7 + 0.07\sin(\omega t), \tag{42}$$

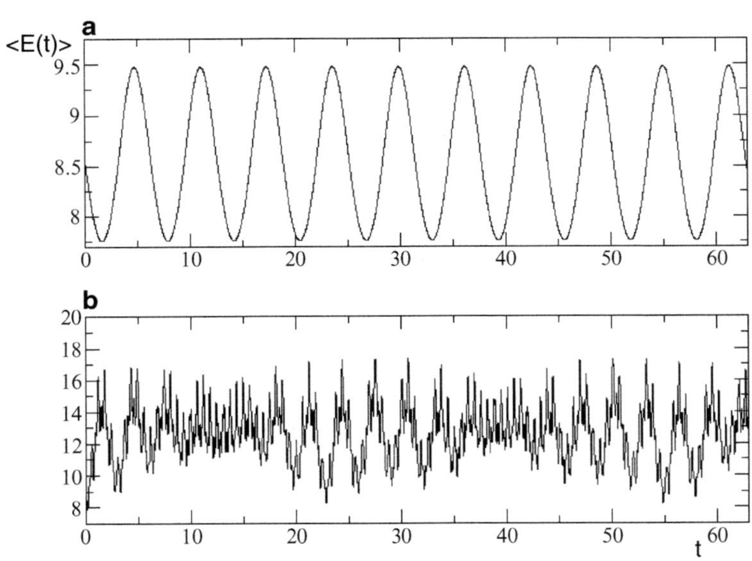


*Figure 3.* Quantum average energy for time-dependent ellipse. Initial state is the ground state: $a(t) = 0.7 + 0.07\sin(\omega t)$, (**a**) $\omega = 1$, (**b**) $\omega = 10$

where $\rho_0 = \tanh^{-1}\sqrt{0.51}$. It can be seen from Fig. 3a, that for $\omega = 1$ the time-dependence of the average energy is periodic. This is due to the fact that for $\omega = 1$ (and for smaller frequencies) the system is in the adiabatic regime where the particle essentially stays in the instantaneous ground state of the corresponding static system. In the non-adiabatic regime of high frequencies ($\omega = 10$) (Glasser et al. 2009 this regime is called "chaotic regime") we observe a completely different behavior. $\langle E(t)\rangle$ is not periodic, but the energy is still bounded. The boundary between adiabatic and non-adiabatic regimes is defined by the frequency of the stationary state (Glasser et al. 2009). It should be noticed that the above prescription can be used only for the case when the eccentricity of the ellipse remains constant.

## 5. Other Billiard Geometries

Now we consider a more general case when the boundary conditions are given for a domain $\{(x, y, t) : \; t > 0, \; (x, y) \in \Omega_{a(t)}\}$, whose boundaries can be defined by the equation

$$g\left(\frac{x}{a(t)}, \frac{y}{a(t)}\right) = 0, \tag{43}$$

with $g(x, y)$ being a continuous function with a smooth first derivative ($a(t) > 0$). Then the boundary condition for (30) can be written as

$$\Psi|_{g\left(\frac{x}{a(t)}, \frac{y}{a(t)}\right)=0} = 0. \tag{44}$$

Some special cases of (43) (for $a(t) = 1$), which reproduces integrable geometries, are:

1. For $g(x, y) = |x + y| - |x - y| = 2$ we have square billiard $[-1; 1] \times [-1; 1]$
2. For $g(x, y) = x^2 + y^2 - 1$ we have a circle
3. For $g(x, y) = \frac{x^2}{\cosh^2 \rho_0} + \frac{y^2}{\sinh^2 \rho_0} = 1$ we get an ellipse with the focus distance of 2

Furthermore, we consider the special case of a time-dependence $a(t)$ which allows the separation of time and coordinate variables of (30). Again, using the transformations given by (32) and (33) we obtain the Schrödinger equation for a time-dependent harmonic oscillator confined in the domain $g(\xi, \eta) = 0$:

$$ia^2 \frac{\partial \phi}{\partial t} = -\frac{1}{2}\left(\frac{\partial^2}{\partial \xi^2} + \frac{\partial^2}{\partial \eta^2}\right)\phi + \frac{1}{2}a^3 \ddot{a}(\xi^2 + \eta^2)\phi, \tag{45}$$

with the following initial and boundary conditions, respectively:

$$\phi(\xi,\eta,t)|_{t=0} = \frac{a(0)}{2}\exp\left(-\frac{i}{2}a(0)\dot{a}(0)\left(\xi^2+\eta^2\right)\right)\Psi_0(\xi,\eta), \qquad \Psi|_{g(\xi,\eta)=0} = 0. \tag{46}$$

Time variable can be separated, if $a^3\ddot{a} = const$. Consider for example the case when the time-dependence of $a(t)$ is given by $a(t) = At + 1$. Then, (45) and (46) can be written as

$$i\frac{\partial\phi}{\partial\tau} = -\frac{1}{2}\left(\frac{\partial^2}{\partial\xi^2} + \frac{\partial^2}{\partial\eta^2}\right)\phi, \tag{47}$$

$$\phi(\xi,\eta,t)|_{\tau=0} = \frac{1}{2}\exp\left(-\frac{i}{2}A\left(\xi^2+\eta^2\right)\right)\Psi_0(\xi,\eta), \qquad \Psi|_{g(\xi,\eta)=0} = 0, \tag{48}$$

where

$$\tau = \frac{t}{At+1}.$$

The solution of the problem is then given by (30) and (44) and can be written as

$$\Psi(x,y,t) = 2\exp\left(\frac{i}{2}A\left(\xi^2+\eta^2\right)\right)\phi\left(\frac{x}{At+1}, \frac{y}{At+1}, \frac{t}{At+1}\right), \tag{49}$$

where $\phi(x, y, t)$ denotes the solution of the corresponding static billiard problem, for which the boundary conditions are given on the curve $g(x, y) = 0$. In general cases, when time and coordinate variables cannot be separated, the problem has to be solved numerically. The most convenient way is by expanding the wave function in terms of the complete set of eigenfunctions of the corresponding static problem.

## 6. Conclusion

In this work we studied the quantum dynamics of billiards with moving boundaries. The problem is reduced to a two-dimensional Schrödinger equation with time-independent boundary conditions but additional time-dependent terms. We explored the time-dependence of the average energy of the particle moving in such billiards. Billiards geometries with integrable static counterparts, such as rectangles and ellipses are considered. An exact analytical solution of the time-dependent Schrödinger equation is obtained in the case of a monotonically expanding (contracting) rectangle. In the rectangular billiard with two harmonically oscillating walls the average energy can be a periodic function of time for some values of the frequency and the amplitude, while for other values, $\langle E(t)\rangle$ is not periodic. The comparison of the average energy for the classical and quantum time-dependent rectangular billiards shows that the average energy is much stronger suppressed in the quantum system than in the classical one. For the elliptical billiard, we considered the case where the eccentricity remains constant. We found that for harmonically oscillating boundaries the average energy is time-periodic. It is shown that the method used for rectangular and elliptical billiards can be used for more general billiard geometries, which fulfill certain boundary conditions.

## Acknowledgements

This work is supported in part by a grant of Volkswagen Foundation (Nr. I/ 82 136).

## References

Abramowitz M. and Stegun I.A.: 1964. Handbook of Mathematical Functions. Dover, New York
Badrinarayanan R., Jose J.V. and Chu G.: 1995. *Physica D*, **83**, 1
Cohen D. and Wisniacki D.A.: 2003. *Phys. Rev. E*, **67**, 026206
Cohen D., Lepore N. and Heller E.J.: 2004. *J. Phys. A*, **37**, 2139
de Carvalho R.E., de Souza F.C. and Leonel E.D.: 2006. *J. Phys. A*, **39**, 3561
Doescher S.W. and Rice M.H.: 1969. *Am. J. Phys.*, **37**, 1246
Glasser M.L., Mateo J., Negro J. and Nieto L.M.: 2009. *Chaos, Solit. Fract.*, **41**, 2067
Heller E.J.: 1984. *Phys. Rev. Lett.*, **53**, 1515
Jose J.V. and Gordery R.: 1986. *Phys. Rev. Lett.*, **56**, 290
Kamphorst S.O. and de Carvalho S.P.: 1999. *Nonlinearity*, **12**, 1363
Kamphorst S.O., Leonel E.D. and da Silva J.K.L: 2007. *J. Phys. A*, **40**, F887
Karner G.: 1989. *Math. Phys. A*, **17**, 329
Koiller J., Markarian R., Kamphorst S.O. and de Carvalho S.P.: 1995. *Nonlinearity*, **8**, 983
Koiller J., Markarian R., Kamphorst S.O. and de Carvalho S.P.: 1996. *J. Stat. Phys.*, **83**, 127
Lenz F., Diakonos F.K. and Schmelcher P.: 2007. *Phys. Rev. E*, **76**, 066213
Lenz F., Diakonos F.K. and Schmelcher P.: 2008. *Phys. Rev. Lett.*, **100**, 014103
Loskutov A.Yu. and Ryabov A.B.: 2002. *J. Stat. Phys.*, **108**, 995
Loskutov A.Yu., Ryabov A.B. and Akinshin L.G.: 2000. *J. Phys. A*, **33**, 7973
Makowski A.J.: 1992. *J. Phys. A*, **25**, 3419
Makowski A.J. and Dembinski S.T.: 1991. *Phys. Lett.*, **154**, 217
Makowski A.J. and Peplowski P.: 1992. *Phys. Lett. A*, **163**, 142
Moralez D.A., Parra Z. and Almeida R.: 1994. *Phys. Lett. A*, **185**, 273
Munier A., Burgan J.R., Feix M. and Fijalkow E.: 1981. *J. Math. Phys.*, **22**, 1219
Pereshogin P. and Pronin P.: 1991. *Phys. Lett. A*, **156**, 12
Pinder D.N.: 1990. *Am. J. Phys.*, **58**, 54
Razavy M.: 1991. *Phys. Rev. A*, **44**, 2384
Scheininger C. and Kleber M.: 1991. *Physica D*, **50**, 391
Seba P.: 1990. *Phys. Rev. A*, **41**, 2306

Traiber A.J.S., Fendrik A.J. and Bernath M.: 1989. *J. Phys. A*, **22**, L365
Waalkens H., Wiersig J. and Dullin H.R.: 1997. *Ann. Phys.*, **260**, 50
Willemsen J.E.: 1994. *Phys. Rev. E*, **50**, 3116
Yuce C.: 2004. *Phys. Lett. A*, **321**, 291

# SCREENING OF THE COULOMB INTERACTION IN A GENERIC BALLISTIC QUANTUM DOT

D. Ullmo
*Université Paris-Sud, LPTMS, Orsay Cedex, France; denis.ullmo@u-psud.fr*

**Abstract.** In semiconductor quantum dots, electrons behave as Fermi–Landau quasi-particles interacting through a week screened interaction. For confined systems a genuine "microscopic" derivation of how this screening takes place is made more complicated than in the bulk case because it has to be treated concurrently with the reorganization of charges generating the smooth electrostatic confining potential in which the electrons are evolving. This contribution gives a discussion of this problem in a semiclassical framework.

**Key words:** Quantum dots; Semiclassical framework; Screened interaction

## 1. Introduction

Mesoscopic physics, in the sense of the study of small, fully coherent, electronic systems on the micron or sub-micron scale, is a field already a few decades old, with a significant number of achievements both experimental and theoretical (Aleiner et al. 2002; Sohn et al. 1997; Grabert and Devoret 1992; Richter 2000; Ullmo 2008). The first experimental realizations of such mesoscopic systems were small metallic grains, for which the motion of the electrons within the sample is diffusive. Progress made in the patterning of two dimensional electron gas, in GaAs/AlGaAs or other kind of heterostructure, made it possible however since the early to produce and study ballistic quantum dots, for which the electrons motion is governed by the electrostatic confining potential.

In such systems, the interplay between Coulomb interaction $V_{\mathrm{Coul}}(\mathbf{r}, \mathbf{r}')$ and interference effects due to confinement plays, in many circumstances, a fundamental role. It is therefore somewhat surprising that, even at this time, there does not exist a proper theory of screening of the Coulomb interaction in ballistic quantum dots. For bulk (possibly disordered) electron gas, the mechanisms leading to screening are extremely well understood and have since long found their way into textbooks. For finite systems, however an extra complication arises from the fact that screening takes place "together" with the reorganization of charges required to form the self consistent one

G. Casati and D. Matrasulov (eds.), *Complex Phenomena in Nanoscale Systems*,
NATO Science for Peace and Security Series B: Physics and Biophysics,

© Springer Science+Business Media B.V. 2009

body potential in which the electrons are evolving. Both processes (screening and formation of the self consistent electrostatic potential) need to be treated concurrently, making significantly more involved a microscopic description.

In the semiclassical limit, and more precisely whenever the screening length $\kappa^{-1}$ is much smaller than the typical size $L$ of the system, a common wisdom approach is usually followed. This common wisdom simply states that since the characteristic scales of variation of the screened potential $V_{sc}$ and of the electrostatic mean field potential $U_{mf}$ are parametrically different (the former $\kappa^{-1}$ is a quantum scale, when the latter $L$ is classical), one could nevertheless use the same screened interaction as for the bulk, and furthermore assume that $U_{mf}(\mathbf{r})$ is correctly approximated by a Thomas–Fermi approximation, which amount to solve the self consistent equation

$$U_{mf}(\mathbf{r}) = U_{ext}(\mathbf{r}) + \int \mathbf{dr'n(r')V_{Coul}(r, r')} \tag{1}$$

$$n(\mathbf{r}) = \int \frac{d\mathbf{p}}{(2\pi\hbar)^d} \Theta(\mu - U_{mf}(\mathbf{r}) - \mathbf{p}^2/2\mathbf{m_e}). \tag{2}$$

Interestingly enough, there is no general microscopic derivation of the above picture. More precisely, our confidence in having the Thomas–Fermi approximation as a correct starting point for the computation of $U_{mf}$ is due to the fact that this approximation can be derived in a quite general framework starting from a density functional description (in e.g. the local density approximation) and neglecting the effect of interferences (Ullmo et al. 2001, 2004). The "common wisdom" prescription given above therefore essentially amounts to trusting the density functional approach on the classical scale $L$ (although it might be less reliable on the quantum scale $\lambda_F$; cf. for instance the discussion in Ullmo et al. 2004), keeping the usual (bulk) form of the screened interaction on the quantum scale, and assuming that the two scales are not going to interfere in any significant way.

There is a class of systems (namely billiards with weak disorder) for which it is possible to perform a renormalization procedure (Aleiner et al. 2002; Blanter et al. 1997) where the fast modes are integrated out so that only the interesting low-energy physics remains. It is then possible to see how both the mean field and the screened interaction emerge in that case from this procedure.

Ideally, the aim of this contribution should have been to generalize this renormalization approach to a generic ballistic quantum dot. This appears out of reach at the present time, and I will limit myself here to a significantly more modest goal, namely to investigate how far the approach introduced in Sect. 2.3.2 of Aleiner et al. (2002) can be generalized to the case of a generic quantum dot (i.e. one for which $U_{mf}(\mathbf{r})$ is not well approximated by a constant), but also to identify the main difficulty preventing a full solution of the problem. I hope this will provide some useful milestone in the solution of this problem in the ballistic case, as well as indicate some possible limitations existing already in the diffusive one.

## 2. Renormalization Scheme

In the bulk, the RPA screened interaction, is obtained by considering the Dyson equation for the dressed interaction (see the discussion in Sect. 9 of Fetter and Walecka 1971)

$$V_{\text{dressed}}(\mathbf{r_1}, \mathbf{r_2}, \omega) = \mathbf{V}_{\text{Coul}}(\mathbf{r_1} - \mathbf{r_2}) + \int d\mathbf{r} \int \mathbf{dr'} \mathbf{V}_{\text{Coul}}(\mathbf{r_1} - \mathbf{r}) \mathbf{\Pi}(\mathbf{r}, \mathbf{r'}, \omega) \mathbf{V}_{\text{dressed}}(\mathbf{r'}, \mathbf{r_2}, \omega), \quad (3)$$

which is exact if all the one-particle irreducible diagrams are included for the polarization operator $\Pi$ but gives the RPA approximation if only the (lowest order) bubble diagram

$$\Pi^0(\mathbf{r}, \mathbf{r'}, \omega) = g_s \int_{-\infty}^{+\infty} \frac{\mathbf{d}\omega'}{\mathbf{2i}\pi} \mathbf{G}(\mathbf{r}, \mathbf{r'}, \omega + \omega') \mathbf{G}(\mathbf{r'}, \mathbf{r}, \omega') \quad (4)$$

is kept. $G(\mathbf{r}, \mathbf{r'}, \omega) = \mathbf{\Theta}(\omega)\mathbf{G^R}(\mathbf{r}, \mathbf{r'}, \omega) + \mathbf{\Theta}(-\omega)\mathbf{G^A}(\mathbf{r}, \mathbf{r'}, \omega)$ is the unperturbed time ordered Green's function, with $\Theta(x)$ the Heaviside function, and $g_s = 2$ is the spin degeneracy factor. In the zero frequency low momentum limit one gets (in the bulk) $\Pi^0(\mathbf{r}, \mathbf{r'}, \omega = \mathbf{0}) \simeq -g_s \nu_0 \delta(\mathbf{r} - \mathbf{r'})$, with $\nu_0$ the local density of states per spin. Inserting this expression for $\Pi^0$ in (3) gives the low momentum zero frequency RPA approximation of the screened interaction.

Let us consider now a mesoscopic systems, and assume that its typical dimensions are much larger than the screening length. One then expects that the residual screened Coulomb interaction should be very similar to the one in the bulk, and it is therefore natural to approach the question from the same viewpoint. In that case however the Green's function are not known exactly, so one needs to resort to semiclassical approximations of $G^{R,A}$ in the expression of $\Pi^0$. The difficulty encountered then is that semiclassical approximations are valid for high energies (high $\omega$), and in particular one cannot expect the semiclassical expressions for $G(\mathbf{r}, \mathbf{r'}, \omega)$ to be accurate if $\omega$ is not much larger than the mean level spacing $\Delta$ of the system.

Following Aleiner et al. (2002), the idea is then, in the spirit of the renormalization group approach, to integrate out only the "fast variable" (high-energy part) for which a semiclassical approximations can be used, and to deal with the low energy physics by some other methods (based for instance on a random-matrix description, Murthy and Mathur 2002; Murthy and Shankar 2003). Using the exact expression for the polarization bubble

$$\Pi^0(\mathbf{r},\mathbf{r}',\omega) = g_s \sum_{\mathbf{nn}'} \mathbf{\Theta}(-\epsilon_{\mathbf{n}}\epsilon_{\mathbf{n}'}) \frac{\varphi^*_{\mathbf{n}}(\mathbf{r}')\varphi_{\mathbf{n}}(\mathbf{r})\varphi^*_{\mathbf{n}'}(\mathbf{r})\varphi_{\mathbf{n}'}(\mathbf{r}')}{\omega + \epsilon_{\mathbf{n}'} - \epsilon_{\mathbf{n}}} (-\mathrm{sgn}(\epsilon_{\mathbf{n}})) \tag{5}$$

with $(\epsilon_n, \varphi_n(\mathbf{r}))$ the one-particle energies and eigenstates, we see that this can be achieved by restricting the sum in the above expression to pair $(n, n')$ such that at least one energy is outside a band centered at the Fermi energy $\epsilon_F$ (taken as the origin of energies) and of width $\epsilon^*$ chosen such that $\Delta \ll \epsilon^* \ll E_{\mathrm{Th}}$, and which precise value (once in this range) is expected to be irrelevant. Up to an unimportant boundary term, this is equivalent to restricting the sum to particle-hole energies $\epsilon_{n'} - \epsilon_n$ larger (in absolute value) than $\epsilon^*$. Introducing $\Pi^{R,A}(\mathbf{r},\mathbf{r}',\omega) \stackrel{\mathrm{def}}{=} \lim_{\eta\to 0+} \mathbf{\Pi^0}(\mathbf{r},\mathbf{r}',\omega \pm \mathbf{i}\eta)$ the retarded and advance polarization bubbles, one can therefore write the polarization operator in which only the fast modes are integrated out as

$$\hat{\Pi}_{\epsilon^*}(\mathbf{r}',\mathbf{r},\tilde{\omega}=\mathbf{0}) = \frac{\mathbf{1}}{\mathbf{2i}\pi} \int \frac{\mathbf{d}\omega}{\omega} \left[ \mathbf{\Pi^R}(\mathbf{r},\mathbf{r}',\omega) - \mathbf{\Pi^A}(\mathbf{r},\mathbf{r}',\omega) \right] \mathbf{\Theta}(|\omega| - \epsilon^*). \tag{6}$$

The insertion of $\hat{\Pi}_{\epsilon^*}$ in (3) will then give the effective interaction describing the low energy ($\leq \epsilon^*$) physics of the quantum dot.

## 3. Calculation of the Polarization Loop

Let us first consider positive energies $\omega > 0$. Noting that phase cancellation is possible only for the product $G^A G^R$, but not for $G^R G^R$ or $G^A G^A$, one has

$$\Pi^R(\mathbf{r},\mathbf{r}',\omega) = g_s \int_{-\omega}^{\mathbf{0}} \frac{\mathbf{d}\omega'}{\mathbf{2i}\pi} \mathbf{G^R}(\mathbf{r},\mathbf{r}',\omega' + \omega + \mathbf{i}\eta)\mathbf{G^A}(\mathbf{r}',\mathbf{r},\omega') \tag{7}$$

and

$$\Pi^A(\mathbf{r},\mathbf{r}',\omega) = \mathbf{\Pi^R}(\mathbf{r}',\mathbf{r},\omega)^*. \tag{8}$$

The Green's functions in the l.h.s. of (7) can be evaluated semiclassically as a sum over classical trajectories (Gutzwiller 1990)

$$\begin{aligned} G^R(\mathbf{r},\mathbf{r}';\epsilon) &\simeq \sum_{j:\mathbf{r}'\to\mathbf{r}} G^R_j(\mathbf{r},\mathbf{r}';\epsilon) \\ G^R_j(\mathbf{r},\mathbf{r}';\epsilon) &\stackrel{\mathrm{def}}{=} \frac{2\pi}{(2i\pi\hbar)^{(d+1)/2}} D_j(\epsilon) \exp\left(iS_j(\epsilon)/\hbar - i\zeta_j\pi/2\right), \end{aligned} \tag{9}$$

with $S_j$ the classical action of the trajectory $j$ going from $\mathbf{r}'$ to $\mathbf{r}$, $\zeta_j$ is a Maslov index, and the determinant $D_j$ is a measure of the stability of the orbit. Keeping only the diagonal approximation in which a trajectory $j$ is paired with itself to cancel the oscillating phases, one gets

$$[G^R(\mathbf{r},\mathbf{r}',\omega+\omega')\mathbf{G}^{\mathbf{A}}(\mathbf{r}',\mathbf{r},\omega')]_{\text{diag}} = \sum_{j:\mathbf{r}\to\mathbf{r}'} \frac{4\pi^2}{(2\pi\hbar)^{d+1}} |D_j|^2 \exp\left[i(S_j(\omega+\omega') - S_j(\omega'))/\hbar\right]. \quad (10)$$

In this equation, one would like then to perform a Taylor expansion of the action

$$(S_j(\omega+\omega') - S_j(\omega')) = (\partial S_j/\partial\epsilon)\omega = t_j\omega. \quad (11)$$

Inserting the unity $\int_0^\infty \delta(t-t_j)$ and making use of the sum rule (Argaman 1996)

$$\sum_{j:\mathbf{r}'\to\mathbf{r}} \frac{|D_j(\epsilon)|^2}{(2\pi\hbar)^d}\delta(t-t_j) = \nu_0^{(d)}(\mathbf{r}')\mathbf{P}^{\epsilon}_{\text{cl}}(\mathbf{r},\mathbf{r}',\mathbf{t}), \quad (12)$$

where $\nu_0^{(d)}(\mathbf{r}')$ is the bulk density of states per unit area (and spin) for the local value of $k_F$ and $P^{\epsilon}_{\text{cl}}(\mathbf{r},\mathbf{r}',\mathbf{t})$ is the classical (density of) probability that a trajectory launched in $\mathbf{r}'$ is in the neighborhood of $\mathbf{r}$ at time $t$, we obtain

$$\begin{aligned}[G^R(\mathbf{r},\mathbf{r}',\omega+\omega')G^A(\mathbf{r}',\mathbf{r},\omega')]_{\text{diag}} &= \frac{4\pi^2}{(2\pi\hbar)^{d+1}}\int_0^\infty dt \sum_{j:\mathbf{r}\to\mathbf{r}'} |D_j|^2\delta(t-t_j)\exp[it\omega/\hbar] \quad (13)\\ &= \frac{2\pi\nu_0(\mathbf{r}')}{\hbar}\int_0^\infty dt P^{\epsilon}_{\text{cl}}(\mathbf{r},\mathbf{r}',\mathbf{t})\exp[\mathbf{i}\mathbf{t}\omega/\hbar] \quad (14)\\ &= 2\pi\nu_0(\mathbf{r})\hat{\mathbf{P}}^{\epsilon}_{\text{cl}}(\mathbf{r},\mathbf{r}',\omega)\,, \quad (15)\end{aligned}$$

with $P^{\epsilon}_{\text{cl}}$ the Fourier transform of the classical probability $P^{\epsilon}_{\text{cl}}$. Interestingly enough $[G^R(\omega+\omega')G^A(\omega')]_{\text{diag}}$ is independent of $\omega'$, so that finally

$$\Pi^R(\mathbf{r},\mathbf{r}',\omega) = -\mathbf{i}\omega g_s\nu_0(\mathbf{r}')\hat{\mathbf{P}}^{\epsilon}_{\text{cl}}(\mathbf{r},\mathbf{r}',\omega). \quad (16)$$

Note that the fact that we have computed $\Pi^R$, i.e. that $\omega \equiv \omega + i\eta$, is what is making the Fourier transform in (14) convergent. If we had computed $\Pi^A$ the above approach would have lead to divergences. $\Pi^A$ should therefore be derived from $\Pi^R$ using (8), giving

$$\Pi^A(\mathbf{r},\mathbf{r}',\omega) = \mathbf{i}\omega g_s\nu_0(\mathbf{r})\hat{\mathbf{P}}^{\epsilon}_{\text{cl}}(\mathbf{r}',\mathbf{r},\omega). \quad (17)$$

For negative $\omega$, $\Pi^A$ should be calculated first and $\Pi^R$ derived from it with (8), leading to the same result.

Here, one rather important remark is in order. Expression (11) assumes obviously that $\omega$ is small. This is usually not a significant constraint since the actions $S_i$ are classical quantities, so that the relevant scale is the Fermi energy (or bandwidth) $\epsilon_F$. It is therefore enough that $\omega \ll \epsilon_F$ to apply (11). However the integral in the left hand side of (6) is not limited to the neighborhood of the Fermi surface. Replacing $\Pi^{R,A}$ by the approximations (16) and (17) will be incorrect on the edge of the energy band, which will be associated to short distances $|\mathbf{r}-\mathbf{r}'| < \lambda_F$. This will be the cause of the problems we shall encounter later on. Let us ignore this issue for the time being, and come back to this discussion when it will become obvious that the results obtained in this way are unphysical.

Then, inserting (16) and (17) into (6) and writing $\Theta(x) = 1-\Theta(-x)$ we get

$$\begin{aligned}\hat{\Pi}_{\epsilon^*}(\mathbf{r}',\mathbf{r},\tilde{\omega}=\mathbf{0}) &= -g_s \int_{-\infty}^{+\infty} \frac{d\omega}{2\pi} \left[\nu_0(\mathbf{r}')\tilde{\mathbf{P}}^{\epsilon}_{\mathrm{cl}}(\mathbf{r},\mathbf{r}',\omega) + \nu_0(\mathbf{r})\tilde{\mathbf{P}}^{\epsilon}_{\mathrm{cl}}(\mathbf{r},\mathbf{r}',\omega)\right] \\ &\quad \times (1-\Theta(\epsilon^*-|\omega|)) \ . \end{aligned} \tag{18}$$

The term proportional to one in the integrand of (18) gives rise to $\int(d\omega/2\pi)$ $\tilde{P}^{\epsilon}_{\mathrm{cl}}(\mathbf{r},\mathbf{r}',\omega) = \mathbf{P}^{\epsilon}_{\mathrm{cl}}(\mathbf{r},\mathbf{r}',\mathbf{t}=\mathbf{0})$. To evaluate the remaining term, it is useful to discuss the weight function $\Theta(\epsilon^* - |\omega|)$. Its precise form is irrelevant here, and, rather than the actual Heaviside step function, I shall assume that $\Theta(\epsilon^* - |\omega|)$ is actually a smooth function $\Theta_{\epsilon^*}(\omega)$ which is zero for $|\omega| \gg \epsilon^*$ and one for $|\omega| \ll \epsilon^*$. To fix the idea one can think for instance of $\Theta_{\epsilon^*}(\omega) = \exp(-(1/2)(\omega/\epsilon^*)^2)$, but this precise form will not play any particular role. If one introduces $\tilde{\Theta}_{\epsilon^*}(t)$ the Fourier transform of $\Theta_{\epsilon^*}(\omega)$, one has, with $t^* = \hbar/\epsilon^*$

$$\tilde{\Theta}_{\epsilon^*}(t) \simeq 1/t^* \qquad \text{for } t \ll t^* \tag{19}$$

$$= 0 \qquad \text{for } t \gg t^* \tag{20}$$

$$\int_0^{\infty} dt\, \tilde{\Theta}_{\epsilon^*}(t) = \Theta_{\epsilon^*}(\omega = 0) = 1. \tag{21}$$

Assuming furthermore that $\tilde{\Theta}_{\epsilon^*}(t)$ is a positive function (this hypothesis can be easily relaxed), we see that $\tilde{\Theta}_{\epsilon^*}(t)$ is a density probability (since it is positive and normalized to one) which selects trajectory shorter than $t^*$. We thus can write

$$\begin{aligned}\hat{\Pi}_{\epsilon^*}(\mathbf{r}',\mathbf{r},\tilde{\omega}=\mathbf{0}) = -g_s \Bigg[ & \nu_0(\mathbf{r}')\mathbf{P}^{\epsilon}_{\mathrm{cl}}(\mathbf{r},\mathbf{r}',\mathbf{t}=\mathbf{0}) \\ & -\frac{1}{2}\left(\nu_0(\mathbf{r}')\langle \mathbf{P}^{\epsilon}_{\mathrm{cl}}(\mathbf{r},\mathbf{r}',\mathbf{t})\rangle_{\mathbf{t}\leq\mathbf{t}^*} + \nu_0(\mathbf{r})\langle \mathbf{P}^{\epsilon}_{\mathrm{cl}}(\mathbf{r}',\mathbf{r},\mathbf{t})\rangle_{\mathbf{t}\leq\mathbf{t}^*}\right)\Bigg], \end{aligned} \tag{22}$$

where $\langle f(t)\rangle_{t\leq t^*} \stackrel{\text{def}}{=} \int_0^{\infty} dt\, f(t)\tilde{\Theta}_{\epsilon^*}(t)$ is the average over time $t$ lesser than $t^*$ of the function $f(t)$.

Now $P^{\epsilon}_{\rm cl}(\mathbf{r}, \mathbf{r}', \mathbf{t} = \mathbf{0}) = \delta(\mathbf{r} - \mathbf{r}')$. Furthermore, the condition $\Delta \ll \epsilon^* \ll E_{\rm Th}$ is equivalent to $t_f \ll t^* \ll t_H$, with $t_H = \hbar/\Delta$ the Heisenberg time and $t_f$ the time of flight across the system (for ballistic systems) or time needed to diffuse to the boundary (for diffusive systems). We see that the choice of $\epsilon^*$ is made precisely so that (1) semiclassical approximations are valid, but also (2) that most of the range $[0, t^*]$ is such that *for diffusive or chaotic systems* (the case of integrable or mixed system should be investigated in this respect), the motion can be assumed randomized. Assuming ergodicity we can therefore write

$$\langle P^{\epsilon}_{\rm cl}(\mathbf{r}, \mathbf{r}', \mathbf{t}) \rangle_{\mathbf{t} \leq \mathbf{t}^*} \simeq \frac{\int \mathbf{dp} \delta(\epsilon_{\mathbf{F}} - \mathbf{H}(\mathbf{r}, \mathbf{p}))}{\int \mathbf{dr}'' \mathbf{dp}'' \delta(\epsilon_{\mathbf{F}} - \mathbf{H}(\mathbf{r}'', \mathbf{p}''))} = \mathbf{\Delta} \nu_{\mathbf{0}}(\mathbf{r}). \tag{23}$$

This eventually leads to

$$\hat{\Pi}_{\epsilon^*}(\mathbf{r}', \mathbf{r}, \tilde{\omega} = \mathbf{0}) = -g_s \left[ \nu_{\mathbf{0}}(\mathbf{r}) \delta(\mathbf{r} - \mathbf{r}') - \nu_{\mathbf{0}}(\mathbf{r}) \nu_{\mathbf{0}}(\mathbf{r}') \mathbf{\Delta} \right], \tag{24}$$

where one recognize the first term as the zero frequency low momentum bulk polarization $\Pi^0_{\rm bulk}(\mathbf{r}', \mathbf{r}, \tilde{\omega} = \mathbf{0}) = -g_s \nu_{\mathbf{0}}(\mathbf{r}) \delta(\mathbf{r} - \mathbf{r}')$, and I will denote by

$$\Pi_{l.r.} \stackrel{\rm def}{=} g_s \nu_0(\mathbf{r}) \nu_{\mathbf{0}}(\mathbf{r}') \mathbf{\Delta} \tag{25}$$

the remaining long range part. For billiard systems for which $\nu_0(\mathbf{r}) = (\mathcal{A}\mathbf{\Delta})^{-1} = $ const., with $\mathcal{A}$ the volume of the system, (24) is for instance exactly (60) of Aleiner et al. (2002).

## 4. Self-Consistent Equation

In the bulk, both the Coulomb interaction and the polarization operator $\Pi^0_{\rm bulk}$ are translation invariant and the Dyson equation (3) can be solved in momentum representation as

$$\hat{V}_{\rm dressed}(\mathbf{q}) = \frac{\hat{\mathbf{V}}_{\rm Coul}(\mathbf{q})}{\mathbf{1} - \hat{\mathbf{V}}_{\rm Coul}(\mathbf{q}) \hat{\mathbf{\Pi}}^{\mathbf{0}}_{\rm bulk}(\mathbf{q})}. \tag{26}$$

The resulting interaction is then short range, effectively much weaker than the original Coulomb interaction, and is therefore well adapted for a perturbative treatment.

The difficulty one encounters in the mesoscopic case is twofold. First, lack of translational invariance for $\Pi_{\epsilon^*}$ makes in principle (3) impossible to be solved in closed form for a generic spatial variations of $\nu_0(\mathbf{r})$. Second, we know that even at the level of electrostatics, the effects of the interactions cannot be small since they will at minima rearrange considerably the static charges within the system. Therefore, even if (3) could be solved, there is little chance that the resulting dressed interaction could be effectively used in a perturbative approach starting from the non-interacting electrons Hamiltonian.

For a quantum dot with a fixed number $(N + 1)$ of electrons, one way to solve both of these difficulties is to derive a self-consistent equation following one of the standard derivation of the Hartree Fock approximation (Thouless 1961). For this purpose, let us note that any one-body potential $\tilde{U}(\mathbf{r})$, can be written formally as the two-body interaction

$$\tilde{V}(\mathbf{r}, \mathbf{r}') = \frac{\mathbf{1}}{\mathbf{N}}(\tilde{\mathbf{U}}(\mathbf{r}) + \tilde{\mathbf{U}}(\mathbf{r}')) \tag{27}$$

since, using for instance a second quantization formalism

$$\frac{1}{2}\int d\mathbf{r}\mathbf{dr'}\hat{\mathbf{\Psi}}^\dagger(\mathbf{r})\hat{\mathbf{\Psi}}^\dagger(\mathbf{r}')\tilde{\mathbf{V}}(\mathbf{r},\mathbf{r}')\hat{\mathbf{\Psi}}(\mathbf{r}')\hat{\mathbf{\Psi}}(\mathbf{r}) = \int \mathbf{dr}\hat{\mathbf{\Psi}}^\dagger(\mathbf{r})\tilde{\mathbf{U}}(\mathbf{r})\hat{\mathbf{\Psi}}(\mathbf{r}). \tag{28}$$

As a consequence, the total Hamiltonian, as well as the formalism presented in the first part of this appendix, are unmodified if the confining potential $U_{\text{ext}}(\mathbf{r})$ and the Coulomb potential $V_{\text{Coul}}(\mathbf{r}, \mathbf{r}')$ are respectively replaced by

$$U(\mathbf{r}) = U_{\text{ext}}(\mathbf{r}) + \tilde{\mathbf{U}}_(\mathbf{r}) \tag{29}$$

$$V(\mathbf{r}, \mathbf{r}') = V_{\text{Coul}}(\mathbf{r}, \mathbf{r}') - \tilde{\mathbf{V}}(\mathbf{r}, \mathbf{r}') \tag{30}$$

One can then now use the freedom in the choice of the function $\tilde{U}(\mathbf{r})$ to simplify the Dyson equation. In particular, if we can impose that

$$\int d\mathbf{r} \int \mathbf{dr'}\mathbf{V}(\mathbf{r_1}, \mathbf{r})\mathbf{\Pi}_{\text{l.r.}}(\mathbf{r}, \mathbf{r}')\mathbf{V}_{\text{dressed}}(\mathbf{r}', \mathbf{r_2}) \equiv \mathbf{0}, \tag{31}$$

the Dyson equation (3) would then have the usual "bulk-like" form

$$V_{\text{dressed}}(\mathbf{r_1}, \mathbf{r_2}) = \mathbf{V}(\mathbf{r_1}, \mathbf{r_2}) - \int \mathbf{dr} \int \mathbf{dr'}\mathbf{V}(\mathbf{r_1}, \mathbf{r})\mathbf{\Pi}^{\mathbf{0}}_{\text{bulk}}(\mathbf{r} - \mathbf{r}')\mathbf{V}_{\text{dressed}}(\mathbf{r}', \mathbf{r_2}), \tag{32}$$

which, if $\nu_0(\mathbf{r})$ and $\tilde{U}(\mathbf{r})$ vary slowly on the scale of the bulk screening length $\kappa^{-1}$ has the same solution as in the bulk.

Now, (31) might seem at first sight difficult to solve since it involves the unknown function $V_{\text{dressed}}(\mathbf{r}', \mathbf{r_2})$. However, since $\Pi_{\text{l.r.}}(\mathbf{r}, \mathbf{r}')$ does actually not correlate $\mathbf{r}$ and $\mathbf{r}'$, the two integrals in (31) actually decouple, and a sufficient condition to solve this equation is that $\int d\mathbf{r}\mathbf{V}(\mathbf{r_1} - \mathbf{r})\nu(\mathbf{r}) \equiv \mathbf{0}$, i.e.

$$\int d\mathbf{r}\mathbf{V}_{\text{Coul}}(\mathbf{r_1} - \mathbf{r})\nu(\mathbf{r}) = \frac{\mathbf{1}}{\mathbf{N}}\int \mathbf{dr}\nu(\mathbf{r})(\tilde{\mathbf{U}}(\mathbf{r}) + \tilde{\mathbf{U}}(\mathbf{r_1})). \tag{33}$$

The constant term $\int d\mathbf{r}\nu(\mathbf{r})\tilde{\mathbf{U}}(\mathbf{r})/\mathbf{N}$ is irrelevant here as it can be eliminated by a constant shift of $\tilde{U}$. One therefore obtain in this way the self-consistent equation

$$\tilde{U}(\mathbf{r}_1) = \mathbf{N}\boldsymbol{\Delta} \int \mathbf{dr}\mathbf{V}_{\mathrm{Coul}}(\mathbf{r}_1 - \mathbf{r})\nu(\mathbf{r}). \tag{34}$$

In other words what we have obtained for the self-consistent potential are the equations

$$U_{\mathrm{mf}}(\mathbf{r}) = U_{\mathrm{ext}}(\mathbf{r}) + \tilde{\mathbf{N}}\boldsymbol{\Delta} \int \mathbf{dr}'\nu_0(\mathbf{r}')\mathbf{V}_{\mathrm{Coul}}(\mathbf{r}, \mathbf{r}') \tag{35}$$

$$\nu_0(\mathbf{r}) = \int \frac{d\mathbf{p}}{(2\pi\hbar)^d}\delta(\mu - U_{\mathrm{mf}}(\mathbf{r}) - \mathbf{p}^2/2\mathbf{m}). \tag{36}$$

## 5. Discussion and Conclusion

The above result looks very much like what we expected to obtain as we began this discussion. Indeed (32), (35) and (36), express than integrating out the "fast degree of freedom" allows one to separate in a natural way the screened potential in two parts: a first one which is a one body potential determined self consistently, and a second which is just the usual bulk screened interaction with the local parameters of the electron gas (provided that $\tilde{U}(\mathbf{r})$ (and thus $\tilde{V}(\mathbf{r}, \mathbf{r}')$) varies slowly on the scale of the screening length).

What makes this description less useful however is the self consistent equations (35) and (36) are obviously incorrect. Indeed we know that whatever self-consistent equations we end up writing, they should contain in some approximation the electrostatic equilibrium of the problem. This is not the case here. If the self-consistent potential $U_{\mathrm{mf}}(\mathbf{r}) \stackrel{\mathrm{def}}{=} \mathbf{U}_{\mathrm{ext}}(\mathbf{r}) + \tilde{\mathbf{U}}(\mathbf{r})$ obtained from (34) is well approximated by a constant (giving for instance a billiard system with weak disorder as was considered in Blanter et al. 1997; Aleiner et al. 2002), and assuming $(N + 1) \gg 1$, one can perform the replacement $N\Delta\nu(\mathbf{r}) \to \mathbf{n}(\mathbf{r})$ in (34) and write instead

$$\tilde{U}(\mathbf{r}_1) = \int \mathbf{dr}\mathbf{V}_{\mathrm{Coul}}(\mathbf{r}_1 - \mathbf{r})\mathbf{n}(\mathbf{r}), \tag{37}$$

i.e. (1) and (2), which is just the Thomas Fermi equation, from which plain electrostatic is obtained by neglecting the kinetic energy associated to Pauli exclusion principle. However for a generic confining potential $U_{\mathrm{ext}}(\mathbf{r})$, solutions of (34) will not in general be an approximation of the solution of (37).

To identify why we ended up with an incorrect equation is actually not very difficult. Indeed, the symptom is that the expected density of particle

$n(\mathbf{r})$ has been replaced in (34) by the $N\nu(\mathbf{r})$, i.e. an extrapolation of what is going on near the Fermi energy. What we see is that, in some sense, (34) is "aware" of the properties of the system near the Fermi energy (the density of states $\nu(\mathbf{r})$), but misses the relevant information at large energies, of the order of the bandwidth, which determine the actual density of particles. This is to be expected since our approximation of the polarization operator $\Pi^*_\epsilon$ (24) involves only the local density of states at the Fermi energy $\nu(\mathbf{r})$. Going further back we see that in the end this behavior can be tracked back to the approximation (11) where the action $S(\omega)$ has been linearized near the Fermi energy, eliminating in this way any information relevant to the large (i.e. $\omega \sim \epsilon_F$) energies.

In the end, what makes non trivial the discussion of screening in a finite system is that the renormalization procedure which transform the bare Coulomb interaction into the screened one should also describe the emergence of the electrostatic mean field. When the former is essentially dominated by the neighborhood of the Fermi energy, the latter requires an accurate description of the physics at the band edge.

To conclude, and since obviously this contribution do not provide a satisfactory solution of the issue of screening in closed quantum dot, a few word about its possible usefulness in presumably warranted. The necessity to treat concurrently different energy scales is what is making non-trivial the problem at hand, but the corresponding separation of scale is however what sort of guarantee that the "common wisdom" approach of the screening described in the introduction is at least qualitatively correct. This common wisdom approach is moreover presumably sufficient as far as qualitative or statistical descriptions are concerned. The absence of a proper microscopic understanding of screening for closed system might however become a limitation when accurate simulation tools are required to describe quantitatively the particular behavior of a specific mesoscopic system. This is particularly true in the view of the fact that standard self consistent approaches, such as density functional theory calculations, fail to describe some of their important aspects (Ullmo et al. 2004). I hope this contribution, by identifying an essential conceptual difficulty in an approach which seems to get us extremely close to the expected result, will provide a useful step toward its full resolution. Moreover, as it turns out that it is not the diffusive character of the dynamics, but rather the possibility to extrapolated the local density of state up to the band edge which turn out to be the relevant difficulty here, it might also indicate some limitation of results derived in the more traditional context diffusive quantum dots.

## References

Aleiner I.L., Brouwer P.W. and Glazman L.I.: 2002, Quantum effects in Coulomb blockade. *Phys. Rep.*, **358**, 309–440

Argaman N.: 1996, Semiclassical analysis of the quantum interference corrections to the conductance of mesoscopic systems. *Phys. Rev. B*, **53**(11), 7035–7054

Blanter Y.M., Mirlin A.D. and Muzykantskii B.A.: 1997, Fluctuations of conductance peak spacings in the Coulomb blockade regime: Role of electron-electron interaction. *Phys. Rev. Lett.*, **78**, 2449–2452

Fetter A. and Walecka J.: 1971, *Quantum Theory of Many-Particle Systems*. New York: McGraw-Hill

Grabert H. and Devoret M.H. (eds.): 1992, *Single Charge Tunneling*. London: Plenum Press

Gutzwiller M.C.: 1990, *Chaos in Classical and Quantum Mechanics*. New York: Springer

Murthy G. and Mathur H.: 2002, Interactions and disorder in quantum dots: Instabilities and phase transitions. *Phys. Rev. Lett.*, **89**(12), 126804

Murthy G. and Shankar R.: 2003, Quantum dots with disorder and interactions: A solvable large-$g$ limit. *Phys. Rev. Lett.*, **90**(6), 066801

Richter K.: 2000, *Semiclassical Theory of Mesoscopic Quantum Systems*. New York: Springer

Sohn L.L., Schön G. and Kouwenhoven L.P. (eds.): 1997, *Mesoscopic Electron Transport*. Dordrecht: Kluwer

Thouless D.J.: 1961, *The Quantum Mechanics of Many-Body Systems*. New York: Academic Press

Ullmo D.: 2008, Many body physics and quantum chaos. *Rep. Prog. Phys.*, **71**, 026001

Ullmo D., Nagano T., Tomsovic S. and Baranger H.U.: 2001, Semiclassical density functional theory: Strutinsky energy corrections in quantum dots. *Phys. Rev. B*, **63**, 125339-1–125339-13

Ullmo D., Jiang H., Yang W. and Baranger H.U.: 2004, Landau Fermi-liquid picture of spin density functional theory: Strutinsky approach to quantum dots. *Phys. Rev. B*, **70**(20), 205309

# KICKED BOSE–EINSTEIN CONDENSATES: IN SEARCH OF EXPONENTIAL INSTABILITY *

T.S. Monteiro*, A. Rancon, and J. Ruostekoski
*Department of Physics and Astronomy, University College London, Gower Street, London WC1E 6BT, UK; t.monteiro@theory.phys.ucl.ac.uk*

**Abstract.** Bose–Einstein condensates subject to short pulses ("kicks") from standing waves of light represent a nonlinear analogue of the well-known chaos paradigm, the quantum kicked rotor. We review briefly our current understanding of dynamical or exponential instability in weakly kicked BECs. Previous studies of the onset of dynamical instability associated it with some form of classical chaos. We show it is due to parametric instability: resonant driving of certain collective modes. We map the zones of instability and calculate the Liapunov exponents.

**Key words:** Bose–Einstein condensate; Kicked systems; Dynamical instability

## 1. Introduction

Cold atoms subjected to strong $\delta$-kicks from standing waves of light provide a near-perfect experimental realization of the leading quantum chaos paradigm, the quantum kicked rotor (QKR) (Casati et al. 1982; Fishman et al. 1982). They provided convincing demonstrations of the quantum supression of classical chaotic diffusion (*"Dynamical Localization"*) and further proof that chaos and exponential sensitivity does not persist in quantum dynamics and the linear Schrödinger equation (Moore et al. 1995; Raizen 1999; Szriftgiser et al. 2002; Duffy et al. 2004; Jones et al. 2004, 2007; Chabe et al. 2008).

More recently, kicked cold atom experimental studies have shifted their attention from dynamical localization, to another interesting regime, namely the Quantum Resonance regime: if the kicking period $T$ is a rational fraction of $T = 4\pi$, the so called "Talbot time," absorption of energy by the atomic cloud peaks at a complex series of narrow resonances. These were analysed in Fishman et al. (2004) and Wimberger et al. (2004) in terms of a novel "image"

* Note that many versions of the famous quantum field theory Bogoliubov transformation $\hat{a}_k = U_k\hat{b}_k - V_k\hat{b}^\dagger_{-k}$ differs from (9) by a minus sign. We adopt in this manuscript the sign convention of Castin (in Les Houches Lecture notes, 2001), related simply by taking $V_k \to -V_k$ and $A_k \to A_k^{-1}$ in all equations.

G. Casati and D. Matrasulov (eds.), *Complex Phenomena in Nanoscale Systems*, NATO Science for Peace and Security Series B: Physics and Biophysics,
© Springer Science+Business Media B.V. 2009

classical dynamics. Further theory (Wimberger et al. 2005; Buchleitner et al. 2006; Saunders et al. 2007; Lepers et al. 2005) includes proposed applications such as the realization of a quantum random walk algorithm (Ma et al. 2006).

Many recent experiments employed Bose Einstein Condensates (Ryu et al. 2006; Sadgrove et al. 2007; Behinaenin et al. 2006; Dana et al. 2008; Kanem et al. 2007). This suggests a new possibility: namely the regime where nonlinear dynamics, arising from the many-body nature of the BEC, combine with the $\delta$-kicked quantum dynamics. Elsewhere, solitonic dynamics also provides another arena for the interaction between BEC physics and chaos or more generic nonlinear dynamics (Martin et al. 2007). To date, the conditions for the initial onset of dynamical instability in kicked BECs and the behavior at longer times in these regimes, including the growth in non-condensate particles is not yet well understood.

## 2. Kicked BECs: The Origin of Dynamical Instability

We attempt to review here the current state of understanding, regarding onset of exponential instability in the *weakly* kicked BEC. We consider only kicking strengths $K \leq 1$, and remain within the framework of the Gross Pitaevski equation (GPE) plus linearized perturbations. There is discussion elsewhere of the behavior of the GPE in the strong chaos regime $K \gg 1$ e.g. Shepelyansky (1993) and Mieck and Graham (2004) but it becomes debatable whether the GPE provides an adequate representation of an atomic condensate in a regime where large amounts of energy are driven into the cloud, especially if the occupation number of individual excited modes no longer remains much larger than one. It is likely that, in that case, a quantum treatment of thermal excitations is essential; see, e.g. Isella and Ruostekoski (2006a, b), Ruostekoski and Isella (2005) and Gardiner and Morgan (2007) for discussions of the available options.

Most previous work on instability in kicked BEC resorted to the time-dependent Bogoliubov using the approach introduced by Castin and Dum (1997, 1998). In practical terms, in this method, excitations of the condensate are evolved in time, concurrently with the GPE (in practice this involves simultaneous propagation of about 2–10 equations analogous to the GPE itself). But orthogonality between the condensate mode and the excited modes is at all times enforced. The method provides an estimate of the number of non-condensate atoms. But since its regime of validity lies in the limit of weak condensate depletion and perturbations, in the presence of resonant driving or exponential instability, it can become unreliable within very few kicks for realistic condensates. Hence it provides, in effect, just an initial rate at which the non-condensate population starts growing.

Using the Castin–Dum (CD) method, previous works suggested a link between exponential behavior and chaos. In Ma et al. (2006), the possibility that instability was related to chaos in the one-body limit was investigated for the Kicked Harmonic Oscillator. In Zhang et al. (2004) and Liu et al. (2006) the correlation between chaos in the mean-field dynamics, rather, and the onset of dynamical instability, was investigated. An instability border, determined by the kick strength $K$ and the nonlinearity $g$ was identified for $T = 2\pi$. It was then found (Liu et al. 2006) that the parameter ranges for this instability corresponds closely to a transition from regular to chaotic motion, of an effective classical Hamiltonian derived from the mean-field dynamics. A similar onset of instability was identified by Poletti et al. (2007) in a variant of the kicked BEC system. In Rebuzzini et al. (2007), a study in the Quantum accelerator regime (Quantum Resonances with gravitational acceleration), no exponential instability was found.

However, in Reslen et al. (2008) it was found that the dynamical instability at $T = 2\pi$ was due to parametric resonances, rather than chaos. In other areas of BEC physics, the relationship between parametric resonance and dynamical instability of a BEC has already been well studied. A particular case concerns atomic behavior trap modulated periodically in time, a topic of both theoretical (Kramer et al. 2005a, b; Garcia-Ripoll et al. 1999; Kagan Yu and Manakova 2006) and experimental interest (Gemelke et al. 2005; Campbell et al. 2005; Engels et al. 2007). Here, the typical method in an analysis of the stability of small perturbations of the initial ground state configuration, rather than evaluating the excited modes in a time-dependent formalism (as in the CD method). The two techniques are closely related. Both involve in some way a decomposition of a small perturbation in terms of low-lying collective modes (Bogoliubov modes). Hence both assume relatively weak perturbations. Condensate stability analysis does not enforce orthogonalization between the evolving condensate and excitations at all times. But it has the advantage that explicit stability parameters are obtained at little computational expense by diagonalising a small stability matrix. It thus becomes easier to distiguish stable from unstable resonances; for instance, in a weak instability, the probability of a particular momentum may be growing slowly with time, but still be exponential; in contrast, rapid growth may be found in a strong but stable resonance.

In Reslen et al. (2008) no stability parameters were explicitly calculated so instability was largely identified with rapid growth at short times in non-condensate atoms, using the CD method. It was found that the condensate recovers stability after losing it and in fact that the zones of exponential instability occur over narrow ranges of the nonlinearity $g$ and kick period $T$. These correspondeded to resonant excitations of *pairs* of coupled normal modes. A simple model using Bogoliubov analysis of the initial ground state confirmed this.

In a more recent work (Monteiro et al. 2009), the stability parameters were calculated from the condensate stability matrix and mapped. It was found that in fact only coupled mode resonances give exponential behavior. Resonant driving of single Bogoliubov modes yields a far stronger condensate response (in this case, the CD method would predict much faster growth of excitations measured in terms of non-condensate population in the coupled equations) but growth is not exponential. This seems surprising at first but is well-known in other instances of BEC resonant driving. With increasing $K$ and $g$, the number of resonances which can be strongly excited by the kicking proliferate and overlap. The CD method suggests this is associated with generalized exponential instability; this regime is however beyond the scope of analyses based on decomposition on Bogoliubov modes.

Mapping the zones of stability made it clear that for strong nonlinearity $g > 1$, the Talbot time $T = 4\pi$ (or rational multiples thereof) is no longer of any significance for resonances of kicked BECs. The main resonances are displaced to other values of $T$. This is an important point, since all recent experimental studies (and many theoretical ones) have focused exclusively on values of $T = n\pi$ where $n$ is an integer or rational fraction. Based on the map, we investigate two types of resonances: (a) Linear resonances are single-mode resonances: as $g \to 0$, they evolve into the usual Talbot time (or a rational fraction) resonances of the non-interacting QKR. They are the strongest resonances but are stable. (b) Nonlinear resonances involve a coupled-pair of modes; they vanish as $g \to 0$. They can yield exponential growth, but are relatively weak: the exponential growth ceases after a finite time. Oddly, for $K \sim 0.4$–$0.8$ a regime of exponential oscillations is identified: exponential growth is interspersed with exponential decay; for $K > 0.8$, exponential growth simply saturates after a short time.

In Monteiro et al. (2009) a model including coupling between Bogoliubov modes (Beliaev and Landau terms) gave a quantitative description of novel features of the strongest "Linear" resonances. In particular an extraordinarily sharp "cut-off" in the leading resonances for $g \gtrsim 1$ was identified: this provides one of the clearest and most robust experimental signature of the effect of interactions in kicked BECs.

The suppression of the exponential growth in the "NL" resonances or the presence of exponential oscillations is not satisfactorily accounted for by the models, even with Beliaev/Landau corrections. Below we review briefly the theory of the Bogoliubov decomposition and stability matrix.

## 3. Kicked BEC Systems

We consider a BEC confined in a ring-shaped trap of radius $R$. We assume that the lateral dimension of the trap is much smaller than its circumference, and thus we are dealing with an effectively 1D system. The dynamics are those of a dimensionless 1D Gross-Pitaevskii (GP) (Zhang et al. 2004; Monteiro et al. 2009) Hamiltonian with an additional kicking potential:

$$H = H_{GP} + K \cos\theta\, f(t), \tag{1}$$

yielding a dimensionless GP equation

$$H_{GP} = -\frac{1}{2}\frac{\partial^2}{\partial\theta^2} + g|\psi(\theta,t)|^2. \tag{2}$$

We write the condensate wavefunction in the form $\psi = \psi_0 + \delta\psi$, where $\psi_0$ is the unperturbed condensate and $\delta\psi$ represent the excited components. Inserting this form in the GPE and linearizing with respect to $\delta\psi$, we can write:

$$i\hbar\frac{d}{dt}\begin{pmatrix}\delta\psi\\ \delta\psi^*\end{pmatrix} = \mathcal{L}(t)\begin{pmatrix}\delta\psi\\ \delta\psi^*\end{pmatrix}. \tag{3}$$

where,

$$\mathcal{L}(t) = \begin{pmatrix}H(t) + g|\psi|^2 & g\psi^{*2}\\ -g\psi^{*2} & -H(t) - g|\psi|^2\end{pmatrix}. \tag{4}$$

Insight into the time-evolution of small perturbations of the condensate is best described by a decomposition into a few normal modes. The effect of the nonlinearity is to provide an effective coupling between $\psi$ and $\psi^*$, so the standard way to do this is to work with a "dual vector" in (4) as elaborated in Castin and Dum (1997, 1998). Excluding the kick term for the moment, we recall that the time propagation under $H_{GP}$ can be analyzed in terms of the eigenmodes $(u_k(t), v_k(t))$ and eigenvalues of $\omega_k(t)$ of the 2×2 matrix on the right hand side of (4). Setting $\psi = 1/\sqrt{2\pi}$, the matrix can be diagonalized and there are well-known analytical expressions for the unperturbed eigenmodes:

$$(u_k(t=0), v_k(t=0)) = \begin{pmatrix}U_k\\ V_k\end{pmatrix}\frac{e^{ik\theta}}{\sqrt{2\pi}}, \tag{5}$$

where $U_k + V_k = A_k,\ U_k - V_k = A_k^{-1}$, and $A_k = \left(\frac{k^2}{2}(\frac{k^2}{2} + \frac{g}{\pi})\right)^{-1/4}$.

In order to understand the behavior at the resonances, we introduce below a simple model using the eigenmodes (5) as a basis. Writing the small perturbation in this basis:

$$\begin{pmatrix}\delta\psi(t)\\ \delta\psi^*(t)\end{pmatrix} = \sum_k b_k(t)\begin{pmatrix}U_k\\ V_k\end{pmatrix}\frac{e^{ik\theta}}{\sqrt{2\pi}} + b_k^*(t)\begin{pmatrix}V_k\\ U_k\end{pmatrix}\frac{e^{-ik\theta}}{\sqrt{2\pi}}. \tag{6}$$

Neglecting the kick, evolving the modes from some initial time $t_0$, each eigenmode $(u_k, v_k)$ simply acquires a phase i.e.:

$$b_k(t) = b_k(t_0)e^{-i\omega_k(t-t_0)}, \tag{7}$$

where $\omega_k = \sqrt{\frac{k^2}{2}(\frac{k^2}{2} + \frac{g}{\pi})}$.

After a time interval $T$, a kick is applied which couples the eigenmodes. Its effect is obtained by expressing the perturbation in a momentum basis, $\psi = \sum_l a_l(t)|l\rangle$ where $|l\rangle = \frac{e^{il\theta}}{\sqrt{2\pi}}$, and we can restrict ourselves to the symmetric subspace $a_l = a_{-l}$ of the initial condensate (parity is conserved in our system). Then, we can see by inspection that

$$a_k(t) = U_k b_k(t) + V_k b^*_{-k}(t). \tag{8}$$

Note that $b_k = b_{-k}$ for this system. Conversely, the corresponding amplitude $b_k$ in each eigenmode $k$ is given trivially from (6) using orthonormality of the momentum states and the relation $U_k^2 - V_k^2 = 1$, yielding

$$b_k(t) = U_k a_k(t) - V_k a^*_k(t). \tag{9}$$

This is a classical version of the well-known Bogoliubov transformation of quantum field theoretical Hamiltonian and becomes appropriate when there is macroscopic occupation of low-lying modes.

If the evolving condensate is given in the momentum basis, the effect of a kick operator $U_{kick} = e^{-iK\cos\theta}$ is well-known. The matrix elements:

$$U_{nl} = \langle n|U_{kick}|l\rangle = J_{n-l}(K)i^{\pm(l-n)} \tag{10}$$

The $J_{n-l}$ are Bessel functions.

The amplitudes $a_l(t)$ are given by

$$a_n(t^+) = \sum_l i^{\pm(l-n)} J_{n-l}(K)\ a_l(t^-), \tag{11}$$

where $a_n(t^+)/a_n(t^-)$ denotes the amplitude in state $|n\rangle$ just after/before the kick.

We can now define a "time-evolution" operator

$$\mathbf{a}((n+1)T) = \mathcal{L}'(T)\,\mathbf{a}(t = nT) \tag{12}$$

noting that the **a** vector of momentum amplitudes is in the dual form:

$$\mathbf{a} = (...a_{-l}, a_{-l+1}..., a_0, ..a_{+l}..., ...a^*_{-l}...a^*_0, ..a^*_{+l}...) \tag{13}$$

With the above basis $\mathcal{L}'(T)$ is the product of four simple matrices:

$$\mathcal{L}'(T) = \mathcal{B}^{-1}\mathcal{L}_{free}(T)\mathcal{B}\,U_{kick} \tag{14}$$

$\mathcal{B}$ and $\mathcal{B}^{-1}$ between the Bogoliubov and the usual plane-wave basis:

$$\mathbf{b} = B\,\mathbf{a} \tag{15}$$

Where both **a**, **b** have the dual form of (13). Their elements are easily inferred from (8) and (9) respectively and couple the two parts of the dual vector. $\mathcal{L}_{free}$ gives the free ringing of the eigenmodes and is a diagonal vector containing the phases $e^{-i\omega_l T}$ (or $e^{+i\omega_l T}$ for the lower half of the **b** dual vector). $U_{kick}$ is the action of the kick, with matrix elements from (11) (and obviously do not mix the $a_l$ with the $a_l^*$).

Since we truncate the number of modes at $n$, the matrix $\mathcal{L}'(T)$ has dimension $2n+1$-by-$2n+1$. In the present case, $a_l = a_{-l}$ and we may easily transform to a matrix of size $n+1$-by-$n+1$ using $\cos lx$ states rather than plane waves.

One might hope naively to use $\mathcal{L}'(T)$ for time propagation as in (12). But this is only provides agreement with GPE numerics for very short times, or for long time in the limit of weak driving, $K \lesssim 0.05$. Much better agreement with the long time evolution under the GPE (up to $K < 0.5$ is obtained by allowing the Bogoliubov modes to interact by means of Beliaev–Landau coupling (Monteiro et al. 2009). However the short time behavior suffices for $\mathcal{L}'(T)$ to identify regions of instability.

A more detailed analysis shows that the eigenvalues of $\mathcal{L}'(T)$ in general come in quartets $\lambda, 1/\lambda, \lambda^*, 1/\lambda^*$. Then $|\lambda_{max}| > 1$ (where $\lambda_{max}$ is the largest eigenvalue and the local Lyapunov exponent) imply dynamical instability and exponential growth in the relevant modes (at least for short times).

Parametric resonance is associated with large oscillations in the resonantly driven modes, regardless of whether they are stable or not. For comparison, we plot in Fig. 1 the behavior at the instability identified by Zhang et al. (2004). The CD method indicates rapid growth of non-condensate atoms at $T = 2\pi$ and $g \simeq 2.2$. The lower figure shows the condensate energy before, during and after the resonance. Large oscillations in energy are seen in the instability region. However, exponential behavior is not clear since we are not isolating the unstable modes.

A better indicator is to plot the average probability of mode 2 (averaged over 100 kicks) i.e. $\sum_n \frac{1}{100}|a_2(t = nT)|^2$. We map this for all $T, g$ for $K = 0.5$ in Fig. 2 (left). The right hand side maps regions of dynamical instability $|\lambda_{max}| > 1$. We analyze dynamical stability by mapping the eigenvalues of $\mathcal{L}'(T)$ for all the resonances of the lowest three excited modes. We divide the resonances into (1) the "linear" family $L(n, l)$ (i.e. those which evolve from the linear case and converge at $g = 0$ to a rational fraction of the Talbot time). They are given by the resonance condition $\omega_l T \simeq 2\pi n$. The resonance in Fig. 1a is the $L(1, 1)$ (first resonance of mode $l = 1$). (2) The "nonlinear" resonances $N_n$ and $\nu_n$ which vanish in the absence of interactions, at $g = 0$;

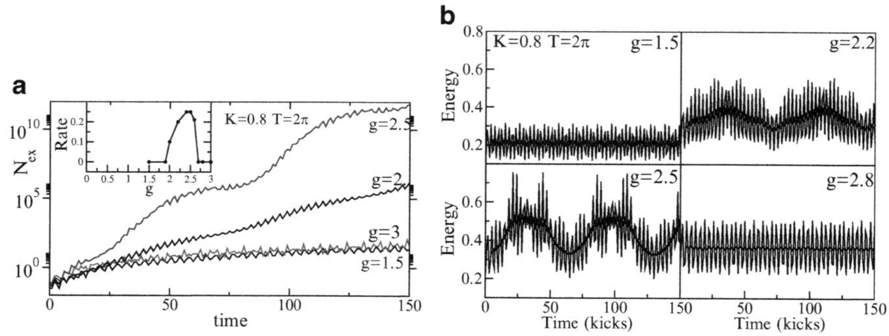


*Figure 1.* (**a**) Behavior near the instability identified by Zhang et al. (2004). Non-condensate particles for kicking period $T = 2\pi$, $K = 0.8$, $g \simeq 2$, obtained by CD method. The inset shows the rate of exponential growth; zero denotes polynomial growth or less. The graph from Reslen et al. (2008) shows that the instability is a resonance: the condensate is unstable for $g = 2$–$2.5$ but is stable for before ($g = 1.5$) and after ($g = 3.0$). (**b**) Energy oscillations as a function of time; smoothed plots are also shown. Before and after the resonance ($g = 1.5$ and $2.8$) the smoothed plots are flat. Near-resonance, ($g = 2.2$ and $2.5$) the energy shows the characteristic large, slow, resonant oscillations. The energy plots does not show exponential behavior too clearly as they does not isolate individual unstable modes

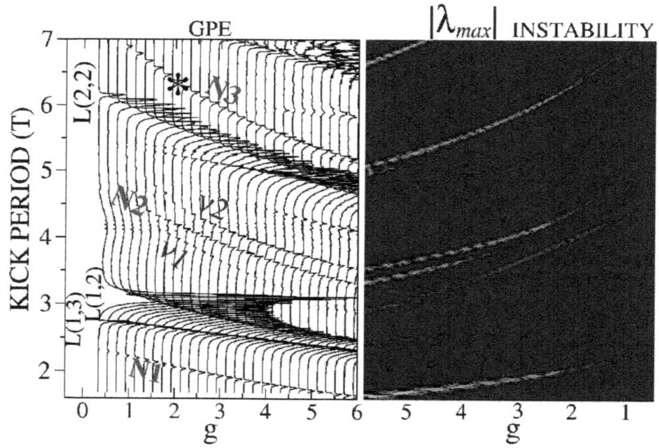


*Figure 2.* Comparison between GP equation numerics and stability analysis. Upper figure (left): Probability for mode 2 (from GPE) averaged over 100 kicks for $K = 0.5$; (right) Shows $|\lambda_{max}|$, largest eigenvalue of matrix $\mathcal{L}'(T)$. Bright regions denote $|\lambda| > 1$ and hence exponential behavior (dynamical instability). The unstable $N_1, N_2, ..N_n$ series of nonlinear resonances (which only appear for $g \gtrsim 1$) correspond to $(\omega_1 + \omega_2)T \simeq 2n\pi$. The asterisk denotes position of the instability found by Zhang et al. (2004), which we show is due to $N_3$. The $L$ series are resonances which evolved from partial or full resonances of the Talbot time at $g = 0$. They are stable, but are *much* stronger than the exponential resonances. $L(n, l)$ denotes the resonance condition $\omega_l T \simeq 2n\pi$

the $N_n$ correspond to $(\omega_1 + \omega_2)T \simeq 2\pi n$, while $\nu_n$ are somewhat analogous to "counter-propagating mode" resonances found in modulated traps (Kramer et al. 2005a, b) and imply $2\omega_n T \simeq 2\pi$.

The resonance of Fig. 1 is thus $N_3$. Contrary to the suggestion of Reslen et al. (2008) where no Liapunov exponents were calculated, we find that none of the $L(n, l)$ resonances have any $|\lambda| > 1$. They are all stable, including $L(1, 1)$, by far the strongest of all. But counter-intuitively, they are associated with a much stronger BEC response, even after a very long-time (100 kicks) than the nonlinear resonances $N_n$ and $\nu_n$ which are unstable.

A peculiar feature seen near the maxima of the unstable resonances $N_n$ series are exponential oscillations (For $K \simeq 0.4$–$0.8$ they are seen for both $N_1$ and $N_3$; however the $N_2$ resonance overlaps strongly with two other resonances so behaves somewhat differently). These are illustrated in Fig. 3.

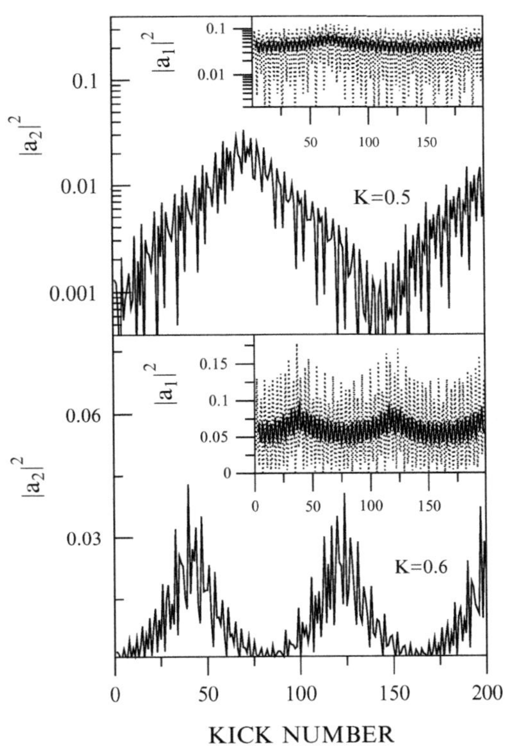


*Figure 3.* Log of Mode 2 probability of $N_3$, near the asterisk of Fig. 1. $T = 6.12$, $g = 2.5$. The exponential growth persists for only a finite time; it is then replaced by exponential decay, leading to *exponential oscillations* (log scale shown in inset). Lower figure shows similar oscillations for $K = 0.6$ (for $T = 6.17$, $g = 2.5$, not on log scale). Inset shows exponential oscillations of mode 1

The rates of exponential increase and decay correspond quite closely with the largest eigenvalues $\lambda_{max}, \lambda_{max}^{-1}$ of $\mathcal{L}'(T)$. Nonetheless, the cause of the oscillatory behavior is not clear. To our knowledge it has not been seen in other related BEC systems.

In conclusion, we have reviewed our present understanding of dynamical instability in kicked BECs.

## References


Behinaenin G., Ramareddy V., Ahmadi P. and Summy G.S.: 2006. *Phys. Rev. Lett.*, **97**, 244101

Buchleitner A., d'Arcy M.B., Fishman V., Gardiner S.A., Guarneri I., Ma Z-Y, Rebuzzini L. and Summy G.S.: 2006. *Phys. Rev. Lett.*, **96**, 164101

Campbell G. et al: 2005. *Phys. Rev. Lett.*, **96**, 020406

Casati G., Chirikov B.V., Izraelev F.M. and Ford J.: 1982, Stochastic behaviour in classical and quantum Hamiltonian systems. In: *Lecture notes in Physics*, Springer, Berlin, 93, 334

Castin Y. and Dum R.: 1997. *Phys. Rev. Lett.*, **79**, 3553

Castin Y. and Dum R.: 1998. *Phys. Rev. A*, **57**, 3008

Chabe J., Lemariä G., Grämaud B., Delande D., Szriftgiser P. and Garreau J.-C.: 2008. *Phys. Rev. Lett.* in press.

Dana I., Ramareddy V., Talukdar I. and Summy G.S.: 2008. *Phys. Rev. Lett.*, **100**, 024103

Duffy G.J., Mellish A.S., Challis K.J. and Wilson A.C.: 2004. *Phys. Rev. A*, **70**, 041602(R)

Engels P., Atherton C. and Hoefer M.A.: 2007. *Phys. Rev. Lett.*, **99** 095301

Fishman S., Grempel D.R. and Prange R.E.: 1982. *Phys. Rev. Lett.*, **49**, 509

Fishman S., Guarneri I. and Rebuzzini L.: 2004. *Phys. Rev. Lett.*, **89**, 0084101

Garcia-Ripoll J.J., Perez-Garcia V.M. and Torres P.: 1999. *Phys. Rev. Lett.*, **83**, 1715

Gardiner S.A. and Morgan S.A.: 2007. **75**, 043621

Gemelke N. et al: 2005. *Phys. Rev. Lett.*, **95**, 170404

Isella L. and Ruostekoski J.: 2006. *Phys. Rev. A*, **74**, 063625

Isella L. and Ruostekoski J.: 2006. *Phys. Rev. A*, **72**, 011601(R)

Jones P.H., Stocklin M., Hur G. and Monteiro T.S.: 2004. *Phys. Rev. Lett.*, **93**, 223002

Jones P.H., Goonasekera M., Meacher D.R., Jonckheere T. and Monteiro T.S.: 2007. *Phys. Rev. Lett.*, **98**, 073002

Kagan Yu and Manakova L.A.: 2006. cond-mat/0609159

Kanem J.F., Maneshi S., Partlow M., Spanner M. and Steinberg A.M.: 2007. *Phys. Rev. Lett.*, **98**, 083004

Kramer M., Tozzo C. and Dalfovo F.: 2005. *Phys. Rev. A*, **71**, 061602(R)

Kramer M., Tozzo C. and Dalfovo F.: 2005. *Phys. Rev. A*, **72**, 023613

Lepers M., Zehnlä V. and Garreau J.-C.: 2008. *Phys. Rev. A*, **77**, 043628

Liu J., Zhang C., Raizen M.G. and Niu Q.: 2006. *Phys. Rev. A.*, **73**, 013601

Ma Z.-Y., Burnett K., d'Arcy M.B. and Gardiner S.A.: 2006. *Phys. Rev. A*, **73**, 013401

Martin A.D., Adams C.S. and Gardiner S.A.: 2007. *Phys. Rev. Lett.*, **98**, 020402

Mieck B. and Graham R.: 2004. *J. Phys. A: Math. Gen.*, **37**, L581

Monteiro T.S., Rancon A., Ruostekoski J.: 2009. *Phys. Rev. Lett.*, in press.

Moore F.L., Robinson J.C., Bharucha C.F., Sundaram B. and Raizen M.G.: 1995. *Phys. Rev. Lett.*, **75**, 4598

Poletti D., Benenti G., Casati G. and Li B.: 2007. *Phys. Rev. A*, **76**, 023421

Poletti D., Alexander T.J., Ostrovskaya E.A., Li B. and Kivshar Y.S.: 2008. *Phys. Rev. Lett.*, **101**, 150403

Raizen M.G.: 1999. *Adv. At. Mol. Opt. Phys.*, **41**, 43

Rebuzzini L., Artuso R., Fishman S. and Guarneri I.: 2007. *Phys. Rev. A*, **76**, 031603

Reslen J., Creffield C.E. and Monteiro T.S.: 2008. *Phys. Rev. A*, **77**, 043621

Ruostekoski J. and Isella L.: 2005. *Phys. Rev. Lett.*, **95**, 110403

Ryu C., Andersen M.F., Vaziri A., d'Arcy M.B., Grossman J.M., Helmerson K. and Phillips W.D.: 2006. *Phys. Rev. Lett.*, **96**, 160403

Sadgrove M., Horikoshi M., Sekimura T. and Nakagawa K.: 2007. *Phys. Rev. Lett.*, **99**, 043002

Saunders M., Halkyard P.L., Challis K.J. and Gardiner S.A.: 2007. *Phys. Rev. A*, **76**, 043415

Shepelyansky D.L.: 1993. *Phys. Rev. Lett.*, **70**, 1787

Szriftgiser P., Ringot J., Delande D. and Garreau J.C.: 2002. *Phys. Rev. Lett.*, **89**, 224101

Wimberger S., Guarneri I. and Fishman S.: 2004. *Phys. Rev. Lett.*, **92**, 084102

Wimberger S., Mannella R., Morsch O. and Arimondo E.: 2005. *Phys. Rev. Lett.*, **94**, 130404

Zhang C., Liu J., Raizen M.G. and Niu Q.: 2004. *Phys. Rev. Lett.*, **92**, 054101

# VORTICES IN THE GROUND STATE OF SPINOR BOSE–EINSTEIN CONDENSATES

A.F. Sadreev
*Kirensky Institute of Physics, 660036 Krasnoyarsk, Russia; almsa@ifm.liu.se*

**Abstract.** We have calculated the $F = 1$ ground state of a spinor Bose–Einstein condensate trapped harmonical potential with an applied Ioffe–Pitchard magnetic field. The vortex phase diagram is found in the plane spanned by perpendicular and longitudinal magnetic fields. The ferromagnetic condensate has two vortex phases which differ by winding number in the spinor components. The two vortices for the $F_z = -1$ antiferromagnetic condensate are separated in space. Moreover we considered an average local spin $|\langle\vec{S}\rangle|$ to testify to what extend it is parallel to magnetic field (the nonadiabatic effects). We have shown that the effects are important at vortex cores.

**Key words:** Vortices; Spinor Bose–Eistein condensates

## 1. Introduction

Vortices appear in many physical systems from tornadoes and bathtub whirlpools to type-II superconductors and rotating Bose–Einstein condensates. Interest to vortices in different media exists long time ago. Apparently the first observation of vortices is related to water flow in tube. As Iwo Bialynicki-Birula et al. (2000) cite, "Vortices have been a source of fascination since time immemorial. Empedocles, Aristotle and Descartes tried to explain the formation of the Earth, its gravity, and the dynamics of the whole solar system as due to primordial cosmic vortices." Berry in his course "Singularities in waves and rays" (Berry 1981) presents another reference to vortices in ocean tides studied by Whewell (1833).

The physical significance of the singularities of the phase of quantum-mechanical wave functions has been recognized by Dirac in his work on magnetic monopoles (Dirac 1931). The hydrodynamic formulation of the Schrödinger theory discovered by Madelung (1926) provided a vivid interpretation of the lines in space where the phase is singular. These are simply the vortex lines in the flow of the probability fluid. Theoretical foundation for the study of these fascinating and ubiquitous objects was provided by Nye and Berry in an important series of papers (Berry 1981; Nye and Berry 1974;

G. Casati and D. Matrasulov (eds.), *Complex Phenomena in Nanoscale Systems*,
NATO Science for Peace and Security Series B: Physics and Biophysics,
© Springer Science+Business Media B.V. 2009

Berry 1977) and by Hirschfelder et al. (1974a); Hirschfelder et al. (1974b). The vortices described in these fundamental papers are the phase singularities of the complex scalar wave field $\psi(x, y)$ which satisfies the Helmholtz wave equation

$$(\nabla^2 + E)\psi = 0. \tag{1}$$

Explicit descriptions of the nodal points as phase singularities or topological charges associated with a complex wave function are given in many articles, for example, Berry (1981), Nye and Berry (1974), Hirschfelder et al. (1974a), Shvartsman and Freund (1994), and Freund and Kessler (1996). There are two separate sets of nodal lines at which either $Re\psi$ or $Im\psi$ vanish. The intersections of the two sets at which $Re\psi = 0, Im\psi = 0$ define the nodal points. Let us write the wave function as $\psi(x, y) = \sqrt{\rho(x, y)}\exp(i\theta(x, y))$, where $\theta(x, y)$ and $\rho(x, y)$ denote the phase and norm. As Dirac demonstrated already in 1931 (Dirac 1931) nodal points give rise to current vortices. This means that when the following loop integral encloses a nodal point one has

$$\oint \mathbf{v} d\mathbf{r} = \oint \nabla\theta d\mathbf{r} = \pm 2\pi. \tag{2}$$

Therefore, the nodal points of complex wave function are the centers of current vortices. Although direct observation of vortices in quantum mechanics is hardly possible, the theory of them shortly reviewed above has been applied in different wave phenomena. That is clear because of universality of the Helmholtz equation (1). First of all, the vortices were studied long time ago for acoustic vibrations (Ebeling 1984; Waterhouse 1985, 1987) whose description is completely equivalent to the quantum vortices. Recently the Marburg group of Stöckmann confirmed a mass of theoretical predictions for vortices and saddles experimentally using the one-to-one correspondence between the Poynting vector in a microwave plane billiard and the probability current density in the corresponding quantum system (Barth and Stöckmann 2002; Kim et al. 2003). Beautiful visualization of phase singularities was done by a highly isotropic microchip laser with a large Fresnel number (Chen and Huang 2003; Chen et al. 2006a, b) and monochromatic light beams (Soskin et al. 1997; Courtial and O'Holleran 2007; Flossmann et al. 2008).

However electromagnetic systems like microwave resonators or elastic systems like elastic membranes are described in general by vectorial fields (electromagnetic and displacement respectively). In particular, in elastic billiards there are two types of elastic eigen modes. The flexural modes with displacement $\psi$ perpendicular to the plane of the plate are well described by the scalar biharmonic Kirchoff–Love equation (Landau and Lifshitz 1959)

$$D\nabla^4\psi = \rho h\Omega^2\psi. \tag{3}$$

Here, $h$ is the thickness of the plate, and $D$ denotes the flexural rigidity, given by $D = Eh^3/12(1-\sigma^2)$, where $E$ is Young's modulus, $\sigma$ is Poisson's ratio, and $\rho$ is the density. Solutions of (3) are characterized by nodal lines near which fine powder is collected under vibrations are visualized as the Chladni patterns (Waller 1952). As dependent on size of fine particles a very thin powder can collect at anti-nodal regions where the amplitude of vibrations is maximal (Thomas and Squires 1998; Aronson and Tsimring 2006; Dorrestijn et al. 2007).

Other modes have vectorial displacement in the plane of the plate. They are described by a two-dimensional Navier-Cauchy equation for the in-plane displacement vector (Landau and Lifshitz 1959; Achenbach 1973)

$$\mu\nabla^2\mathbf{u} + (\lambda+\mu)\nabla(\nabla\mathbf{u}) + \rho\Omega^2\mathbf{u} = 0 \tag{4}$$

where $\mathbf{u}(x, y)$ is the displacement field in the plate, $\lambda, \mu$ are the material dependent Lamé coefficients. Introducing elastic potentials $\psi$ and $\mathbf{A}$ with the help of the Helmholtz (Achenbach 1973) decomposition the displacement field $\mathbf{u}$ could be written as

$$\mathbf{u} = \mathbf{u}_l + \mathbf{u}_t, \quad \mathbf{u}_l = \nabla\psi, \; \mathbf{u}_t = \nabla\times\mathbf{A}. \tag{5}$$

Then (4) reduces to two Helmholtz equations for the elastic potentials

$$\begin{aligned} -\nabla^2\psi &= k_l^2\psi, \\ -\nabla^2\mathbf{A} &= k_t^2\mathbf{A}. \end{aligned} \tag{6}$$

Here $k_l = \omega/c_l$, $k_t = \omega/c_t$ are the wave numbers for the longitudinal and transverse waves, respectively and $\omega^2 = \rho\Omega^2/E$. In the 2D-case potential $\mathbf{A}$ has only one none-zero component $A_z$ and the dimensionless longitudinal and transverse sound velocities $c_{l,t}$ are given by

$$c_l^2 = \frac{1}{1-\sigma^2}, \; c_t^2 = \frac{1}{2(1+\sigma)}, \tag{7}$$

where $\sigma$ is Poisson's ratio (Landau and Lifshitz 1959; Achenbach 1973). $\sigma$ is function of the Lamé coefficients (Landau and Lifshitz 1959; Achenbach 1973).

The flexural modes are characterized by nodal lines while the vectorial in-plane displacements $\mathbf{u}(x, y) = (u(x, y)\ v(x, y))$ in closed elastic plate have nodal points

$$u(x_0, y_0) = 0, \quad v(x_0, y_0) = 0, \tag{8}$$

at the point $\mathbf{x}_0 = (x_0, y_0)$. They are specified by the Poincaré index (topological charge) (Dubrovin et al. 1985; Liu and Mazenko 1992). Here, we consider the statistical properties of NPs, in elastic random waves. In this case, only structurally stable zeros of the Poincaré indexes $q \pm 1$ occur (Dennis 2003)

$$q = sign(\det M_{\mathbf{x}_0}) = sign(\lambda_1 \lambda_2), \qquad M = \begin{pmatrix} \frac{\partial u}{\partial x} & \frac{\partial u}{\partial y} \\ \frac{\partial v}{\partial x} & \frac{\partial v}{\partial y}, \end{pmatrix} \tag{9}$$

where $\lambda_{1,2}$ are eigenvalues of matrix $M$ at NP $\mathbf{x}_0$. Depending on these eigenvalues NPs split on the four types (Dubrovin et al. 1985; Liu and Mazenko 1992): (1) centers for imaginary $\lambda_{1,2}$ with the index $q = 1$; (2) nodes for real $\lambda_{1,2}$ with the same sign and $q = 1$; (3) focuses for complex $\lambda_1 = \lambda_2^*$ with $q = 1$; and (4) saddles for real $\lambda_{1,2}$ with opposite sign and $q = -1$. Eigenvalues of matrix $M$ are $\frac{u_x+v_y}{2} \pm \sqrt{(\frac{u_x+v_y}{2})^2 - J} = \frac{u_x+v_y}{2} \pm \sqrt{D}/4$ where we have introduced

$$S = (u_x + v_y)^2 - 4J, \qquad J = \det(M). \tag{10}$$

If $q = 1,\ S > 0$ NP can be classified as node, while for $q = 1,\ S < 0$ we have NP as focus (center for particular case $u_x + v_y = 0$). At last for $q = -1$ NP is a saddle.

In order to clearly show these types of NP we consider simple superposition (Maksimov and Sadreev 2007, 2008)

$$u(\mathbf{x}) = \sqrt{\frac{2(1-\gamma)}{N}} \sum_{n=1}^{N} \cos\phi_{ln} \exp[i\mathbf{k}_{ln}\mathbf{x}] + \sqrt{\frac{2\gamma}{N}} \sum_{n=1}^{N} \sin\phi_{tn} \exp[i\mathbf{k}_{tn}\mathbf{x}]$$

$$v(\mathbf{x}) = \sqrt{\frac{2(1-\gamma)}{N}} \sum_{n=1}^{N} \sin\phi_{ln} \exp[i\mathbf{k}_{ln}\mathbf{x}] - \sqrt{\frac{2\gamma}{N}} \sum_{n=1}^{N} \cos\phi_{tn} \exp[i\mathbf{k}_{tn}\mathbf{x}] \tag{11}$$

consisted of only three plane waves ($N = 3$) with the wave vectors directed by $2\pi/3$ angles relative each other. Here where $\phi_{ln}$, $\phi_{tn}$ are the angles between $\mathbf{k}_{ln}$, $\mathbf{k}_{tn}$ and the x-axis respectively. The prefactors $\sqrt{\gamma}$, $\sqrt{1-\gamma}$ are chosen from the normalization condition $\langle \mathbf{u}^\dagger \mathbf{u} \rangle = 1$ where $\langle \ldots \rangle$ means integration over the elastic membrane. For $\gamma = 1$ we have the vectorial "electric" field $\mathbf{u} = \nabla\psi$. Then at NP the "scalar potential" $\psi$ achieves maximum or minimum for $q = 1$ or has a saddle point for $q = -2$. Respectively, NP of $\mathbf{u}$ is a node or a saddle as demonstrated in Fig. 1. For $\gamma = 1$ we have the vectorial "magnetic" field $\mathbf{u} = \nabla \times \mathbf{A}$ NPs of which are centers of displacements and saddles as demonstrated in Fig. 2.

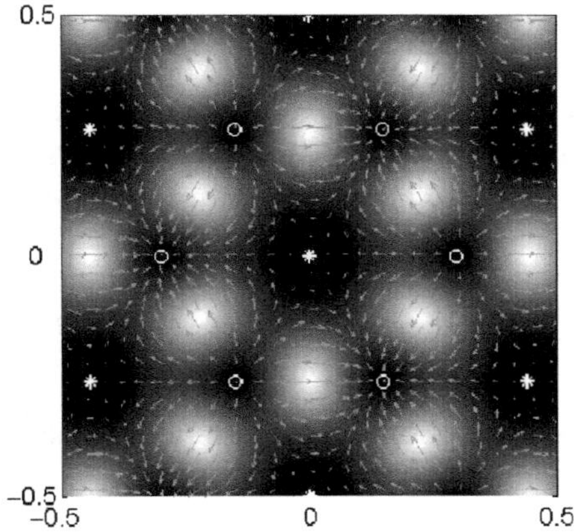


*Figure 1.* (Color online) Intensity of in-plane displacement $u^2 + v^2$ and vectorial field $\mathbf{u} = \nabla\psi$ shown by arrows for superposition of three plane longitudinal waves with the wave vectors directed by $2\pi/3$ angles relative each other and zero random phases $\theta_{ln} = 0$. $k_l = 20, \sigma = 0.5$. NPs are shown by circles (nodes), and by stars (saddles)

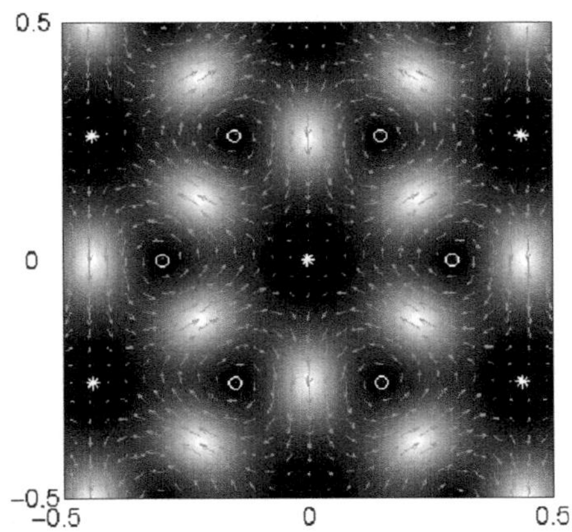


*Figure 2.* (Color online) Intensity of in-plane displacement $u^2 + v^2$ and vectorial field $\mathbf{u} = \nabla \times \mathbf{A}$ for superposition of three plane transverse waves with the wave vectors directed by $2\pi/3$ angles relative each other and zero random phases $\theta_{tn} = 0$. $k = 10, \sigma = 0.5$. Saddles are shown by stars, and centers by circles

## 2. Vortices in the Ground State

The mean energy of two-dimensional quantum system equals

$$\langle E \rangle = \int d^3\mathbf{r} \left[ \frac{\hbar^2}{2m} |\nabla\psi|^2 + U(\mathbf{r})|\psi(\mathbf{r})|^2 \right]. \tag{12}$$

From here we immediately obtain that the ground state $\psi_0(\mathbf{r})$ has no nodal lines (points in the two-dimensional case) because each nodal line increases the gradient term in (12). If there were nodal points we had vortical currents around the nodal point which would supply kinetical energy. The wave systems obeying to the Helmholtz equation (1) have the nodeless ground state too. However this simple argument forbidding nodal points in the ground state can not applicable for charged quantum particles in external magnetic field with the mean energy

$$\langle E\rangle = \int d^3\mathbf{r}\left[\frac{1}{2m}|(i\hbar\nabla + \frac{e}{c}\mathbf{A})\psi|^2 + U(x,y)|\psi(x,y)|^2 + \frac{1}{8\pi}(\nabla\times\mathbf{A})^2\right]. \quad (13)$$

For uniform external magnetic field the solution of the Schrödinger equation is well known (Landau and Lifshitz 1975)

$$\psi(\mathbf{r}) = exp(ik_x x + ik_z z)\phi_m(y), \quad (14)$$

where $\phi_m(y)$ are the eigen functions of the harmonic oscillator. Then for the ground state we have again nodeless state $k_x = 0, k_z = 0, m = 0$.

Firstly, vortices in the ground state were discovered by Abrikosov in 1957 superconductors of the second order for Landau–Ginzburg description (the Abrikosov vortices) (Saint-James and Sarma 1969)

$$F = \tau|\psi|^2 + \frac{\beta}{2}|\psi|^4 + \frac{1}{4m}(i\hbar\nabla + \frac{2e}{c}\mathbf{A})\psi|^2 + \frac{1}{8\pi}(\nabla\times\mathbf{A})^2 \quad (15)$$

where $\tau = \frac{T-T_c}{T_c}$, $\psi$ is a condensate of Cooper pairs. Physical origin of the Abrikosov vortices is related to an existence of two correlation lengths (Saint-James and Sarma 1969). The first one is given by the superconducting order parameter $\psi$

$$\xi = \frac{\hbar}{\sqrt{4m|\tau|}}.$$

The second length, the London penetration depth is given by magnetic field

$$\lambda = \sqrt{\frac{\beta}{8\pi m|\tau|}}\frac{c}{\hbar e}.$$

Then, if $\xi < \lambda$ the loss of energy around the nodal point might be less than a gain of energy by penetration of magnetic field through the vortex core for the field exceeding the low critical magnetic field.

Vortices in the ground state of quantum system of many interacting electrons are the next example. In the fractional quantum Hall effect (Chakraborty and Pietiläinen 1995) the external magnetic field penetrates through the two-dimensional (2D) electron system at the vortex positions. Every vortex corresponds to a single magnetic field flux quantum. For the quantum Hall state of the filling factor $\nu = 1$, a single vortex is on top of each electron. For stronger B, more vortices appear and, e.g. the Laughlin state of $\nu = 1/3$ attaches three vortices on top of each electron. The vortices keep electrons farther apart, reducing the interaction energy and causing strong correlations between the electrons (Saarikoski et al. 2004).

## 3. Vortices in the Ground State of Spinor Bose–Einstein Condensates

Our major part of the paper is to show an existence of vortices in the ground state of the spinor Bose–Einstein condensates (BEC) confined in an external inhomogeneous magnetic field. We argue that the quadratic Zeeman energy gives rise to the vortex states. This energy was ignored in previous theoretical studies, although it exists in experimental systems (Isoshima and Yip 2006). Despite these extensive studies on vortices, studies on simpler systems such as the spinor BEC of atoms in a uniform magnetic have not been conducted till 1996. The existence of different vortices has been predicted in the seminal papers by Ho and Shenoy (1996), Ho (1998), and Ohmi and Machida (1998) and was further studied by several others (Yip 1999; Isoshima et al. 2000; Isoshima and Machida 2002; Mizushima et al. 2002a, 2004; Martikainen et al. 2002; Bulgakov and Sadreev 2003; Mueller 2004; Zhang et al. 2007). In the spinor BEC, a vortex state is inseparably related to the spin texture. By utilizing this relation, it is possible to imprint the vorticity without rotating the cloud, as proved experimentally (Leanhardt et al. 2003; Leanhardt 2003). By applying a quadrupole (nonuniform) magnetic field, the spin texture of the BEC is forced, resulting in the nucleation of the vortex.

One of the recent developments in Bose–Einstein condensates (BEC) in atomic gases is the study of dilute Bose gases with spin degrees of freedom. The first realization of such a system is found in optically trapped $^{23}Na$, which is a spin-1 Bose gas (Anderson et al. 1995). The nature of the spinor condensate depends on the magnetic interaction. In zero magnetic field the spinor condensate can be either ferromagnetic or antiferromagnetic ("polar"). Both have very different properties (Davis et al. 1995; Petrich et al. 1995).

In the present paper we study the ground state structure of BEC described by a constituent atom with the hyperfine state $|F| = 1$ ($F_z = \pm 1, 0$) where the order parameter of the Bose condensate is characterized by three components: $\Psi_\alpha, \alpha = \pm 1, 0$ similarly to the spin part of superfluid $^3He$. However these degrees of freedom bring about a remarkable difference between the BEC of alkali atoms and that of $^4He$. The hyperfine spin aligns along the direction of the local magnetic field when a BEC is magnetically trapped. Then, even though the alkali atoms carry spins, they behave like scalar particles. In contrast, the spin of the alkali atoms is an important degree of freedom in an optical trap formed by the optical dipole force which confines atoms in all hyperfine states $F_z = \pm 1, 0$ (Stamper-Kurn et al. 1998). In order to manipulate by spin states we assume at the same time that the BEC is created in a magnetic Ioffe–Pitchard trap (Miesner et al. 1999).

Following to Ohmi and Machida (1998) and Isoshima and Machida (2002) we introduce the basis set $|x>, |y>, |z>$ defined by $F_i|i>= 0$, $i = x, y, z$. The order parameter is then expressed via a three-dimensional vector $\Psi_i$ where

$$|\Psi>= \Psi_x|x> + \Psi_y|y> + \Psi_z|z> . \tag{16}$$

$\vec{\Psi}$ behaves as a vector under spin space rotation. In what follows the Latin indexes define the XYZ basis (16) while the Greek indexes denote the z-quantized basis with $F_z = \pm 1, 0$.

In particular a mean value of spin is equal to

$$\langle \vec{F}(\mathbf{r})\rangle = \Psi^*(\mathbf{r})_\alpha \vec{F}_{\alpha\beta}\Psi(\mathbf{r})_\beta \tag{17}$$

where $\vec{F}_{\alpha\beta}$ are the matrix elements of the spin operators $F_i$ in the basis (16). We write the order parameter via the Bose condensate density $n$

$$\Psi_i(\mathbf{r}) = \xi_i^*(\mathbf{r})\sqrt{n(\mathbf{r})}, \tag{18}$$

and the average local spin via normalized spinor $\xi_\alpha$

$$\langle \vec{S}(\mathbf{r})\rangle = \xi_\alpha \vec{F}_{\alpha\beta}\xi_\beta. \tag{19}$$

In terms of the order parameter (16) the BEC free energy density has the form (Ho 1998; Ohmi and Machida 1998)

$$\begin{aligned} H = {} & \Psi_i^*\left(-\tfrac{\hbar^2}{2m}\nabla^2 + U(\mathbf{r}) - \mu\right)\Psi_i + \tfrac{1}{2}g_1(\Psi_i^*\Psi_i)^2 \\ & + \tfrac{1}{2}g_2(\Psi_i^*\Psi_i^*)(\Psi_j\Psi_j) + i\gamma_\mu\epsilon_{ijk}B_k\Psi_i^*\Psi_j, \end{aligned} \tag{20}$$

where

$$U(\mathbf{r}) = m(\omega_z^2 z^2 + \omega r^2)/2 \tag{21}$$

is the potential of the optical trap, $\gamma_\mu$ is the gyromagnetic ratio and $\mathbf{B}(\mathbf{r})$ is magnetic field of the Ioffe–Pitchard trap. The two interaction constants $g_1$ and $g_2$ are characteristics of the three-component order parameter which represent the spin degrees of freedom of the condensate. The two nonlinear terms in (20) originate from the interactions (Ohmi and Machida 1998)

$$\frac{1}{2}g_n n^2 + \frac{1}{2}g_s\langle\mathbf{F}\rangle^2 \tag{22}$$

where $g_1 = g_s + g_n, g_2 = -g_s$. As shown by Klausen et al. (2001), the spin interaction of $^{87}Rb$ is ferromagnetic ($g_s < 0$), while for $^{23}Na$ this interaction is antiferromagnetic ($g_s > 0$) (Stenger et al. 1998).

When the system is uniform and infinitely large, the ground state is either ferromagnetic of antiferromagnetic (Ho 1998; Ohmi and Machida 1998). However a rich variety of topological defects have been predicted (Ho 1998; Ohmi and Machida 1998; Ho and Shenoy 1996; Yip 1999; Khawaja and Stoof 2001; Mizushima et al. 2002b). Ho and Shenoy (1996) have shown that the spatial variations of the magnetic field **B** give rise to vortical ground state. In particular these variations are necessary to produce the magnetic trapping. The basic assumption of Ho and Shenoy is that the spin state $\Psi_i(\mathbf{r}) = \xi_i(\mathbf{r})\Phi(\mathbf{r})$, defined by the normalized spinor $\xi_i$ is aligned with the magnetic field. This approach has established local a spin-gauge symmetry of the condensate. It means that a local gauge $U(1)$ transformation is undoing by a local spin rotation. Nodal points of the scalar field $\Phi$ define a vortices. Yip (1999) has considered composite vortices in the spin-1 BEC in a rotating trap. These vortices display interesting internal structure. They may have broken cylindrical symmetry with nodes of the order parameter of individual components appearing at positions other than the trap center.

Topological defects similar to composite vortices, called skyrmion in general, have been proposed in the spinor BEC (Khawaja and Stoof 2001; Mizushima et al. 2002b; Marzlin et al. 2000). However it was shown that in the ferromagnetic spin-1 BEC trapped in a harmonic potential, the skyrmions or composite vortices are not thermodynamically stable without rotation (Khawaja and Stoof 2001; Mizushima et al. 2002b). The skyrmions were shown to be favored over the singular vortices and other non-axis-symmetric vortices. Following (Mizushima et al. 2002b) we introduce a specification of different vortex phase winding numbers as $(m_1, m_0, m_{-1})$ for the condensate wave function $(\Psi_1, \Psi_0, \Psi_{-1})$ with $m_\alpha = 0, \pm 1, \pm 2, \ldots$. We show that the vortices with different winding numbers are stable even without rotation in the Ioffe–Pitchard trap for the ground state.

## 4. Gross–Pitaevskii Equations

We consider the ground state of the spin-1 BEC which is uniform along the z axis. We introduce cylindrical coordinates $\mathbf{r} = (r, \varphi, z)$. Suppose that a Ioffe–Pitchard magnetic field

$$\mathbf{B} = (B_\perp(r)\cos\varphi, -B_\perp(r)\sin\varphi, B_z) \tag{23}$$

is applied to the system. The trapping potential (21) gives rise to a characteristic length $d = \sqrt{\frac{\hbar}{2m\omega}}$ and a characteristic energy $E_0 = \hbar\omega$ which allow to write the dimensionless form of the free energy density (20)

$$\tilde{H} = \frac{g_1}{E_0^2 d^3} H = \psi_i^*(-\nabla^2 + v(\rho) - \tilde{\mu})\psi_i$$
$$+\tfrac{1}{2}(\psi_i^*\psi_i)^2 + \tfrac{\tilde{g}}{2}(\psi_i^*\psi_i^*)(\psi_j\psi_j) + i\epsilon_{ijk}b_k\psi_i^*\psi_j, \tag{24}$$

where

$$\tilde{\mu} = \frac{\mu}{\hbar\omega}, \quad \mathbf{b} = \frac{\gamma_\mu \mathbf{B}}{\hbar\omega}, \quad \vec{\psi} = \sqrt{\frac{g_1}{\hbar\omega}}\vec{\Psi}, \quad \tilde{g} = \frac{g_2}{g_1}. \tag{25}$$

Since the spin-1 BEC is uniform along the z axis we fix the linear density of the Bose gas by condition

$$\int d^2\mathbf{r}|\Psi_j|^2 = \frac{N}{L}.$$

Substituting notations (25) and the coupling constants (Ho 1998)

$$g_1 = \frac{4\pi\hbar^2}{m}a_2, \quad g_2 = \frac{4\pi\hbar^2}{3m}(a_0 - a_2), \tag{26}$$

we obtain

$$\int d^2\vec{\rho}\,|\psi_i|^2 = 8\pi a_2 \frac{N}{L}, \tag{27}$$

where $\vec{\rho}$ is a dimensionless two-dimensional radius vector. For the case of $^{23}Na$ $\tilde{g}$ is negative ($\approx -0.1$) , and an external magnetic field is uniform, then the ground state is antiferromagnetic. For the case of $^{87}Rb$ $\tilde{g}$ is positive ($\approx 0.03$), and the ground state is ferromagnetic. For a strong magnetic field or small BEC density it is reasonable to consider $\tilde{g} = 0$ as the first step. Then the Gross–Pitaevskii equation in the z-quantized basis $\psi_\alpha$ takes the following form

$$(-\nabla^2 + v(\rho) - \tilde{\mu})\psi_\alpha + n(\rho)\psi_\alpha - E_{\alpha\beta}\psi_\beta = 0, \tag{28}$$

where

$$E = \begin{pmatrix} b_z & \frac{b_\perp(\rho)}{\sqrt{2}}e^{i\varphi} & 0 \\ \frac{b_\perp(\rho)}{\sqrt{2}}e^{-i\varphi} & 0 & \frac{b_\perp(\rho)}{\sqrt{2}}e^{i\varphi} \\ 0 & \frac{b_\perp(\rho)}{\sqrt{2}}e^{-i\varphi} & -b_z \end{pmatrix}. \tag{29}$$

One can see that (28) give rise to a separation of variables $\rho$ and $\varphi$. Equations (28) and (29) imply that $\psi_\alpha = \psi_{\alpha-1}e^{i\varphi}$. This equality gives us the simple relation

$$m_\alpha = m_{\alpha-1} + 1 \tag{30}$$

between vortical winding numbers $m_\alpha$ of the spinor component $\psi_\alpha$:

On the one hand, for the strong magnetic field $b_z$ the spinor component $\psi_1$ is prevailing. On the other hand, the kinetic energy prevents this component to have nodes. Therefore the solution for the ground state has the form

$$\begin{pmatrix} h_1(\rho) \\ h_0(\rho)e^{-i\varphi} \\ h_{-1}(\rho)e^{-2i\varphi} \end{pmatrix} \tag{31}$$

for which the Gross–Pitaevskii equation is

$$\begin{aligned}
\frac{d^2h_1}{d\rho^2} + \frac{1}{\rho}\frac{dh_1}{d\rho} - (v + n - \tilde{\mu})h_1 + b_z h_1 + \frac{b_\perp}{\sqrt{2}}h_0 &= 0 \\
\frac{d^2h_{-1}}{d\rho^2} + \frac{1}{\rho}\frac{dh_{-1}}{d\rho} - (v + n + \frac{4}{\rho^2} - \tilde{\mu})h_{-1} - b_z h_{-1} + \frac{b_\perp}{\sqrt{2}}h_0 &= 0 \\
\frac{d^2h_0}{d\rho^2} + \frac{1}{\rho}\frac{dh_0}{d\rho} - (v + n + \frac{1}{\rho^2} - \tilde{\mu})h_0 + \frac{b_\perp}{\sqrt{2}}(h_1 + h_{-1}) &= 0.
\end{aligned}$$

Here

$$v(\rho) = \frac{1}{4}\rho^2, \quad b_\perp = b\rho. \tag{32}$$

with obvious normalization condition $n(\rho) = h_\alpha(\rho)^2$.

For small longitudinal magnetic field $b_z$ a different solution of the Gross–Pitaevskii equation

$$\begin{pmatrix} f_1(\rho)e^{i\varphi} \\ f_0(\rho) \\ f_{-1}(\rho)e^{-i\varphi} \end{pmatrix} \tag{33}$$

could be favorable for the ground state. Substituting (33) into the Gross–Pitaevskii equation (28) we obtain

$$\begin{aligned}
\frac{d^2f_1}{d\rho^2} + \frac{1}{\rho}\frac{df_1}{d\rho} - (v + n + \frac{1}{\rho^2} - \tilde{\mu})f_1 + b_z f_1 + \frac{b_\perp}{\sqrt{2}}f_0 &= 0 \\
\frac{d^2f_{-1}}{d\rho^2} + \frac{1}{\rho}\frac{df_{-1}}{d\rho} - (v + n + \frac{1}{\rho^2} - \tilde{\mu})f_{-1} - b_z f_{-1} + \frac{b_\perp}{\sqrt{2}}f_0 &= 0 \\
\frac{d^2f_0}{d\rho^2} + \frac{1}{\rho}\frac{df_0}{d\rho} - (v + n - \tilde{\mu})f_0 + \frac{b_\perp}{\sqrt{2}}(f_1 + f_{-1}) &= 0.
\end{aligned}$$

## 5. Numerical Results

In order to find the ground state of the spin-1 BEC for $\tilde{g} = 0$ we numerically solved (32) and (34). The ground states corresponed to solutions (31) and (33) were chosen by the minimum of the energy of the BEC. For computations, since in the magnetic field of the Ioffe–Pitchard trap magnetization does not conserve, we fixed the dimensionless linear density (27) but not magnetization similar to Ho (1998), Ho and Shenoy (1996), and Isoshima and Machida (2002). We took $N/L = 1000$ fitting the chemical potential $\tilde{\mu}$ in the Gross–Pitaevskii equations. The vortex phase diagram in the plane spanned by the perpendicular magnetic field $b$ and longitudinal one $b_z$ for is shown in Fig. 3 $\tilde{g} = 0$. As expected the vortex phase $(0\ -1\ -2)$ is substituted by the

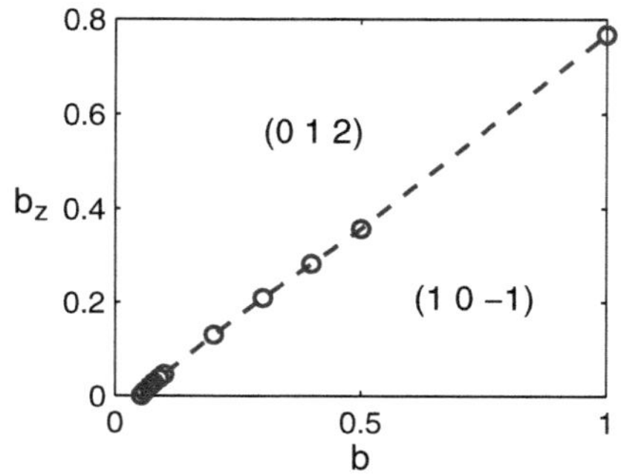


*Figure 3.* Vortex phase diagram of the spin-1 BEC in the Ioffe–Pitchard trap for $\tilde{g} = 0$

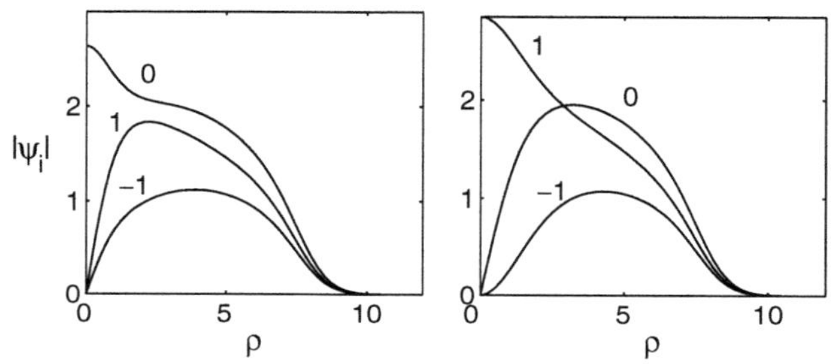


*Figure 4.* The radial behavior of the $(1\ 0 - 1)$ vortex at $b_z = 0.7, b = 1$ and of the $(0 - 1 - 2)$ vortex at $b_z = 0.8, b = 1$

vortex phase $(1\ 0\ -1)$ when the magnetic field $b_z$ is decreased as shown in Fig. 3. One can see that the winding number rule (30) holds for each vortex phase. Figure 4 shows the radial behavior of the spinor components in the vortex phases presented in Fig. 3. We also performed a computation for the ground state of the ferromagnetic case $\tilde{g} = 0.03$ ($^{87}Rb$) using the Metropolis procedure with a total number of sites of the order 50,000. For this case the vortex phase diagram shown in Fig. 3 is slightly deformed. However deviation of the effective constant $\tilde{g}$ from zero gives rise to a solution which violates rotational symmetry of the Bose condensate around the z-axis.

For the antiferromagnetic case $\tilde{g} = -0.1$ ($^{23}Na$) the vortex phase diagram changes. The rule (30) still holds. However a new vortex phase appears in which two nodal points of the component $\psi_{-1}$ are spacely separated as shown in Fig. 5. We denote this kind of vortices as two prime. At the solid line between vortex phases $(0\ -1\ -2')$ and $(0\ -1\ -2)$ shown in Fig. 6 the vortices of the component $\psi_{-1}$ are joining together at $\rho = 0$. As the magnetic field $b$ decreases the distance between the vortices increases. For small $b$ these vortices go to the region $\rho \gg 1$ where the wave function is exponentially small because of optical trapping. As a result the vortices become practically invisible. The dashed line in Fig. 6 shows where this happens.

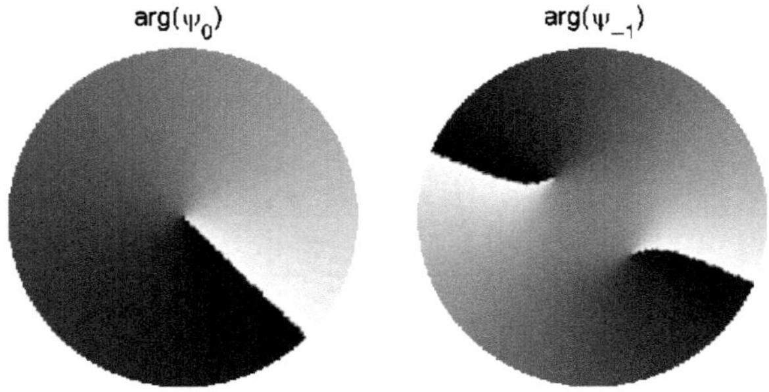


*Figure 5.* Phase images $arg(\psi_i)$ in the vortex phase $(0 \;-1 \;-2')$ for $\tilde{g} = -0.1, b_z = 0.9$, $b = 0.2$, $\tilde{\mu} = 11.44$

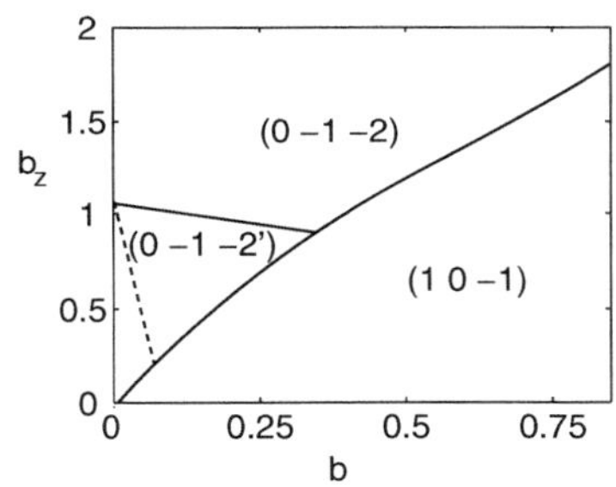


*Figure 6.* Vortex phase diagram of the spin-1 BEC in the Ioffe–Pitchard trap for $\tilde{g} = -0.1$

The ground vortex phase $(1\ 0\ -1)$ is interesting because by that it has non-axis-symmetric vortices for the spinor components $\psi_1$ and $\psi_{-1}$ as it was found by Mizushima et al. for rotating BEC (Mizushima et al. 2002b).

Next, we evaluated $| < \overrightarrow{S}(\rho) > |$ in order to find out to what extend nonadiabatic effects are important (Ho and Shenoy 1996). It is well known that in quantum magnets with anisotropy these effects are important for the ground state. Quantitatively the nonadiabatic effects can be described by a quantum reduction of the spin $| < \overrightarrow{S}(\rho) > | < 1$ and deviation of the direction of average local spin with respect to the magnetic field. In Fig. 7a we show the radial dependence of $| < \overrightarrow{S}(\rho) > |$ in the vortex phases $(0\ -1\ -2)$ and $(1\ 0\ -1)$. As seen from figure the quantum spin reduction is maximal for the vortex phase $(1\ 0\ -1)$. The value of the average local spin $| < \overrightarrow{S}(\rho) > |$ substantially reduces at the vortex core. For the vortex phase $(0\ -1\ -2)$ the quantum spin reduction is almost absent (shown in Fig. 7a by dashed line).

Moreover we calculated the radial behavior of the deviation of the direction $< \overrightarrow{S}(\rho) >$ relative to the direction of the local magnetic field $\overrightarrow{b}$. We found that in the plane perpendicular to the z-axis the local spin completely follows the direction of the magnetic field. In the plane parallel to the z-axis

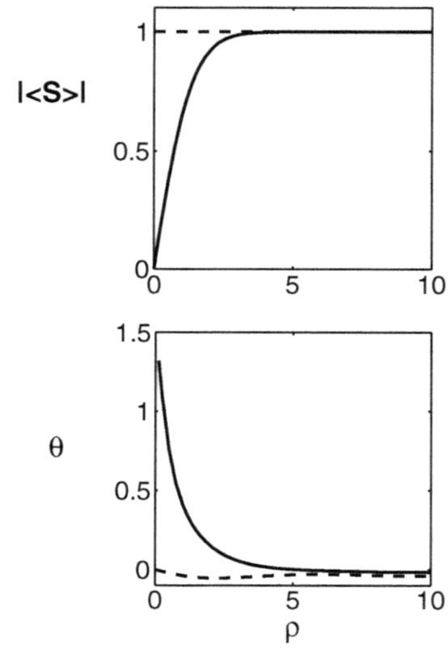


*Figure 7.* Figure shows the radial dependence of the average local spin $< \mathbf{S}(\rho) >$ (**a**) and its projection onto the direction of local magnetic field (**b**) specified by the angle $\theta$ (**b**) for different vortex phases shown in Fig. 3. Solid line refers to the vortex phase $(1\ 0\ -1)$, $\tilde{g} = 0$, $b_z = 0.1$, $b = 0.2$, $\tilde{\mu} = 11.5$, and dashed line refers to the vortex phase $(0\ -1\ -2)$, $\tilde{g} = 0$, $b_z = 1.5$, $b = 0.5$, $\tilde{\mu} = 12$

the situation is different. For the vortex phase $(1\ 0\ -1)$ shown in Fig. 7b the direction of the local spin (19) substantially differs from the direction of the magnetic field while for the phase $(0\ -1\ -2)$ this is not so. Therefore, as seen from Fig. 7 the adiabatic approximation is applicable only for the phase $(0\ -1\ -2)$.

Thus we conclude that the ground state of the spin-1 Bose–Einstein condensate trapped in the harmonic potential and subjected by the Ioffe–Prichard magnetic field is given by a rich variety of winding numbers in different spinor components. At the vortex cores of the ground state the quantum spin reduction is substantially large and the direction of the average local spin can deviate from the magnetic field.

## Acknowledgements

I am very grateful to my permamnent coauthors Karl–Fredrik Berggren, Evgeny Bulgakov and Dmitrii Maksimov without whos the present paper could not appear. I thank Michael Berry for important remarks and discussions. The work was partially supported by the grants 07-02-00694 and 09-02-01211 of Russian Foundation for Basic Research.

## References

Achenbach J.D.: 1973. Wave Propagation in Elastic Solids. North-Holland, Amsterdam

Anderson M.H., Ensher J.R., Matthews M.R., Weiman C.E. and Cornell E.A.: 1995. *Science*, **269**, 198

Aronson I.S. and Tsimring L.S.: 2006. *Rev. Mod. Phys.*, **78**, 641

Barth M. and Stöckmann H.-J.: 2002. *Phys. Rev. E*, **65**, 066208

Berry M.V.: 1977. *J. Phys. A*, **10**, 2083

Berry M.V.: 1981. In: R. Balian, M. Klaeman, and J.-P. Poirier (eds.): Physics of Defects, *Les Houches Lecture Series Session XXXV*. North Holland, Amsterdam

Bialynicki-Birula I., Bialynicka-Birula Z. and Sliwa C.: 2000. *Phys. Rev. A*, **61**, 032110

Bulgakov E.N. and Sadreev A.F.: 2003. *Phys. Rev. Lett.*, **90**, 200401

Chakraborty T. and Pietiläinen P.: 1995. The Quantum Hall Effects: Fractional and Integral. Springer, Berlin

Chen Y.F. and Huang K.F.: 2003. *Phys. Rev. E*, **68**, 066207

Chen Y.F., Lu T.H. and Huang K.F.: 2006a. *Phys. Rev. Lett.*, **96**, 033901

Chen Y.F., Lu T.H. and Huang K.F.: 2006b. *Phys. Rev. Lett.*, **97**, 233903

Courtial J. and O'Holleran K.: 2007. *Eur. Phys. J. Special Top.*, **145**, 35

Davis K.B. et al: 1995. *Phys. Rev. Lett.*, **75**, 3969

Dennis M.R.: 2003. *J. Phys. A: Math. Gen.*, **36**, 6611

Dirac P.A.M.: 1931. *Proc. R. Soc. London, Ser. A*, **133**, 60

Dorrestijn M. et al: 2007. *Phys. Rev. Lett.*, **98**, 26102

Dubrovin B.A., Fomenko A.T., and Novikov S.P.: 1985. Modern geometry-methods and applications. Part II, The geometry and topology of manifolds. Graduate Texts in Mathematics, vol. 104, Springer, New York

Ebeling K.J.: 1984. Statistical Properties of Random Wave Fields in Physical Acoustics: Principles and Methods. Academic, New York

Flossmann F., O'Holleran K., Dennis M.D. and Padgett M.J.: 2008. *Phys. Rev. Lett.*, **100**, 203902

Freund I. and Kessler D.A.: 1996. *Opt. Commun.*, **124**, 321

Hirschfelder J.O, Christoph A.C. and Palke W.E.: 1974a. *J. Chem. Phys.*, **61**, 5435

Hirschfelder J.O, Goebel C.J. and Bruch L.W.: 1974b. *J. Chem. Phys.*, **61**, 5456

Ho T.-L.: 1998. *Phys. Rev. Lett.*, **81**, 742

Ho T.-L. and Shenoy V.B.: 1996. *Phys. Rev. Lett.*, **77**, 2595

Isoshima T. and Machida K.: 2002. *Phys. Rev. A*, **66**, 023602

Isoshima T. and Yip S.: 2006. *J. Phys. Soc. Jpn.*, **75**, 074605

Isoshima T., Nakahara M., Ohmi T. and Machida K.: 2000. *Phys. Rev. A*, **61**, 063610

Khawaja U.A. and Stoof H.T.C.: 2001. *Phys. Rev. A*, **64**, 043612

Kim Y.-H., Barth M., Kuhl U. and Stöckmann H.-J.: 2003. *Progr. Theor. Phys. Suppl.*, **150**, 105

Klausen N.N. et al: 2001. *Phys. Rev. A*, **64**, 053602

Landau L.D. and Lifshitz E.M.: 1959. Theory of Elasticity. Pergamon, Oxford

Landau L.D. and Lifshitz E.M.: 1975. Quantum Mechanics. Non-Relativistic Theory. Pergamon, New York

Leanhardt A.E.: 2003. Microtraps and Waveguides for Bose-Einstein Condensates. PhD Thesis, Massachusetts Institute of Technology

Leanhardt A.E., Shin Y., Kielpinski D., Pritchard D.E. and Ketterle W.: 2003. *Phys. Rev. Lett.*, **90**, 140403

Liu F. and Mazenko G.F.: 1992. *Phys. Rev. B*, **46**, 5963

Madelung E.: 1926. *Z. Phys.*, **40**, 322
Maksimov D.N. and Sadreev A.F.: 2007. *JETP Lett.*, **86**, 670
Maksimov D.N. and Sadreev A.F.: 2008. *Phys. Rev. E*, **77**, 056204
Martikainen J.-P., Collin A. and Suominen K.-A.: 2002. *Phys. Rev. A*, **66**, 043613
Marzlin K.-P., Zhang W. and Sanders B.C.: 2000. *Phys. Rev. A*, **62**, 013602
Miesner H.-J. et al: 1999. *Phys. Rev. Lett.*, **82**, 2228
Mizushima T., Machida K. and Kita T.: 2002a. *Phys. Rev. A*, **66**, 053610.
Mizushima T., Machida K. and Kita T.: 2002b. *Phys. Rev. Lett.*, **89**, 030401
Mizushima T., Kobayashi N. and Machida K.: 2004. *Phys. Rev. A*, **70**, 043613
Mueller E.J.: 2004. *Phys. Rev. A*, **69**, 033606
Nye J.F. and Berry M.V.: 1974. *Proc. R. Soc. London, Ser. A*, **336**, 165
Ohmi T. and Machida K.: 1998. *J. Phys. Soc. Jpn.*, **67**, 1822
Petrich W., Anderson M.H., Ensher J.R. and Cornell E.A.: 1995. *Phys. Rev. Lett.*, **74**, 3352
Saarikoski H., Harju A., Puska M.J. and Nieminen R.M.: 2004. *Phys. Rev. Lett.*, **93**, 116802
Saint-James D. and Sarma G.: 1969. Type II Superconductivity. Pergamon Press, Oxford
Shvartsman N. and Freund I.: 1994. *Phys. Rev. Lett.*, **72**, 1008
Soskin M.S. et al: 1997. *Phys. Rev. A*, **56**, 4064
Stamper-Kurn D.M. et al: 1998. *Phys. Rev. Lett.*, **80**, 2027
Stenger J. et al: 1998. *Nature* (London), **369**, 345
Thomas B. and Squires A.M.: 1998. *Phys. Rev. Lett.*, **81**, 574
Waller M.D.: 1952. *Proc. R. Soc. London, Ser. A*, **211**, 265
Waterhouse R.V.: 1985. *J. Acoust. Soc. Am.*, **76**, 758
Waterhouse R.V.: 1987. *J. Acoust. Soc. Am.*, **82**, 1782
Whewell W.: 1833. *Phil. Trans. Roy. Soc. London*, **123**, 147
Yip S.-K.: 1999. *Phys. Rev. Lett.*, **83**, 4677
Zhang P., Jen H.H., Sun C.P. and You L.: 2007. *Phys. Rev. Lett.*, **98**, 030403

# NONLINEAR DYNAMICS OF WAVE PACKETS AND VORTICES IN BOSE–EINSTEIN CONDENSATES

*Chaos and Billiard Motion*

K. Nakamura
*Heat Physics Department of the Uzbek Academy of Sciences, 28 Katartal Street, 100135 Tashkent, Uzbekistan; nakamura@a-phys.eng.osaka-cu.ac.jp*

**Abstract.** We study the dynamics of single and multi-component Bose–Einstein condensates (BECs) in two dimensions with and without a harmonic trap by using various variants of nonlinear Schrödinger (or Gross–Pitaevskii) equation. Firstly, we examine the three-component repulsive BEC with cubic nonlinearity in a harmonic trap, and see the conservative chaos based on a picture of vortex molecules. We obtain an effective nonlinear dynamics for three vortex cores, which are equivalent to three charged particles under the uniform magnetic field with the repulsive inter-particle potential quadratic in the inter-vortex distance $r_{ij}$ on short length scale and logarithmic in $r_{ij}$ on large length scale. The vortices here acquire the inertia in marked contrast to the standard theory of point vortices since Onsager. We then explore chaos in the three-body problem in the context of vortices with inertia. Secondly, by choosing the nonlinear Schrödinger equation with saturable nonlinearity, we investigate the single and multi-component WP dynamics within the hard-walled square and stadium billiards with neither a harmonic trap nor driving field. We analyze the stability of WPs by using the variational (collective-coordinate) method. By emitting the radiation the Gaussian WP becomes deformed to a bell-shaped one and then stabilized. As the velocity increases, WPs tend to be stable against many collisions with billiard walls.

**Key words:** Bose–Einstein condensate; Vortices; Wave packet dynamics

## 1. Introduction

As well as single-component Bose–Einstein condensates (BECs), the trapping techniques can create multi-component condensates which involve inter-component nonlinear interactions. The multi-component BEC, far from being a trivial extension of the single-component one, presents novel and fundamentally different scenarios for its ground state and excitations. In particular, it has been observed that BEC can reach an equilibrium state characterized by the separation of the species in different domains (Hall et al. 1998a, b; Matthews et al. 1998; Stenger et al. 1998; Miesner et al. 1999; Modugno et al. 2001a, b; Brazhnyi et al. 2006). BEC provides a nice stage when the nonlinear

G. Casati and D. Matrasulov (eds.), *Complex Phenomena in Nanoscale Systems*,
NATO Science for Peace and Security Series B: Physics and Biophysics,

© Springer Science+Business Media B.V. 2009

Schrödinger equation plays a vital role. We study the dynamics of single- and multi-component BECs in two dimensions (2-d) with and without a harmonic trap by using the nonlinear Schrödinger (or Gross–Pitaevskii) equation.

In Sect. 2 we examine the three-component repulsive BEC in a harmonic trap in the absence of magnetic field, and construct a model of conservative chaos based on a picture of vortex molecules. Assuming the vortex solution with a Gaussian tail for each component and applying the collective-coordinate method, we obtain an effective nonlinear dynamics for three vortex cores. This dynamics represents three charged particles under the uniform magnetic field with the repulsive inter-particle potential. We show the novel feature of point vortices in the multi-component BEC unseen in the single-component version, and explore "the chaos in the three-body problem."

In Sect. 3, by choosing the nonlinear Schrödinger equation with saturable nonlinearity, we investigate the single- and multi-component wave packet (WP) dynamics within the hard-walled square and stadium billiards with neither a harmonic trap nor driving field. We study the stability of WPs by using the variational (collective-coordinate) method. We show how the Gaussian WP becomes deformed and stabilized and how the stability of a billiard motion depends on the velocity of WP.

## 2. Chaos in 3-Body Problem in Vortices with Inertia

BEC has a dual aspect of waves and particles. The wave nature is highlighted in the phenomenon of interference leading to fringe patterns (Andrews et al. 1997a, b, 1998). On the other hand, the particle nature of BECs can be seen in typical localized states like vortices and solitons. In fact solitons were observed in the quasi-one dimensional BEC (Strecker et al. 2002; Khaykovich et al. 2002). Solitons are non-diffusive and localized wave packets and behave like particles described in classical mechanics. Recently, Kinoshita et al. (2004) measured the collisional dynamics of two bright solitons in a trapped quasi one-dimensional $^{87}$Rb BEC. Two bright solitons with collisional interaction are integrable, and their dynamics is similar to "*Newton's cradles*."

On the other hand, Martin et al. (2007) theoretically predicted that three bright solitons in a one-dimensional BEC were non-integrable and showed the change of their behavior from regular to chaos, being related to the three-body problem in classical astrophysics. They further did a direct numerical simulation of the Gross–Pitaevskii equation (GPE) which describes the dynamics of the macroscopic wave function, and also confirmed such a change in the behavior of bright soliton dynamics.

In multi-component BECs, there are not only intra-component particle interaction but also inter-component particle interaction which is another origin of nonlinearity, so we expect novel soliton dynamics unseen in single-component BECs. Motivated by this, Pérez-García et al. (1997), Montesinos et al. (2004), and Yamasaki et al. (2005) investigated bright soliton dynamics in two or three component BECs in two dimensions, and discussed their particle-like behavior. However, in two and higher dimensional systems, bright solitons are unstable unless intra-component interaction oscillates between attraction and repulsion or intra-component three-body interaction is strong enough, so experimental realization of bright soliton in two-dimension seems to be very difficult. On the other hand, topological vortex solitons as quantized vortices of the macroscopic wave function can be stable even in two dimensions. Contrary to a bright soliton, vortex solitons in multi-component BECs have already been experimentally realized, and can become a good candidate to study solitons in 2-d multi-component BECs.

In this Section, we consider the vortex soliton in multi-component BECs in two dimensions. To consider the effective dynamics of point-like dynamics of vortices, we extract some degrees of freedom by using the variational approach, and derive an effective dynamics with finite degrees of freedom. For the case of an one-component BEC, it is well known that effective hamiltonian with finite degrees of freedom becomes (Onsager 1949; Neu 1990; Aftalion 2006)

$$H = -8\pi \sum \sum_{j>i} n_i n_j \log r_{ij}, \tag{1}$$

in the limit of infinitesimal vortex cores in the vortex point-like picture. Here $n_i$ is the winding number of the $i$th vortex, and $r_{ij} = \sqrt{(x_i - x_j)^2 + (y_i - y_j)^2}$ is the distance between cores of $i$th and $j$th vortices. Equation (1) shows that there is no momentum degree of freedom and $x_i$ and $y_i$ are conjugate coordinates each other. On the other hand, in the case of multi-component BECs, we can successfully show that the system recovers momentum degrees of freedom and is described as particle-like dynamics rather than vortex point-like dynamics.

Analyzing the multi-component GPE with vortex soliton, we now extract some degrees of freedom of soliton by using the variational principle, and obtain the effective Hamiltonian with finite degrees of freedom for vortex solitons. BEC at zero temperature is described by the GPE. A two-dimensional system of trapped $n$-component macroscopic wave function $\Phi_1(t, x, y)$, $\Phi_2(t, x, y)$, $\cdots$, $\Phi_n(t, x, y)$, obeys

$$i\tfrac{\partial}{\partial t}\Phi_i(t, x, y) =$$
$$\left[ -\nabla^2 + V(x, y) + g_{ii}|\Phi_i(t, x, y)|^2 + \textstyle\sum_{j\neq i} g_{ij}|\Phi_j(t, x, y)|^2 \right]\Phi_i(t, x, y), \tag{2}$$

for $i = 1, \dots, n$. $V(x, y) = (x^2 + y^2)$ stands for the effect of trapping. $g_{ii}$ and $g_{ij}$ are the coupling constant of intra-component and inter-component interaction, respectively. We set the situation of repulsive interactions and assume the same inter-component and the same intra-component interactions, i.e. $g_{11} = g_{22} = \cdots = g_1 > 0$ and $g_{12} = g_{13} = \cdots = g_{1n} = g_{23} = g_{24} = \cdots = g_2 > 0$. In the context of (2), we here made the following replacement: $\frac{t}{2} \to t, 2U_{ij} \to g_{ij}$.

In the absence of the inter-component interaction, each component has stationary states of a vortex. So, we consider the case in which each component has one vortex and vortices interact with each other through the inter-component interaction of BECs. Our goal is to derive from (2) the evolution equation for the collective coordinates of trial vortices. The collective coordinates for a vortex are phase variables besides the center of mass. We Taylor-expand the phase with respect to space coordinates relative to the center of mass. We choose a trial function for the vortex with unit winding number ($n_i = \pm 1$) as (Aranson and Steinberg 1996)

$$\begin{aligned}\Phi_i(t, x, y) &\equiv f_i(t, x, y) \exp[i\phi_i(t, x, y)] \\ &= \frac{1}{\sqrt{\pi\Delta}} \exp\left[-\frac{x^2 + y^2}{2\Delta}\right] \sqrt{\frac{(x - x_i)^2 + (y - y_i)^2}{2\xi^2 + (x - x_i)^2 + (y - y_i)^2}} \\ &\quad \times \exp i\left[n_i \tan^{-1}\left(\frac{y - y_i}{x - x_i}\right) + \alpha_i(x - x_i) + \beta_i(y - y_i)\right], \end{aligned} \tag{3}$$

which is the solution of (2). The meaning of collective coordinates is as follows: $(x_i, y_i)$ is the center of mass and $(\alpha_i, \beta_i)$ are the first-order coefficients of Taylor-expansion of the phase $\phi_i(t, x, y)$ with respect to $(x - x_i, y - y_i)$ rather than $(x, y)$. A trivial constant phase has been suppressed. $\Delta$ is the width of trapped condensates and $\xi$ is the healing length related to vortex core size. For the case of a single-component BEC, $\Delta$ and $\xi$ are approximately obtained as $\Delta = \sqrt{1 + g_1/4\pi}$ and $\xi = \sqrt{2\pi\Delta/g_1}$.

First of all we note: (2) can be derived from the variational principle that minimizes the action obtained from Lagrangian density for field variables,

$$\begin{aligned}\mathcal{L} &= \sum_i \left[\frac{i}{2}(\Phi_i \dot{\Phi}_i^* - \Phi_i^* \dot{\Phi}_i) + |\nabla\Phi_i|^2 + (x^2 + y^2)|\Phi_i|^2 + \frac{g_1}{2}|\Phi_i|^4\right] \\ &\quad + \sum \sum_{j>i} g_2 |\Phi_i|^2 |\Phi_j|^2. \end{aligned} \tag{4}$$

In fact, the multi-component GPE is obtained from Lagrange equation:

$$\frac{\partial}{\partial t}\frac{\partial\mathcal{L}}{\partial\dot{\Phi}_i^*} - \frac{\partial\mathcal{L}}{\partial\Phi_i^*} + \nabla\frac{\partial\mathcal{L}}{\partial\nabla\Phi_i^*} = 0. \tag{5}$$

By regarding $(x_i, y_i)$ and $(\alpha_i, \beta_i)$ as variational parameters, the Lagrangian density (4) becomes

$$\mathcal{L} = \sum_i \left[ f_i^2 \left( \dot{x}_i \frac{\partial \phi_i}{\partial x_i} + \dot{y}_i \frac{\partial \phi_i}{\partial y_i} + \dot{\alpha}_i \frac{\partial \phi_i}{\partial \alpha_i} + \dot{\beta}_i \frac{\partial \phi_i}{\partial \beta_i} \right) + \left( \frac{\partial f_i}{\partial x} \right)^2 + \left( \frac{\partial f_i}{\partial y} \right)^2 \right.$$
$$\left. + \left\{ \left( \frac{\partial \phi_i}{\partial x} \right)^2 + \left( \frac{\partial \phi_i}{\partial y} \right)^2 \right\} f_i^2 + \frac{g_1}{2} f_i^4 + (x^2 + y^2) f_i^2 \right] + g_2 \sum_{j>i} f_i^2 f_j^2. \tag{6}$$

We now insert the trial Gaussian function (3) into (6) and obtain the effective Lagrangian $L$ for their collective coordinates by integrating $\mathcal{L}$ over space coordinates:

$$L = \int dxdy \mathcal{L}. \tag{7}$$

In the limit of $X = \xi^2/\Delta \ll 1$, the Lagrangian for collective coordinates is expressed by

$$\begin{aligned} L(x_i, y_i, \alpha_i, \beta_i) = \sum_i \Big[ & (\alpha_i^2 + \beta_i^2 - \alpha_i \dot{x}_i - \beta_i \dot{y}_i) \Big( c_1 - \frac{c_2 r_i^2}{\Delta} \Big) \\ & + \frac{n_i c_3}{\Delta} (x_i \dot{y}_i - y_i \dot{x}_i + 2\alpha_i y_i - 2\beta_i x_i) - c_4 (x_i \dot{\alpha}_i + y_i \dot{\beta}_i) \\ & + \frac{1}{\Delta} \Big\{ c_5 + c_7 \Big( 1 + \frac{2g_2}{g_1} \Big) + c_9 \Delta^2 \Big\} + \frac{r_i^2}{\Delta^2} \Big\{ c_6 - c_8 \Big( 1 + \frac{g_2}{g_1} \Big) \\ & + c_{10} \Delta^2 \Big\} + \frac{c_{11}}{\Delta^2} \frac{g_2}{g_1} \sum_{j>i} V(x_{ij}, y_{ij}) \Big]. \end{aligned} \tag{8}$$

Here, $r_i = \sqrt{x_i^2 + y_i^2}$, $x_{ij} = x_i - x_j$, and $y_{ij} = y_i - y_j$. Coefficients $c_1 \sim c_{11}$ are expressed in terms of the imperfect Gamma function $\Gamma(\nu, x)$. (Nakamura 2007) Asymptotic behaviors for the interactions $V(x_{ij}, y_{ij})$ between solitons becomes

$$V(x_{ij}, y_{ij}) = \begin{cases} -r_{ij}^2 + V_0 & (r_{ij} \ll 1) \\ -24\Delta^2 X \log r_{ij} + V_0' & (r_{ij} \gg 1), \end{cases} \tag{9}$$

with $r_{ij} = \sqrt{x_{ij}^2 + y_{ij}^2}$. Equation (9) conveys that the repulsive inter-particle potential is quadratic in the inter-vortex distance $r_{ij}$ on short scale and logarithmic in $r_{ij}$ on large scale.

Lagrange equations of motion for the phase variables $\alpha_i$ and $\beta_i$ lead to

$$\begin{aligned} \alpha_i &\simeq B_1 \dot{x}_i - n_i B_2 y_i \\ \beta_i &\simeq B_1 \dot{y}_i - n_i B_2 x_i. \end{aligned} \tag{10}$$

Equation (10) shows that $(\alpha_i, \beta_i)$ correspond to generalized momentum conjugate to $(x_i, y_i)$ under the vector potential. Therefore we can rewrite other two equations by eliminating $\alpha_i$ and $\beta_i$. Another Lagrange equation of motion for $(x_i, y_i)$ gives

$$\ddot{x}_i \simeq -n_i B_3 \dot{y}_i - B_4 x_i - B_5 \textstyle\sum_{j\neq i} \frac{\partial V(x_{ij},y_{ij})}{\partial x_i}$$
$$\ddot{y}_i \simeq n_i B_3 \dot{x}_i - B_4 y_i - B_5 \textstyle\sum_{j\neq i} \frac{\partial V(x_{ij},y_{ij})}{\partial y_i}, \tag{11}$$

Coefficients $B_1 \sim B_5$ are simple functions of $c_1 - c_{11}$.

Equation (11) shows that the dynamics of vortices is very similar to charged particles under the magnetic field in the harmonic potential. The vortices here acquire the inertia in marked contrast to the standard theory of point vortices since Onsager (1949). This is one of the most important discoveries in the present work. The corresponding Hamiltonian to (10) and (11) can be given as

$$H = \sum_i \Big[\frac{1}{2}[(\alpha_i, \beta_i) - n_i \vec{A}_i]^2 + W_i + \sum_{j>i} U(x_{ij}, y_{ij})\Big], \tag{12}$$

with the vector potential $\vec{A}_i = -D_1(-y_i, x_i)$, the harmonic potential $W_i = D_2(x_i^2 + y_i^2)/2$, and interaction $U(x_{ij}, y_{ij}) = D_3 V(x_{ij}, y_{ij})/2$, where $D_1 = B_3/(1 + B_1)$, $D_2 = B_4 - B_2 D_1 - D_1^2$, and $D_3 = 2B_5$. Compared to (1), it is clear that the system has momentum degrees of freedom, i.e. the inertia, and vortices have particle-like behavior rather than vortex point-like behavior in conventional many vortices system. The inertia of the present vortices has appeared due to the multi-component nature of BEC. Parameters $D_1 \sim D_4$ depend on $\xi$, $\Delta$, and $g$, and can be controlled by changing the number of particles, tightness of trapping and strength of interaction which is tunable by Feshbach resonance.

We shall focus on the system of three vortices with equal winding numbers, and find that chaos appears even in three vortices system. This feature is completely different from that of point vortices system in a single component BEC in which chaos appears in the case of more than three vortices.

For three vortices with the same winding numbers ($n_1 = n_2 = n_3 = 1$), Hamiltonian (12) becomes

$$\begin{aligned} H = & \frac{1}{2}(\alpha_1^2 + \alpha_2^2 + \alpha_3^3 + \beta_1^2 + \beta_2^2 + \beta_3^3) \\ & + D_1(x_1\beta_1 + x_2\beta_2 + x_3\beta_3 - y_1\alpha_1 - y_2\alpha_2 - y_3\alpha_3) \\ & + \frac{1}{2}(D_1^2 + D_2)(x_1^2 + x_2^2 + x_3^3 + y_1^2 + y_2^2 + y_3^3) \\ & + U(x_{12}, y_{12}) + U(x_{23}, y_{23}) + U(x_{31}, y_{31}). \end{aligned} \tag{13}$$

With use of Jacobi coordinates

$$
\begin{aligned}
(x_T, y_T) &= \frac{1}{3}(x_1 + x_2 + x_3, y_1 + y_2 + y_3) \\
(x_C, y_C) &= \frac{1}{2}(x_1 - x_3, y_1 - y_3), \\
(x_R, y_R) &= (x_1 + x_3 - 2x_2, y_1 + y_3 - 2y_2) \\
(\alpha_R, \beta_R) &= \frac{1}{2\sqrt{3}}(\alpha_1 + \alpha_3 - 2\alpha_2, \beta_1 + \beta_3 - 2\beta_2) \\
(\alpha_T, \beta_T) &= \frac{1}{\sqrt{6}}(\alpha_1 + \alpha_2 + \alpha_3, \beta_1 + \beta_2 + \beta_3) \\
(\alpha_C, \beta_C) &= \frac{1}{2}(\alpha_1 - \alpha_3, \beta_1 - \beta_3),
\end{aligned}
\tag{14}
$$

Equation (13) can be rewritten as

$$
\begin{aligned}
H = \alpha_T^2 + \beta_T^2 + \sqrt{6}D_1(x_T\beta_T - y_T\alpha_T) + \tfrac{3(D_1^2+D_2)}{2}(x_T^2 + y_T^2) \\
+\alpha_C^2 + \beta_C^2 + 2D_1(x_C\beta_C - y_C\alpha_C) + (D_1^2 + D_2)(x_C^2 + y_C^2) \\
+\alpha_R^2 + \beta_R^2 + \tfrac{D_1}{\sqrt{3}}(x_R\beta_R - y_R\alpha_R) + \tfrac{D_1^2+D_2}{12}(x_R^2 + y_R^2) \\
+U\left(\tfrac{2x_C+x_R}{2}, \tfrac{2y_C+y_R}{2}\right) + U\left(\tfrac{2x_C-x_R}{2}, \tfrac{2y_C-y_R}{2}\right) + U(-2x_C, -2y_C).
\end{aligned}
\tag{15}
$$

$(x_T, y_T), (x_C, y_C)$, and $(x_R, y_R)$ represent the center of mass of all three components, the relative displacement, and the bisector of the vertex $(x_2, y_2)$, respectively. Aside from the separable center-of mass degrees of freedom, the system of three vortices with the same winding numbers has 4 degrees of freedom, but the independent constants of motion is 2 (energy and angular momentum). The system is therefore nonintegrable and can be chaotic. For three vortices with the different winding numbers ($n_1 = n_2 = -n_3 = 1$), on the other hand, even the angular momentum is not conserved, and one can expect chaos more easily.

We now carry out a simulation for time evolution of three vortices with the identical winding number. Under the fixed center-of-mass coordinates, we construct from (15) the equations of motion for $x_C, y_C, x_R, y_R$ and their canonical-conjugate variables, which is solved numerically. Poincaré cross section and power spectra in Fig. 1 show the transition from high-dimensional tori to chaos, as the system's energy is increased. In fact the lowest panels show a collapse of torus and the broad power spectrum, which are manifestations of chaos.

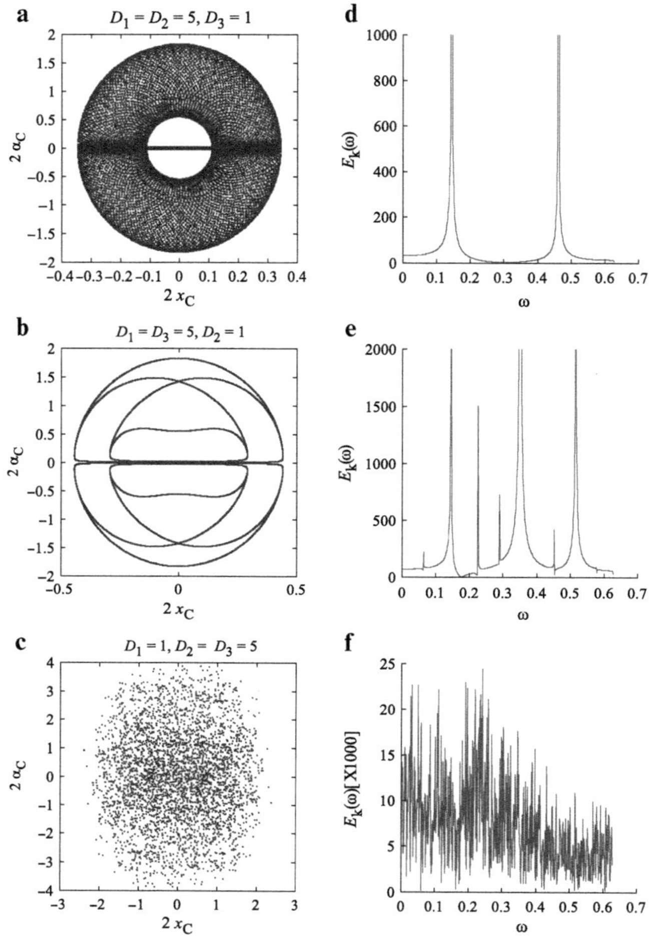


*Figure 1.* [Color online] Poincaré cross section for $\beta_R = 0$ (**a**)-(**c**), and power spectra (**d**)–(**f**). (**a**), (**d**): $D_1 = D_2 = 5, D_3 = 1$. (**b**), (**e**): $D_1 = D_3 = 5, D_2 = 1$. (**c**), (**f**): $D_1 = 1, D_2 = D_3 = 5$

## 3. Billiard Motion of Wave Packets

In nonlinear optics, it is well-known that dynamics of the optical pulse propagating in nonlinear media is described by nonlinear Schrödinger equation. In particular, the dynamics of the optical pulse propagating in the nonlinear media with the nonlinear refractive index $n(I) = n_0 I/(1 + n_1 I)$ is described by saturable nonlinear Schrödinger equation,where $I$ is the intensity of pulse and $n_0$, $n_1$ are positive constants (Jovanoski and Sammut 1994; Schjodt-Erisken et al. 1998).

In this Section we study the billiard motion of macroscopic wave packet (WP) in two dimensions which obeys Gross–Pitaevskii equation with the saturable nonlinearity. As billiards we choose a square and stadium.

In the billiard-free case, the nonlinear Schrödinger equation for the single-componrnt wave function is given by

$$i\hbar\frac{\partial\psi}{\partial t} = -\frac{\hbar^2}{2m}\nabla^2\psi - \frac{c|\psi|^2}{1+\gamma|\psi|^2}\psi, \tag{16}$$

where $\gamma$ and $c$ are positive constans. The last term on the right hand side of (16) stands for the saturable nonlinearity. When this term will be expanded with respect to a small but finite $\gamma$, its first and second terms give a cubic and quintic nonlinearities, respectively. On the other hand, when $\gamma \gg 1$, (16) becomes the linear Schrödinger equation. In this sense, we regard the saturable nonlinearity as the most generic among others. We examine the stability of WP which obeys (16) by using the collective coordinate method. Equation (16) can be derived from the Lagrangian density,

$$\mathcal{L}_{free} = \frac{i}{2}(\psi\dot{\psi}^* - \psi^*\dot{\psi}) + \frac{1}{2}|\nabla\psi|^2 + \frac{c}{\gamma^2}(\log(1+\gamma|\psi|^2) - \gamma|\psi|^2). \tag{17}$$

We take Gaussian function as a trial function for WP:

$$\begin{aligned}\psi &= \sqrt{\frac{1}{\pi w_x w_y}}\exp[-\frac{(x-x_0)^2}{2w_x^2} - \frac{(y-y_0)^2}{2w_y^2} \\ &+ i\alpha_x(x-x_0) + i\alpha_y(y-y_0) + i\beta_x(x-x_0)^2 + i\beta_y(y-y_0)^2],\end{aligned} \tag{18}$$

where $\mathbf{q} \equiv \{x_0, y_0; w_j, \alpha_j, \beta_j\}$ $(j = x, y)$ are collective-coordinates which characterize the WP. Variables $\alpha_j, \beta_j$ $(j = x, y)$ are canonical-conjugate to the center-of-mass $j_0$ and the width $w_j$, respectively. Inserting (18) into (17) and integrating over space variables, the effective Lagrangian $L_{free}$ for $\mathbf{q}$ is obtained. The Lagrange equation for $\mathbf{q}$ is

$$\frac{d}{dt}(\frac{\partial L_{free}}{\partial\dot{\mathbf{q}}}) - \frac{\partial L_{free}}{\partial\mathbf{q}} = 0. \tag{19}$$

Besides the complementary relations $\alpha_j = \frac{d}{dt}j_0$ and $\beta_j = \frac{1}{2w_j}\frac{d}{dt}w_j$ $(j = x, y)$, equation of motion for $w_j$ $(j = x, y)$ leads to the effective potential for the widths:

$$\begin{aligned}U_{free}(w_x, w_y) &= \tfrac{1}{2w_x^2} + \tfrac{1}{2w_y^2} + \tfrac{c\pi^3}{6\gamma^2}w_x w_y + \tfrac{c}{\gamma^2}\pi w_x w_y\{\tfrac{1+e^a}{2}\Phi(\tfrac{1+e^a}{4}, 2, 1) \\ &-\Phi(\tfrac{1}{2}, 2, 1) + 2(\log\tfrac{1+e^a}{2} - \log 2)\log\tfrac{3-e^a}{2} - 2\log\tfrac{1+e^a}{2}\} - \tfrac{2c}{\gamma},\end{aligned} \tag{20}$$

where $\Phi(z, s, v) = \sum_{n=0}^{\infty}(v+n)^{-s}z^n$ Learch function and $a = \frac{\gamma}{\pi w_x w_y}$. The profile of this potential is shown in Fig. 2, which has a stable minimum. Since (20) is not dependent on the center-of-mass coordinates, one might expect a stable propagation of the weakly-breathing WP. But it is not true and the fact is more intricate.

We numerically solved (16) with use of the initial Gaussian WP in (18). Snapshots of WP dynamics are shown in Figs. 3 and 4. Figure 3 represents the cross-section of the WP on the vertical plane passing through the cente-of-mass. We find that the radiation (i.e. short wave length fluctuations) emerges as the WP propagates. Figure 4 shows the analogous dynamics of the WP on longer time scale. Interestingly, there appears a deformation of the WP profile from Gaussian to bell-shaped. It should be noted: the bell-shaped WP can

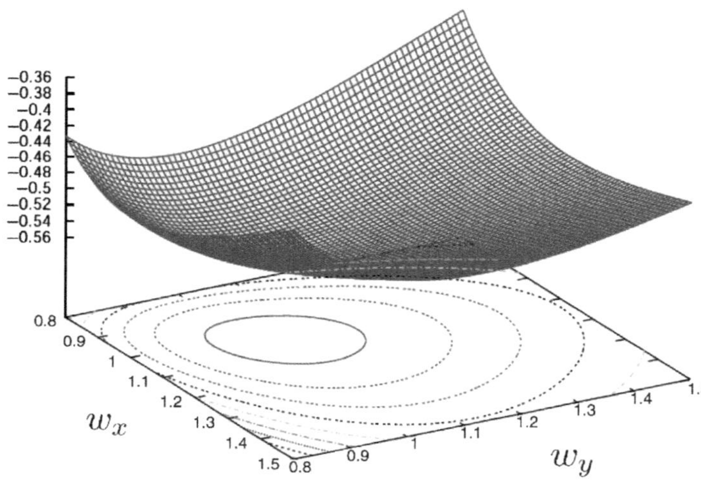


*Figure 2.* Width potential $U_{free}(w_x, w_y)$

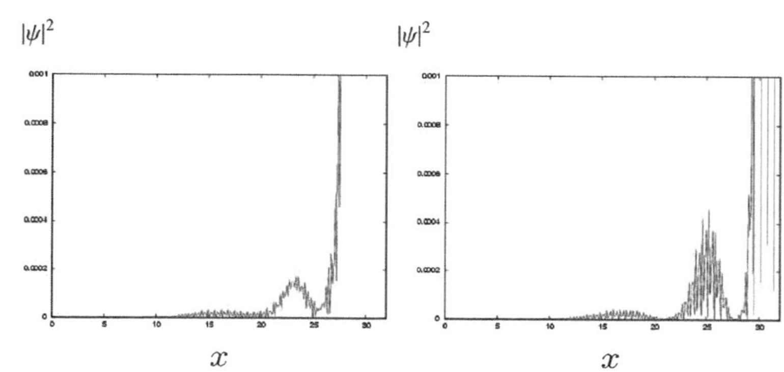


*Figure 3.* Emission of the radiatin from wave packet. Time elapses from left to right

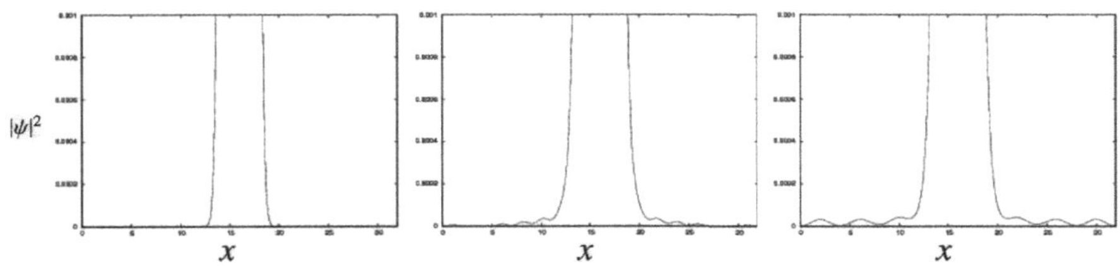


*Figure 4.* Deformation of wave packet from Gaussian to bell-shaped. Time elapses from left to right

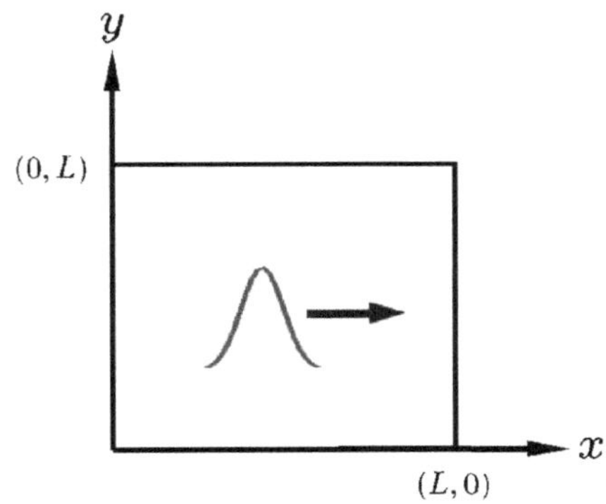


*Figure 5.* Dynamics of wave packet in a square billiard. Initially, the wave packet is located at the center of the billiard, i.e. $x = y = L/2$ with $L = 32$

coexist with the accompanying radiation and the bell-shaped WP keeps its profile for a long time.

We shall proceed to a study of the collisions of WP with square walls (see Fig. 5). Again we apply the variational method (Montesinos et al. 2004; Pérez-Gacía et al. 1997; Montesinos et al. 2005). The nonlinear Schrödinger equation with a hard-walled square boundary can be obtained from the Lagrangian density,

$$\mathcal{L}_{billiard} = \mathcal{L}_{free} + V(\mathbf{r})|\psi|^2, \tag{21}$$

where $V(\mathbf{r})$ represents the confining potential: $V(\mathbf{r}) = 0$ inside the billiard and $V(\mathbf{r}) = V_0$ with $V_0 = \infty$ outside the billiard, although we shall choose a large but finite value for $V_0$ in the numerical analysis below. Inserting Gaussian function (18) into (21) and integrating over all space coordinates, we obtain the effective Lagrangian for the collective-coordinates $\mathbf{q} = \{x_0, y_0; w_j, \alpha_j, \beta_j\}$ $(j = x, y)$. Then Lagrange equation yields equations motion for its center-of-mass and width as

$$\ddot{x}_0 = -\frac{\partial U_{billiard}(x_0, y_0, w_x, w_y)}{\partial x_0}, \tag{22a}$$

$$\ddot{y}_0 = -\frac{\partial U_{billiard}(x_0, y_0, w_x, w_y)}{\partial y_0}, \tag{22b}$$

$$\ddot{w}_x = -\frac{\partial U_{billiard}(x_0, y_0, w_x, w_y)}{\partial w_x}, \tag{22c}$$

$$\ddot{w}_y = -\frac{\partial U_{billiard}(x_0, y_0, w_x, w_y)}{\partial w_y}, \tag{22d}$$

with the potential given by,

$$U_{billiard}(x_0, y_0, w_x, w_y) = U_{free}(w_x, w_y)$$
$$+V_0[A_1(x_0, w_x) + A_2(y_0, w_y) + A_1(x_0, w_x)A_2(y_0, w_y)], \tag{23}$$

where $A_1(x_0, w_x) = \frac{1}{2}[\text{erfc}(\frac{x_0}{w_x} + \text{erfc}(\frac{L-x_0}{w_x})]$ and $A_2(y_0, w_y) = \frac{1}{2}[\text{erfc}(\frac{y_0}{w_y}) + \text{erfc}(\frac{L-y_0}{w_y})]$, and $\text{erfc}(x)$ stands for the error function defined by $\text{erfc}(x) = \frac{2}{\sqrt{\pi}}\int_x^\infty e^{-t^2}dt$. The terms including $A$ functions come from the interaction of WP with the hard-walled square billiard. Equation (23) depends on both the center-of-mass ($x_0$,$y_0$) and widths ($w_x$,$w_y$), which radically changes the argument in the case of the billiard-free motion.

Figure 6 shows the profile of the effective width potential $U_{billiard}(x_0, y_0, w_x, w_y)$ when the center-of mass of WP is located in the vicinity of the wall. $U_{billiard}$ has still an isolated minimum as in the case of the center-of-mass away from the wall, but the potential profile becomes to be elongated along the wall and to have infinitely-degenerated minima when the WP approaches very close to the walls (compare Fig. 6a and b). Because of this, when WP collides with and is reflected from the wall, the high-speed WP has little time to diffuse along the wall, while the low-speed WP has enough time to diffuse along the wall. Therefore the former WP is expected to show a stability against its collision with the wall.

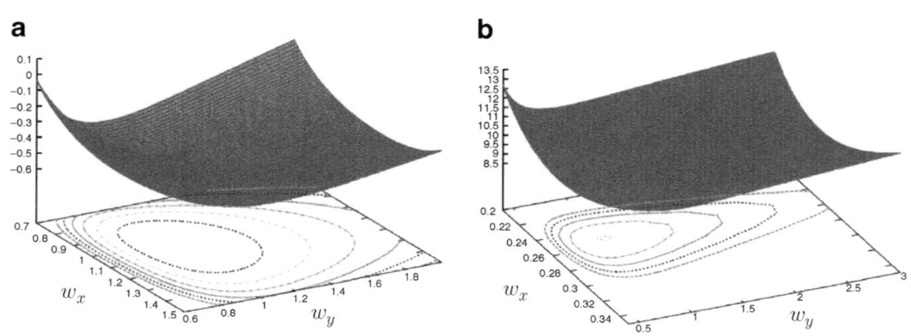


*Figure 6.* The profile of $U_{billiard}$ with $V_0 = 100$ when wave packet is located in the vicinity of the wall at $y_0 = L/2$: (**a**) $x_0 = 0.9L$; (**b**) $x_0 = 0.99L$

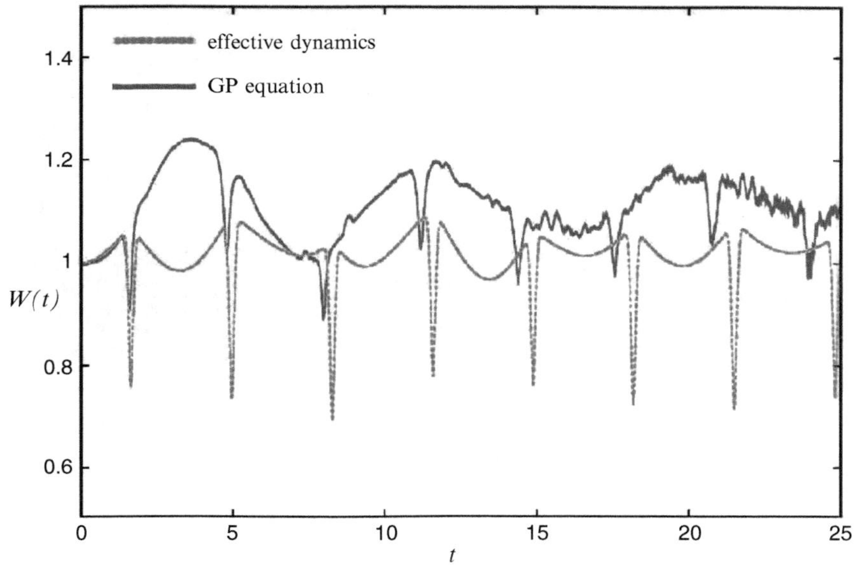


*Figure 7.* Time evolution of the width. Velocity is $v = 10$

To see the influence of collisions of WP with the walls, we show in Fig. 7 the numerical result for the time evolution of the width defined by $W(t) = [\int((x- < x >)^2 + (y- < y >)^2)|\psi|^2 d^2\mathbf{r}]^{1/2}$ in the case of high-velocity WP ($v = 10$). Here $W$ is evaluated with use of the wave function regularized by removing the effect of radiation. The width shows spikes when WP collides with wall boundaries, but is stable between successive collisions with the walls. We also find in Fig. 7 that the billiard motion of WP can also be well interpreted in terms of the effective dynamics of the collective coordinates. It should be noted the stability of width is not guaranteed in the case of low-velocity WP (e.g. $v \leq 1$).

We now investigate the WP dynamics in the stadium billiard. Figure 8 shows wave packet profiles during the collision with a wall. We see the stability of WP in the vicinity of the wall. Figure 9 shows the trajectory of the center of mass of WP. As in the case of the square billiard, the WP keeps its robustness against collisions with the stadium billiard.

In the classical dynamics, the particle moving in the stadium billiard shows a chaotic trajectory. However, in quantum dynamics, WP is well-known to collapse because of the absence of nonlinearity. When a suitable (e.g. saturable) nonlinearity is included, our numerical simulation strongly indicates that the WP does keep its particle picture for a long time.

The feature of WP dynamics in the two-component BEC is much more interesting. We studied the dynamics of the two-component WP described by

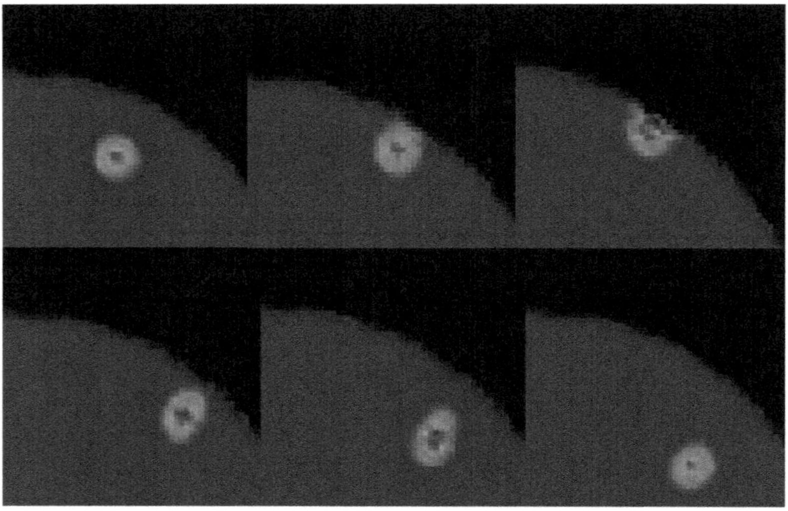

*Figure 8.* Wave packet profiles during the collision with a curved wall

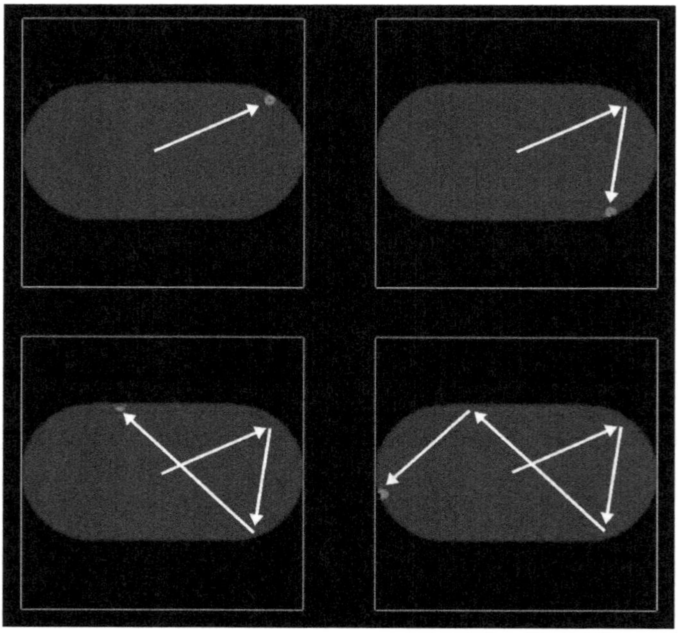

*Figure 9.* Billiard motion of the wave packet in the stadium. Snapshots are taken at the moments that the wave packet collides with the wall

the two-component saturable nonlinear Schrödinger equation. In this case, there exist three types of collisions between WPs (transmission, reflection, and formation of a molecule) depending on the initial relative velocity. Because of the space limitation, however, the concrete results will be described elsewhere.

## 4. Summary and Discussions

We explored vortex dynamics of the multi-component BEC in the harmonic trap in the case that each component has a single vortex. With use of collective coordinates for the vortex core and phase gradients, we have obtained an effective nonlinear dynamics for vortex cores, which represents three charged particles with inertia under the uniform magnetic field in the harmonic potential. The repulsive inter-particle potential is quadratic in the inter-vortex distance $r_{ij}$ on short scale and logarithmic in $r_{ij}$ on large scale. The inertia of vortices is the most novel aspect that has not been found in the conventional theory of point vortices since Onsager. As the energy is increased, the system of three vortex cores shows the transition from regular high-dimensional torus to chaos, which is a manifestation of "chaos in the three-body problem."

We also investigated the single and multi-component WP dynamics within the hard-walled square and stadium billiards with neither a harmonic trap nor driving field. We chose a saturable nonlinearity in the nonlinear Schrödinger equation. The stability of WPs can be analyzed by using the variational (collective-coordinate) method. The initial Gaussian WP is deformed to a bell-shaped one, which coexists with the radiation around the WP. As the velocity increases, WPs tend to be stable against many collisions, whose mechanism is elucidated with use of the effective potential for widths of WP.

## Acknowledgements

The author owes much to T. Ogura, A. Kohi, and D. Matrasulov for their valuable contributions and to M. Kobayashi for innumerable comments and discussions on numerical works. The full details will be published in a separate paper. The work is partly supported through a project of the Uzbek Academy of Sciences (FA-F2-084).

## References

Aftalion A.: 2006. *Vortices in Bose-Einstein Condensates*, Birkhäuser, Boston

Andrews M.R., Townsend C.G., Miesner H.-J., Durfee D.S., Kurn D.M. and Ketterle W.: 1997a. *Science*, **275**, 637

Andrews M.R., Kurn D.M., Miesner H.-J., Durfee D.S., Townsend C.G., Inouye S. and Ketterle W.: 1997b. *Phys. Rev. Lett.*, **79**, 553

Andrews M.R., Kurn D.M., Miesner H.-J., Durfee D.S., Townsend C.G., Inouye S. and Ketterle W.: 1998. *Phys. Rev. Lett.*, **80**, 2967

Aranson I. and Steinberg V.: 1996. *Phys. Rev. B*, **53**, 75

Brazhnyi V.A., Konotop V.V. and Kuzmiak V.: 2006. *Phys. Rev. Lett.*, **96**, 150402

Hall D. et al: 1998a. *Phys. Rev. Lett.*, **81**, 1539

Hall D. et al: 1998b. *Phys. Rev. Lett.*, **81**, 1543

Jovanoski Z. and Sammut R.A.: 1994. *Phys. Rev. E*, **50**, 4087

Khaykovich L., Schreck F., Ferrari G., Bourdel T., Cubizolles J., Carr L.D., Castin Y. and Salomon C.: 2002. *Science*, **296**, 1290

Kinoshita T., Wenger T. and Weiss D.S.: 2004. *Science*, **305**, 1125

Martin A.D., Adams C.S. and Gardiner S.A.: 2007. *Phys. Rev. Lett.*, **98**, 020402

Matthews M.R. et al: 1998. *Phys. Rev. Lett.*, **81**, 243

Miesner H.-J. et al: 1999. *Phys. Rev. Lett.*, **82**, 2228

Modugno G. et al: 2001a. *Science*, **294**, 1320

Modugno G. et al: 2001b. *Phys. Rev. A*, **72**, 023611

Montesinos G.D., Pérez-García V.M. and Michinel H.: 2004. *Phys. Rev. Lett.*, **92**, 133901-1

Montesinos G.D., Rodas-Verde M.I., Pérez-García V.M. and Michinel H.: 2005. *CHAOS*, **15**, 033501

Nakamura K.: 2007. *Prog. Theor. Phys. Suppl.*, **166**, 179

Neu J.C.: 1990. *Physica D*, **43**, 385

Onsager L.: 1949. *Nuovo Cimento Suppl.*, **6**, 279

Pérez-Gacía V.M. et al: 1997. *Phys. Rev. A.*, **56**, 1424

Schjodt-Erisken J. et al: 1998. *Phys. Lett. A.*, **246**, 423

Stenger J. et al: 1998. *Nature* (London), **396**, 345

Strecker K.E., Partridge G.B., Truscott A.G. and Hulet R.G.: 2002. *Nature* (London), **417**, 150

Yamasaki H., Natsume Y. and Nakamura K.: 2005. *J. Phys. Soc. Jpn.*, **74**, 1887

# OPTOMECHANICS

B. Kubala, M. Ludwig, and F. Marquardt*
*Department of Physics, Arnold Sommerfeld Center for Theoretical Physics, Center for NanoScience, Ludwig-Maximilians-Universität, Theresienstrasse 37, 80333 Munich, Germany; Florian.Marquardt@physik.lmu.de*

**Abstract.** We review recent progress in the field of optomechanics, where one studies the effects of radiation on mechanical motion. The paradigmatic example is an optical cavity with a movable mirror, where the radiation pressure can induce cooling, amplification and nonlinear dynamics of the mirror.

**Key words:** Optomechanics; Radiation pressure; Nonlinear dynamics

## 1. Introduction

Optomechanics is an emerging research topic that is concerned with mechanical effects caused by light, particularly in connection with micro- and nanomechanical structures that are deflected by radiation pressure. Thoughts about the mechanical effects of light can be traced back as far as Johannes Kepler. Observing the tails of comets always pointing away from the sun, he speculated that this might be due to the force exerted by the solar radiation. Ever since the first measurements of such radiation forces more than 100 years ago, optomechanical effects have been observed in various areas of physics and engineering: Spacecraft with solar sails are indeed being developed, radiation forces are setting fundamental limits for the precision of laser interferometers used in detecting gravitational waves, and these forces are also used to manipulate cold atoms. A recent addition is the use of optomechanical forces to drive, cool and read out micro- and nanomechanical devices (see a recent review in Kippenberg and Vahala 2008, and other recent developments in Marquardt 2008). To reach the ground state of a mechanical oscillator with a frequency of 100 MHz, it would have to be cooled down to about 1 mK. Achieving such ground state cooling would "put back mechanics into quantum mechanics" (Schwab and Roukes 2005), and quantum effects would become observable in a massive object consisting of roughly $10^{15}$ atoms.

G. Casati and D. Matrasulov (eds.), *Complex Phenomena in Nanoscale Systems*,
NATO Science for Peace and Security Series B: Physics and Biophysics,

© Springer Science+Business Media B.V. 2009

This brief review is organized as follows. In Sect. 2 we introduce the basic setup, an optical cavity, driven by a laser with one mirror placed on an oscillating cantilever. We explain the classical effects of retarded radiation forces. Similar physics was investigated in a variety of other system, like driven LC-circuits coupled to cantilevers (Brown et al. 2007) or single-electron transistors and microwave cavities coupled to nanobeams (Naik et al. 2006; Rodrigues and Armour 2007; Regal et al. 2008). Light-induced forces can not only cool the cantilever, but can also enhance the mechanical motion leading to an instability. In Sect. 3 we show how one can derive an intricate attractor diagram for the resulting self-induced oscillations (Marquardt et al. 2006), which have also been seen in experiment. Section 4 is devoted to a quantum description of the coupled cavity-cantilever system (Ludwig et al. 2008). A new optomechanical setup (Jayich et al. 2008; Thompson et al. 2008), which aims at Fock state detection, is discussed in Sect. 5.

## 2. The Basic Optomechanical Setup

The standard setup of optomechanics is shown in Fig. 1. It consists of an optical cavity driven by a laser impinging on the cavity through a fixed mirror. The other mirror of the cavity is movable. For example, it may be attached to a micro-cantilever as used in atomic force spectroscopy. In such a setup the mechanical effects of light are enhanced, as the light field is resonantly increased in the cavity and each photon will transfer momentum to the mirror in each of the numerous reflections it undergoes, until finally leaving the cavity.

The coupled cavity-cantilever system is described by a Hamiltonian of the form

$$\hat{H}_{\text{cav+cant}} = \hbar\left(\omega_{\text{cav}} - g\frac{\hat{x}_M}{x_{\text{ZPF}}}\right)\hat{a}^\dagger\hat{a} + \hbar\omega_M\hat{c}^\dagger\hat{c}\,. \tag{1}$$

Additional terms in the Hamiltonian describe the driving of the cavity by the laser beam, decay of photons out of the cavity and the mechanical damping of the cantilever. Here, $\omega_M$ denotes the oscillation frequency of a mechanical

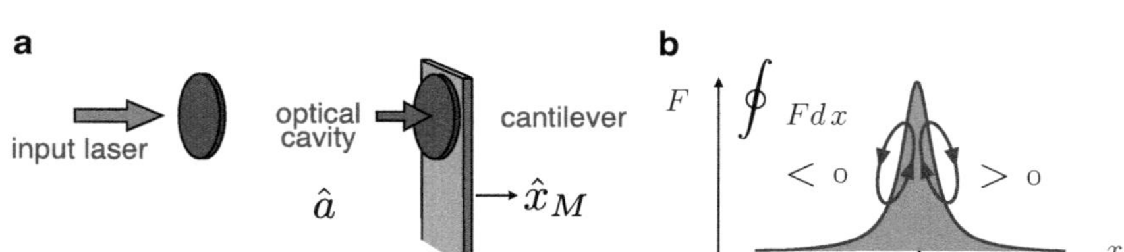


*Figure 1.* (**a**) The standard setup of optomechanics. (**b**) The dependence of the radiation pressure force (circulating intensity) on the cantilever position

oscillator, whose displacement can be expressed as $\hat{x}_M = (\hat{c}^\dagger + \hat{c})x_{ZPF}$ in terms of ladder operators and the oscillator's zero point fluctuations $x_{ZPF} = (\hbar/2m\omega_M)^{-1/2}$. The optical cavity, described by operators $\hat{a}^\dagger$ and $\hat{a}$, has a resonance frequency $\omega_{cav}$ if the cantilever is fixed at position $x_M = 0$.

The coupling term $\propto \hat{x}_M\hat{a}^\dagger\hat{a}$ with a strength depending on the coupling constant $g$ can be understood by two equivalent ways of reasoning: The radiation pressure force should give rise to a term of the form $-\hat{F}_{rad}\hat{x}_M = -\frac{c}{L}\hat{a}^\dagger\hat{a}\,\hbar k_{cav}\,\hat{x}_M$, which leads to (1) with $g = \omega_{cav}x_{ZPF}/L$. Alternatively, we can understand the same term as stemming from the dependence of the cavity's resonance frequency on the cavity length, $L+x_M$, given by $d\omega_{res}/dx_M = -\omega_{cav}x_M/L$.

Two crucial new ingredients are added to the physics of radiation pressure by considering a cavity setup. First, the radiation pressure becomes strongly position dependent due to its proportionality to the total light intensity in the cavity $\propto \hat{a}^\dagger\hat{a}$. The light intensity shows resonances when the cavity length $L + x_M$ is varied. Their full width at half maximum (FWHM) depends on the decay time $\kappa^{-1}$ of the cavity, $x_{FWHM} = \kappa L/\omega_{cav}$ .The resulting dependence of the radiation pressure force on the cantilever position in the stationary state is sketched in Fig. 1. Secondly, the decay time $\kappa^{-1}$ introduces a delay between the mirror motion and the response of the light intensity.

To understand the effects of such a retarded response of the radiation pressure force, let us consider a cantilever at a position $x_M > 0$ to the right of the resonance (see Fig. 1) moving towards the resonance position, $\dot{x}_M < 0$. We consider small delay times and small excursions of the cantilever only. Moving leftwards the cantilever acts against the radiation pressure, which grows as the cantilever moves closer to resonance and the light intensity in the cavity increases. This increase, however, lags behind the movement of the cantilever, so that at any instance the force acting on the cantilever is smaller than its stationary value at the same position would be (see Fig. 1). Moving into the opposite, positive direction the delayed decrease of the intensity leads to an accelerating force on the cantilever, larger than the stationary one. Overall, there is a net input of work into the mechanical motion during one oscillation, given by the enclosed area in the force-position diagram in Fig. 1. Thus, for $x_M > 0$ (where the laser light is blue detuned with respect to the cavity resonance) the cantilever motion gets enhanced, while for $x_M < 0$ the same physics causes an additional damping. In the next section, we will extend these qualitative statements to a detailed description of the classical dynamics of the coupled cavity-cantilever system.

Retarded radiation forces were first investigated in pioneering studies by Braginsky, both experimentally and theoretically (Braginsky and Manukin 1967; Braginsky et al. 1970).

## 3. Nonlinear Classical Dynamics

Operating on the red detuned side of the resonance, any small thermal oscillation amplitude will be damped away more quickly than in the absence of radiation. On the opposite, blue detuned side, damping is effectively reduced. If this effect overcomes intrinsic friction, an arbitrary thermal fluctuation will be amplified into an oscillation with increasing amplitude, driving the coupled system into a nonlinear regime (Aguirregabiria and Bel 1987; Fabre et al. 1994; Braginsky et al. 2001; Marquardt et al. 2006). Finally, the system will settle into a stable, self-sustained oscillation, where radiation power input and dissipation are in balance. This will be the subject of the present section. These effects have already been observed in experiments (Höhberger and Karrai 2004; Carmon et al. 2005; Kippenberg et al. 2005; Metzger et al. 2008).

To derive classical equations of motion, we replace the operator $\hat{a}$ by the complex light amplitude $\alpha$ and the position operator $\hat{x}_M$ by the cantilever's classical displacement $x_M$. From the Hamiltonian equation (1) we then derive

$$\begin{aligned} \dot{\alpha} &= \left[ i \left( \Delta + g \frac{x_M}{x_{\mathrm{ZPF}}} \right) - \frac{\kappa}{2} \right] \alpha - i\alpha_L \\ \ddot{x}_M &= -\omega_M^2 x_M + |\alpha|^2 \hbar g/(m x_{\mathrm{ZPF}}) - \Gamma_M \dot{x}_M , \end{aligned}$$

where $\alpha_L$ is the amplitude of the driving laser field, $\Gamma_M$ describes the mechanical damping of the cantilever, and $\Delta = \omega_L - \omega_{\mathrm{cav}}$ is the detuning of the laser light with respect to the cavity resonance.

Beside a static solution $x_M(t) = \mathrm{const.}$, the system can exhibit self-induced oscillations. The cantilever will then conduct approximately sinusoidal oscillations, $x_M(t) \approx \bar{x} + A\cos(\omega_M t)$, at its unperturbed frequency $\omega_M$. Since radiation pressure effects are small, the amplitude $A$ of the oscillations will change slowly over many oscillation periods only.

From this ansatz, an analytical solution for the coupled dynamics of $x_M(t)$ and $\alpha(t)$ can be found (Marquardt et al. 2006; see also Ludwig et al. 2008). The two parameters of the solution, the amplitude $A$ and the average displacement $\bar{x}$, can be determined from two balance conditions: For any periodic solution the total force should average to zero during one cycle,

$$\langle \ddot{x}_M \rangle \equiv 0 \quad \Leftrightarrow \quad m\omega_M^2 \bar{x} = \langle F_{\mathrm{rad}} \rangle = \frac{\hbar g}{m x_{\mathrm{ZPF}}} \left\langle |\alpha(t)|^2 \right\rangle . \tag{2}$$

This yields an implicit equation for $\bar{x}$, since $\langle F_{\mathrm{rad}} \rangle$ is a function of $\bar{x}$ and $A$. Furthermore, the work performed by the radiation pressure balances on average the frictional losses,

$$\langle F_{\mathrm{rad}} \dot{x} \rangle = \Gamma_M \left\langle \dot{x}^2 \right\rangle . \tag{3}$$

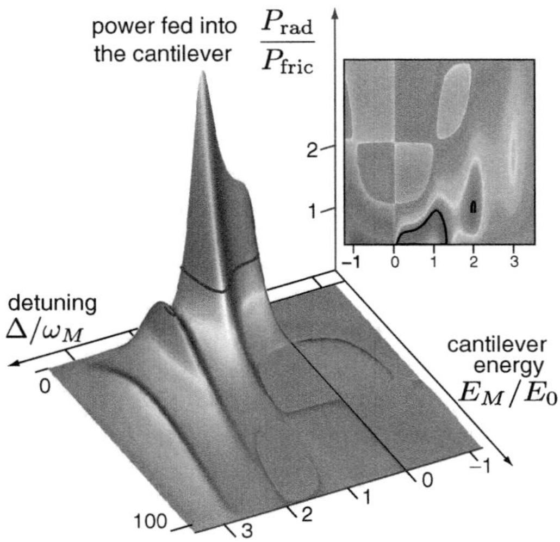


*Figure 2.* The power fed into the cantilever motion by the radiation force, as a function of oscillation amplitude and laser detuning. This can be used to construct the possible attractors for the elf-induced oscillations (indicated by thick lines)

Eliminating $\bar{x}$ by use of (2) we can plot the ratio between radiation power input and frictional loss, the two sides of the last equation, as a function of the oscillation amplitude $A$. Such a plot is shown in Fig. 2, where we chose the detuning $\Delta$ as a second variable, while other parameters are fixed. The condition of (3) is fulfilled if the ratio $P_{\text{rad}}/P_{\text{fric}} = 1$, as indicated by the horizontal cut in Fig. 2. A solution will be stable only if an increase of the amplitude is accompanied by a decrease of $P_{\text{rad}}/P_{\text{fric}}$. By that reasoning the final attractor diagram is constructed, as indicated by the thick black lines in Fig. 2.

Important general features of the dynamics of the coupled system can be seen in Fig. 2. Self-induced oscillations appear for sufficiently strong driving around integer multiples of the cantilever frequency, $\Delta \approx n\omega_M$. Such oscillations appear for a positive detuning $\Delta$, while for red detuned laser light ($\Delta < 0$) the stationary solution, $x_M(t) = \text{const.}$, is stable. Note that stable solutions with large amplitude do exist even for $\Delta < 0$.

The most striking feature, however, is the coexistence of several stable solutions with different finite oscillation amplitudes for a fixed set of system parameters. This dynamical multi-stability, first discussed in this context in Marquardt et al. (2006) and also seen in similar systems (Rodrigues and Armour 2007), is visible in Fig. 2b, while for the parameters of Fig. 2a we find coexistence of a stationary and a finite amplitude solution around $\Delta \approx 2\omega_M$.

These multi-stabilities could be utilized for ultra-sensitive "latching" measurements, as argued in Marquardt et al. (2006).

Self-induced oscillations in an optomechanical system have already been observed in experiments with bolometric forces (Höhberger and Karrai 2004; Metzger et al. 2008) and in microtoroidal structures where radiation pressure dominates (Carmon et al. 2005). Recently, a more detailed comparison of theory and experiment revealed interesting effects due to higher order mechanical modes that get involved in the nonlinear dynamics (Metzger et al. 2008).

## 4. Quantum Theory of Optomechanical Systems

The prospect of reaching the quantum mechanical ground state of a "macroscopic" mechanical object is currently one of the main goals in the field of micro- and nanomechanics. Impressive progress has been made in a series of experiments (Cohadon et al. 1999; Höhberger-Metzger and Karrai 2004; Arcizet et al. 2006; Gigan et al. 2006; Schliesser et al. 2006; Kleckner and Bouwmeester 2006; Corbitt et al. 2007; Thompson et al. 2008), though the ground state has not yet been reached at the time of writing. In the classical picture derived above, we found that a properly detuned laser beam will cool the cantilever by providing extra damping. According to the classical theory, the cantilever can be cooled down to an effective temperature $T_{\text{eff}} = T\Gamma_M/(\Gamma_{\text{opt}} + \Gamma_M)$, apparently arbitrarily close to absolute zero for sufficient drive power and low mechanical damping. However, quantum mechanics sets the ultimate limit for optomechanical cooling.

Starting from an intuitive quantum picture of the cooling process, we will present in the next subsection a quantum noise approach to cooling. Quantum effects on the self-induced oscillations can be described numerically within a quantum master equation discussed in the following subsection, which allows studying the classical-to-quantum crossover.

### 4.1. QUANTUM NOISE APPROACH TO COOLING

In the quantum description, a photon impinging on the cavity will emit or absorb a phonon of the mechanical cantilever motion and change its frequency accordingly, in a Raman-like process. A photon that is red detuned from the resonance will absorb a phonon of energy $\hbar\omega_M$ from the cantilever, so that it is scattered into the cavity resonance, leading to cooling. Detuning to a "sideband" of the cavity at a frequency $\omega_{\text{cav}} - \omega_M$ will be particularly effective.

For a quantitative approach the radiation field of the cavity will be considered as a "bath" acting upon the "system," the cantilever degree of freedom $\hat{x}_M$, via the coupling term, $-\hat{x}_M\hat{F}$, in the Hamiltonian. The influence of the

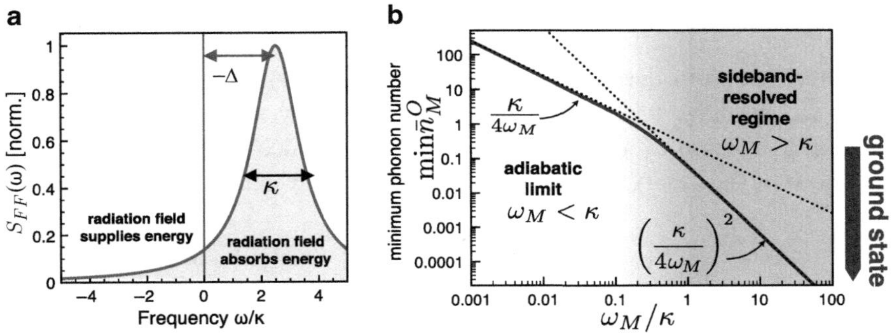


*Figure 3.* (**a**) Power spectrum for the radiation pressure force. (**b**) Quantum-mechanical cooling limit

bath is then characterized by the power spectrum of the force, $\hat{S}_{FF}(\omega) = \int dt\, \exp(i\omega t)\langle \hat{F}(t)\hat{F}(0)\rangle$ . In particular, Fermi's golden rule links the net optical damping rate of the cantilever to the possibility of the cavity to absorb/emit a quantum of energy $\hbar\omega_M$ from/to the bath, $S_{FF}(\pm\omega_M)$, as

$$\Gamma_{\text{opt}} = (x_{\text{ZPF}}/\hbar)^2 \left[S_{FF}(\omega_M) - S_{FF}(-\omega_M)\right] . \tag{4}$$

The power spectrum $S_{FF}$ is directly related (Marquardt et al. 2007) to the spectrum of photon number fluctuations due to shot-noise (see Fig. 3). Crucially, the asymmetry of the power spectrum (which is set by the laser detuning) determines whether the cavity will more readily absorb or emit energy, setting the sign of the net optical damping rate $\Gamma_{\text{opt}}$ [cf. (4)].

One finds (Marquardt et al. 2007; Wilson-Rae et al. 2007) a simple limit on the minimal occupation number, $\bar{n}_M^O = [\kappa/(4\omega_M)]^2$, which can be reached for optimal detuning $\Delta = -\omega_M$ in the resolved-sideband limit $\omega_M \gg \kappa$, for $\Gamma_{\text{opt}} \gg \Gamma_M$. In general, the reachable occupation number $\bar{n}_M$ of the mechanical mode will depend on the initial occupation $\bar{n}_M^T$ (hence, starting from cryogenically precooled samples is advantageous) and the mechanical and optical damping rates, as $\bar{n}_M = (\Gamma_{\text{opt}}\bar{n}_M^O + \Gamma_M\bar{n}_M^T)/(\Gamma_{\text{opt}} + \Gamma_M)$, which reduces to the simple classical expression for the effective temperature given above for $\bar{n}_M^T \gg 1$. As shown in Fig. 3 ground state cooling is most advantageously pursued in the resolved-sideband regime with high finesse cavities and high frequency resonators. With various groups working on a variety of setups further progress and final success in approaching the quantum limit is expected in the very near future.

The strong coupling regime, where $\Gamma_{\text{opt}} > \kappa$, needs a more sophisticated analysis and gives rise to new features (Marquardt et al. 2007, 2008).

## 4.2. QUANTUM DESCRIPTION OF SELF-INDUCED OSCILLATIONS

For a full quantum description (Ludwig et al. 2008) of the self-induced oscillations, we have to consider the reduced density matrix $\hat{\rho}$ of the system consisting of cantilever and cavity mode. Mechanical damping and photon decay out of the cavity are treated using a Lindblad master equation,

$$\frac{d}{dt}\hat{\rho} = \mathcal{L}\hat{\rho} = -\frac{i}{\hbar}\left[\hat{H}_{\text{cav+cant+drive}}, \hat{\rho}\right] + \Gamma_M \mathcal{D}[\hat{c}] + \kappa\mathcal{D}[\hat{a}] \quad (\text{for} \quad T = 0), \quad (5)$$

where $\mathcal{D}[\hat{a}] = \hat{a}\hat{\rho}\hat{a}^\dagger - \frac{1}{2}\hat{a}^\dagger\hat{a}\hat{\rho} - \frac{1}{2}\hat{\rho}\hat{a}^\dagger\hat{a}$ is of the standard Lindblad form.

The stationary state of the system is found as the eigenvector of the Liouvillian $\mathcal{L}$ for eigenvalue zero. This problem can be solved numerically for a restricted, but sufficiently large number of cavity and cantilever states. From the eigenvector, the density matrix $\hat{\rho}_f$, all quantities of interest, for instance, the average kinetic energy of the cantilever motion, can then be calculated.

Before comparing the results of this quantum mechanical description to the classical approach, it is instructive to quantify the degree of "quantumness" of the system. Using the dimensionless parameters $\mathcal{P} = 8|\alpha_L|^2 g^2/\omega_M^4$, characterizing the driving strength, and $\zeta = g/\kappa$, the Hamiltonian is written as

$$\hat{H}_{\text{cav+cant+drive}} = \hbar\left\{\left[-\Delta - \kappa\zeta(\hat{c} + \hat{c}^\dagger)\right]\hat{a}^\dagger\hat{a} + \omega_M\hat{c}^\dagger\hat{c} + \frac{\sqrt{2\mathcal{P}}\omega_M^2}{4\kappa\zeta}(\hat{a} + \hat{a}^\dagger)\right\}. \quad (6)$$

The master (5) then contains only dimensionless quantities, if time and the remaining energy/frequency variables are written in terms of the mechanical oscillation frequency $\omega_M$. Four of the dimensionless parameters in this equation, $\Gamma_M/\omega_M$, $\kappa/\omega_M$, $\Delta/\omega_M$ and $\mathcal{P}$ do also appear in the classical equations of motion, while

$$\zeta = \frac{g}{\kappa} = \frac{x_{\text{ZPF}}}{x_{\text{FWHM}}} \propto \sqrt{\hbar} \quad (7)$$

does not. The so-defined "quantum parameter" $\zeta$ constitutes a measure of the quantum nature of the system and vanishes in the classical limit $\hbar \to 0$. It is defined as the ratio of the quantum mechanical zero point fluctuations of the cantilever to a classical length scale, namely the resonance width $x_{\text{FWHM}}$ of the cavity.

The quantum master equation allows studying the quantum-to-classical crossover of the system dynamics by changing the numerical value of the quantum parameter $\zeta$. Classical results are recovered for small $\zeta$, while for $\zeta \gtrsim 1$ quantum fluctuations tend to smear out the sharp features of the

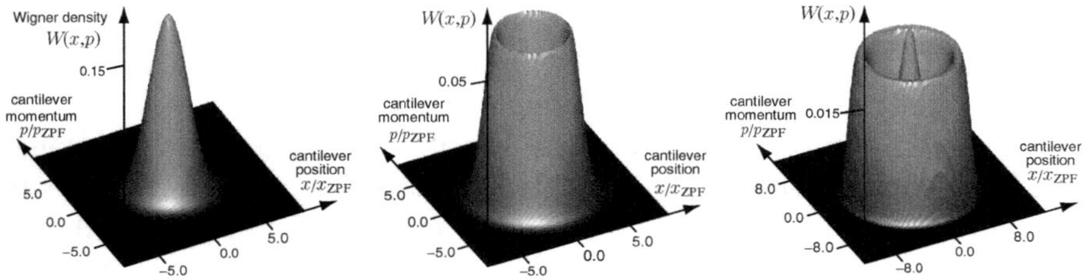


*Figure 4.* Wigner densities for the cantilever under the influence of the radiation force, for varying detuning, displaying the optomechanical instability (middle and right panels)

classical result and favour the occurrence of self-induced oscillations below the classical onset, a feature which can also be deduced from the quantum noise approach introduced above (see Ludwig et al. 2008 for details and figures). Note that to some extent the effects of quantum fluctuations can be mimicked by introducing quantum zero-point fluctuations into the classical equations of motion (Ludwig et al. 2008).

The existence of classical bi- or multistable solutions can be seen by considering the Wigner density of the cantilever. As illustrated in Fig. 4, the Wigner density shows characteristic features corresponding to (a) a single stationary classical solution (broad peak in phase space), (b) a single finite amplitude classical solution (ring structure – the phase of the oscillatory solution is undetermined), or (c) the coexistence of a classical stationary and finite amplitude solution (peak with superimposed ring structure).

We find that most optomechanical experiments are well in the classical regime, in the sense that the quantum parameter remains small (e.g. $\zeta \approx 10^{-3} \ll 1$ in the Bouwmeester setup, Kleckner and Bouwmeester 2006). In two recent setups, however, combining standard optomechanics with cold-atom physics (Gupta et al. 2007; Murch et al. 2008; Brennecke et al. 2008), $\zeta$ is of the order of one. In these experiments a cloud of cold atoms is placed in an optical cavity, so that the collective motion of the cloud couples to an optical mode of the cavity, replacing the cantilever motion.

## 5. Towards Fock-State Detection

Linked inextricably to the race towards ground state cooling is the question how to confirm the quantum nature of the final state. Measurement of the displacement quadratures is possible via optical readout (Clerk et al. 2008). However, probably the most straightforward demonstration would be to observe the quantum jumps from the ground state to progressively higher energy eigenstates (Fock states), as the system heats up again. Such quantum jumps between different Fock states have been observed in the mechanical motion of an electron in a Penning trap (Peil and Gabrielse 1999). In optomechanics such quantum jumps might eventually be observed for the mechanical motion of a truly macroscopic object, consisting of billions of atoms.

Recently, the Yale group of Jack Harris introduced a novel optomechanical setup (Thompson et al. 2008; Jayich et al. 2008), where a thin dielectric membrane is placed in the middle of a cavity with two fixed, high finesse mirrors. Beside the technological advances offered by this setup, it also leads to a different coupling of the mechanical displacement of the oscillating membrane to the cavity, which is advantageous for the aim of Fock state detection. To find the structure of the coupling term in the Hamiltonian, consider first the limit of a perfectly reflecting membrane at some position $x$ in the middle of the cavity. Moving the membrane will change the frequencies of resonances in the left and right halves of the cavity in opposite directions, which would lead to a resonance crossing at some displacement $x_{\text{cross}}$. A finite transmission of the membrane, however, produces an anti-crossing, with $\omega(x) - \omega(x_{\text{cross}}) \propto x^2$ near the degeneracy point. In rotating wave approximation the coupling is then of the form $\propto (\hat{c}^\dagger \hat{c} + \frac{1}{2})\hat{a}^\dagger \hat{a}$, so that $\left[\hat{H}_{\text{cant+cav+drive}}, \hat{c}^\dagger \hat{c}\right] = 0$, allowing non-destructive measurement of the phonon number. Detecting the phase of the transmitted beam driving the cavity at resonance frequency then constitutes a direct quantum non-demolition (QND) measurement of the phonon number.

Shot noise in the transmitted beam can be overcome by time averaging, which, however, is restricted by the life time of Fock states due to finite damping and temperature. Optimal averaging times and strategies, how best to distinguish classical from quantum fluctuations, even when the QND readout time is comparable to the state's life time, have been explored in Jayich et al. (2008).

## 6. Conclusions

Optomechanics is a new research topic that has been established in the past four years, with strong progress being made through a tight interplay of theory and experiment. Even the classical nonlinear dynamics of these systems is far from being fully explored: For example, chaotic motion has been observed at strong drive (Carmon et al. 2005), but not yet analyzed systematically. In the quantum regime, ground-state cooling and creation of nonclassical states (e.g. entanglement) are interesting challenges. New setups expand the applicability of these concepts, e.g. in superconducting microwave circuits or with cold atoms.

## Acknowledgements

F.M. acknowledges the inspiring atmosphere of the NATO workshop in Tashkent, Usbekistan, where this overview was presented, as well as J. Harris, S. Girvin, K. Karrai, C. Metzger, A. Clerk, and C. Neuenhahn for collaboration on this topic. This research is funded by the German Science Foundation (DFG) through NIM, SFB631, and the Emmy-Noether program.

## References

Aguirregabiria J.M. and Bel L.: 1987, Delay-induced instability in a pendular Fabry–Perot cavity. *Phys. Rev. A* **36**, 3768–3770

Arcizet O., Cohadon P.F., Briant T., Pinard M. and Heidmann A.: 2006, Radiation-pressure cooling and optomechanical instability of a micro-mirror. *Nature* **444**, 71

Braginsky V. and Manukin A.: 1967, Ponderomotive effects of electromagnetic radiation. *Sov. Phys. JETP* **25**, 653

Braginsky V.B., Manukin, A.B. and Tikhonov, M.Y.: 1970, Investigation of dissipative ponderomotove effects of electromagnetic radiation. *Sov. Phys. JETP* **31**, 829

Braginsky V.B., Strigin S.E. and Vyatchanin S.P.: 2001, Parametric oscillatory instability in Fabry–Perot interferometer. *Phys. Lett. A* **287**, 331–338

Brennecke F., Ritter S., Donner T. and Esslinger T.: 2008, Cavity optomechanics with a Bose–Einstein condensate. *Science* **322**, 235–238

Brown K.R., Britton J., Epstein R.J., Chiaverini J., Leibfried D. and Wineland D.J.: 2007, Passive cooling of a micromechanical oscillator with a resonant electric circuit. *Phys. Rev. Lett.* **99**, 137205

Carmon T., Rokhsari H., Yang L., Kippenberg T.J. and Vahala K.J.: 2005, Temporal behavior of radiation-pressure-induced vibrations of an optical microcavity phonon mode. *Phys. Rev. Lett.* **94**, 223902

Clerk A.A., Marquardt F. and Jacobs K.: 2008, Back-action evasion and squeezing of a mechanical resonator using a cavity detector. *New J. Phys.* **10**, 095010

Cohadon, P.F., Heidmann A. and Pinard M.: 1999, Cooling of a mirror by radiation pressure, *Phys. Rev. Lett.* **83**, 3174–3177

Corbitt T., Chen Y., Innerhofer E., Muller-Ebhardt H., Ottaway D., Rehbein H., Sigg D., Whitcomb S., Wipf C. and Mavalvala N.: 2007, An all-optical trap for a gram-scale mirror. *Phys. Rev. Lett.* **98**, 150802

Fabre C., Pinard M., Bourzeix S., Heidmann A., Giacobino E. and Reynaud S.: 1994, Quantum-noise reduction using a cavity with a movable mirror. *Phys. Rev. A* **49**, 1337–1343

Gigan S., Bohm H.R., Paternostro M., Blaser F., Langer G., Hertzberg J.B., Schwab K.C., Bauerle D., Aspelmeyer M. and Zeilinger A.: 2006, Self-cooling of a micromirror by radiation pressure. *Nature* **444**, 67–70

Gupta S., Moore K.L., Murch K.W. and Stamper-Kurn D.M.: 2007, Cavity nonlinear optics at low photon numbers from collective atomic motion. *Phys. Rev. Lett.* **99**, 213601

Höhberger C. and Karrai K.: 2004, Self-oscillation of micromechanical resonators. In: *Nanotechnology 2004, Proceedings of the 4th IEEE conference on nanotechnology*, IEEE, New York, p. 419

Höhberger-Metzger C. and Karrai K.: 2004, Cavity cooling of a microlever. *Nature* **432**, 1002

Jayich, A.M., Sankey J.C., Zwickl B.M., Yang C., Thompson J.D., Girvin S.M., Clerk A.A., Marquardt F. and Harris J.G.E.: 2008, Dispersive optomechanics: a membrane inside a cavity. *New J. Phys.* **10**, 095008 (28pp)

Kippenberg T.J. and Vahala K.J.: 2008, Cavity optomechanics: Back-action at the mesoscale. *Science* **321**, 5893.

Kippenberg T.J., Rokhsari H., Carmon T., Scherer A. and Vahala K.J.: 2005, Analysis of radiation-pressure induced mechanical oscillation of an optical microcavity. *Phys. Rev. Lett.* **95**, 033901

Kleckner D. and Bouwmeester D.: 2006, Sub-kelvin optical cooling of a micromechanical resonator. *Nature* **444**, 75

Ludwig M., Kubala B. and Marquardt F.: 2008, The optomechanical instability in the quantum regime. *New J. Phys.* **10**, 095013 (19pp)

Marquardt F.: 2008, Optomechanics: Push towards the quantum limit, *Nat. Phys.* **4**, 513–514

Marquardt F., Harris J.G.E. and Girvin S.M.: 2006, Dynamical multistability induced by radiation pressure in high-finesse micromechanical optical cavities. *Phys. Rev. Lett.* **96**, 103901

Marquardt F., Chen J.P., Clerk A.A. and Girvin S.M.: 2007, Quantum theory of cavity-assisted sideband cooling of mechanical motion. *Phys. Rev. Lett.* **99**, 093902

Marquardt F., Clerk A.A. and Girvin S.M.: 2008, Quantum theory of optomechanical cooling, *J. Mod. Optic.* **55**, 3329

Metzger C., Ludwig M., Neuenhahn C., Ortlieb A., Favero I., Karrai K. and Marquardt F.: 2008, Self-induced oscillations in an optomechanical system driven by Bolometric backaction. *Phys. Rev. Lett.* **101**, 133903

Murch K.W., Moore K.L., Gupta S. and Stamper-Kurn D.M.: 2008, Observation of quantum-measurement backaction with an ultracold atomic gas. *Nat. Phys.* **4**, 561–564

Naik A., Buu O., LaHaye M.D., Armour A.D., Clerk A.A., Blencowe M.P. and Schwab K.C.: 2006, Cooling a nanomechanical resonator with quantum back-action. *Nature* **443**, 193–196

Peil S. and Gabrielse G.: 1999, Observing the quantum limit of an electron cyclotron: QND measurements of quantum jumps between fock states. *Phys. Rev. Lett.* **83**, 1287–1290

Regal C.A., Teufel J.D. and Lehnert K.W.: 2008, Measuring nanomechanical motion with a microwave cavity interferometer. *Nat. Phys.* **4**, 555–560

Rodrigues D.A., Imbers J., Harvey T.J. and Armour A.D.: 2007, Dynamical instabilities of a resonator driven by a superconducting single-electron transistor. *New J. Phys.* **9**, 84

Schliesser A., Del'Haye P., Nooshi N., Vahala K.J. and Kippenberg T.J.: 2006, Radiation pressure cooling of a micromechanical oscillator using dynamical backaction. *Phys. Rev. Lett.* **97**, 243905

Schwab K.C. and Roukes M.L.: 2005, Putting mechanics into quantum mechanics. *Phys. Today* **July**, 36

Thompson J.D., Zwickl B.M., Jayich A.M., Marquardt F., Girvin S.M. and Harris J.G.E.: 2008, Strong dispersive coupling of a high-finesse cavity to a micromechanical membrane. *Nature* **452**, 900–900

Wilson-Rae I., Nooshi N., Zwerger W. and Kippenberg T.J.: 2007, Theory of ground state cooling of a mechanical oscillator using dynamical backaction. *Phys. Rev. Lett.* **99**, 093901

# HOHENBERG–MARTIN DILEMMA FOR BOSE CONDENSED SYSTEMS AND ITS SOLUTION

A. Rakhimov* and Z. Narzikulov
*Institute of Nuclear Physics, Tashkent 102132, Uzbekistan;*
*rakhimovabd@yandex.ru*

**Abstract.** The properties of the uniform Bose gas is studied within the optimized variational perturbation theory (Gaussian approximation) in a self-consistent way. It is shown that the atomic BEC with a repulsive interaction becomes unstable when the gas parameter $\gamma = \rho a^3$ exceeds a critical value $\gamma_{crit} \approx 0.01$. The quantum corrections beyond the Bogoliubov–Popov approximation to the energy density, chemical potential and pressure in powers of $\sqrt{\gamma}$ expansions are presented.

**Key words:** Bose condensate; Critical density; Field theoretical methods; Nonperturbative approach

## 1. Introduction

The outline of the present talk is following:

- What is Hohenberg–Martin dilemma? Illustration on the one loop level (Hohenberg and Martin 1965)
- Yukalov–Kleinert prescription (Yukalov and Kleinert 2006)
- Hartree Fock–Bogolubov approximation and unstability of uniform BEC

## 2. Hohenberg–Martin Dilemma for Bose Condensed Systems

1. $\mathcal{L}_{\text{BOSE}}(\psi, \psi^*)$ is invariant under $U(1)$ globe gauge symbol. $\exists\ T = T_c$, $T \langle T_c\ U(1)$ is broken.
2. The theory of Bose-condensed systems is based on Bogolubov idea of breaking the global gauges symmetry by means of the Bogolubov shift for field operators.

$$\psi(r,t) \rightarrow v(r,t) + \tilde{\psi}(r,t),$$
$$\int v^*(r,t)\tilde{\psi}(r,t)d\mathbf{r} = \mathbf{0}, \tag{1}$$

G. Casati and D. Matrasulov (eds.), *Complex Phenomena in Nanoscale Systems*,
NATO Science for Peace and Security Series B: Physics and Biophysics,
© Springer Science+Business Media B.V. 2009

$$N_0 = \langle \int \vartheta^* \vartheta(r,t) d\mathbf{r} \rangle,$$
$$N_1 = \langle \int \tilde{\psi}^* \psi(r,t) d\mathbf{r} \rangle. \quad (2)$$

statistical $N = N_0 + N_1$

3. Statistical condition stability

- $$\frac{d\Omega(\mu, v)}{dv} = 0, \quad (3)$$

  MIN of thermodynamic potential
- Particle spectrum, under the broka gauge symmetry must be gapless:

  $$\lim_{k \to \infty} E(k) = 0. \quad (4)$$

  This is equivalent to Hugaholtz–Bines theorem: $\mu = \Sigma_n(0,0) - \Sigma_{a_n}(0,0)$ $\Sigma_{\alpha_\beta}$-self energies.
- Conservation of quantum number (spin, momentum)

  $$\langle \tilde{\psi}(r,t) \rangle = 0 \quad (5)$$

Hohenberg–Martin dilemma: if you fix $\mu$ by (4) then you can not satisfy (3) and vice verso.

## 3. Illustration of Hohenberg–Marin Contradiction (One Loop Approximation) UNIFORM Bose System

Partitution function:

$$Z(\beta) = \int \mathcal{D}\psi \mathcal{D}\psi^* e^{-s} \quad (6)$$

action

$$S = \int_0^T d\tau d\mathbf{r} \left\{ \psi^+ [\partial_\tau - \frac{\nabla^2}{2m} - \mu] \psi(x) + \frac{g}{2} (\psi^+ \psi)^2 \right\}, \quad (7)$$

Thermodynamic potential:

$$\Omega = -T \ln Z(\beta) \quad (8)$$

Thermodynamic average:

$$\langle \hat{A} \rangle = \frac{1}{Z(\beta)} \int \mathcal{D}\psi^+ \mathcal{D}\psi \mathcal{D}(\psi, \psi^+) \, e^{-s(\psi,\psi^+)} \quad (9)$$

Bogolubov shift

$$\begin{aligned} \psi &= v + \tilde{\psi} \\ \tilde{\psi} &= \frac{1}{\sqrt{2}}(\psi_1 + i\psi_2), \\ \tilde{\psi^+} &= \frac{1}{\sqrt{2}}(\psi_1 - i\psi_2) \qquad (10) \\ \psi_a(r,\tau) &= \frac{1}{\sqrt{V\beta}} \sum_{n,k} \psi_a(\omega_n, k) e^{i\omega_n \tau + i\mathbf{kr}}, \qquad (a,b = 1,2) \qquad (11) \end{aligned}$$

$\omega_n = 2\pi n T$ – Matsubara frequency

$$\sum_k \to \forall \int \mathbf{k}/(2\pi)^3 \qquad (12)$$

then

$$\begin{aligned} S &= S^0 + S^{(1)} + S^{(2)} + S^{(3)} + S^{(4)}, \\ S^0 &= \int d\tau di[\frac{gv^4}{2} - \mu v^2], \\ S^{(1)} &= di \int d\tau \psi_1 \sqrt{2} v[\mu + gv^2], \\ S^{(2)} &= \frac{1}{2} \sum_{\substack{n,n' \\ k,k'}} \psi_a(\omega_n, \mathbf{k}) M_{ab}(\omega_n, \mathbf{k}; \omega_n, \mathbf{k}')\psi_b(\omega_n; \mathbf{k}') \qquad (13) \end{aligned}$$

$$M_{ab} = \frac{(2\pi)^4}{\sqrt{\beta}}(S(\mathbf{k}+\mathbf{k}')S(\omega_n + \omega_{n'})\begin{pmatrix} \varepsilon_k + x_1 & \omega_n \\ -\omega_n & \varepsilon_k + x_2 \end{pmatrix} \qquad (14)$$

where $\varepsilon_k = \mathbf{k}^2/2m$,

$$\begin{aligned} x_1 &= 2gv^2 - \mu; \qquad (15) \\ x_2 &= gv^2 - \mu. \end{aligned}$$

One loop:

$$\begin{aligned} s^{(3)} &\to 0 \\ s^{(4)} &\to 0 \qquad (16) \end{aligned}$$

Then $\int \mathcal{D}\psi \mathcal{D}\psi^+ e^{-S}$ is Gaussian. Green function:

$$G_{ab}^{(i,i')} = \langle T_\tau \psi_a(r)\psi_b(r')\rangle = \frac{1}{Z(\beta)} \frac{\delta^2 Z(\beta, j_1, j_2)}{\delta j_a(2)\delta j_b(2')}\Big|_{j_1=j_2=0} \qquad (17)$$

$$G_{ab}(k,\omega_n) = \frac{1}{\omega_n^2 + E_k^2}\begin{pmatrix} \varepsilon_k + x_2 & \omega_n \\ -\omega_n & \varepsilon_k + x_1 \end{pmatrix} \tag{18}$$

$$E_k = \sqrt{(\varepsilon_k + k_1)(\varepsilon_k + k_2)}, \tag{19}$$

$$\varepsilon_k = \mathbf{k} =^2 /2m; \quad x_1 = 3gv^2 - \mu,$$
$$x_2 = gv^2 - \mu \tag{20}$$

$$\Sigma_n = \frac{1}{2}[x_1 + x_2 + 2\mu], \quad \Sigma_{an} = \frac{1}{2}[x_1 - x_2], \tag{21}$$

Check condition:

- $\langle\psi_1\rangle = 0 \Rightarrow$

$$S^{(1)} = \int d\tau d\tau \psi_1 \sqrt{2} v[\mu + dv^2] = 0 \tag{22}$$

  that is $\mu = -g\vartheta^2$ must be.

- H.P. theorem

$$\Sigma_n - \Sigma_{an} = x_2 + \mu = \mu \qquad \text{mute} \Rightarrow x_2 = 0; \tag{23}$$

that is $E_k = \sqrt{k^2/2m + x_1\varepsilon_k} \approx ck + 0(11)$ that is $x_2 = g\vartheta^2 - \mu; \mu = gv^2$
One can cheek that $\mu = g\vartheta^2$ not appropriate with $\frac{d\Omega}{dv} = 0$ mins $\Omega_i$.

## 4. Yukalov–Kleinert Prescription

For each three conditions for equilibrium

1. $\langle\tilde{\psi}\rangle = 0 \; \psi \to v + \tilde{\psi}$: quantum conservation
2. $\langle\hat{N}_0\rangle = \langle\int v^2 d\mathbf{r}\rangle = N_0$
3. $\langle N\rangle = N_0 + N_2; N_! = \hat{N}_1 = \langle\int(\tilde{\psi})^2 d\mathbf{r}\rangle$

One should introduce three Lagrange multipliers to satisfy these conditions

$$S[v,\psi] = \int \left[\hat{\mathcal{L}} + \mu_0\hat{N}_0 + \hat{\Lambda}\right] dt \tag{24}$$

$$\Lambda \equiv \int \left[\mathcal{L}(r,t)\hat{\psi}^+ + \mathcal{L}^*\hat{\psi}\right] d\mathbf{r} \tag{25}$$

$\mu_0$ – chemical potential for condensed particles: $N_0 = -\frac{\partial\Omega}{\partial\mu_0}$
$\mu_1$ – chemical potential for uncondensed particles: $N_1 = -\frac{\partial\Omega}{\partial\mu_1}$
$\mathcal{L}$ – is chosen such that $\langle\hat{\Lambda}\rangle = 0$
All and all grand Hamiltonian

$$H[v,\hat{\psi}] = \hat{H} - \mu_0\hat{N}_0 - \mu_1\hat{N}_1 - \hat{\Lambda} \;\leftrightarrow\quad \hat{H} - \mu N \tag{26}$$

$\mu = -\left(\frac{\partial\Omega}{\partial N}\right)$ is $\mu = (\mu_0 N_0 + \mu_1 N_1)/N$. This prescription leads to Illowis action for interacting (content potential) uniform Bose gases:

$$S[\psi,\psi^*] = \int d\tau d\mathbf{r}\{\psi^*[\partial_\tau - \frac{\nabla^2}{2m}]\psi - \mu_1\hat{\psi}^*\hat{\psi}$$

$$-\mu_0 v^2 - (\mathcal{L}\hat{\psi}^+ + \mathcal{L}^*\hat{\psi}) + \frac{g}{2}[\psi^*\psi]^2\} \tag{27}$$

## 5. Illustration for One Loop

After $\psi \to v + \hat{\psi}$, $\quad \hat{\psi} = \frac{1}{\sqrt{2}}(\psi_1 + i\psi_2)$, $\quad \hat{\psi^+} = \frac{1}{\sqrt{2}}(\psi_1 - i\psi_2)$

$$S = S^0 + S^{(1)} + S^{(2)} + 0(\psi^3); \tag{28}$$

$$S^0 = V\beta[-\mu_0 v^2 + g v^4/2] \tag{29}$$

$$S^{(1)} = \int d\tau dr\{g v^2[\psi^+ \hat{+} \hat{\psi}] - \mathcal{L}^*\hat{\psi} - \mathcal{L}\hat{\psi}^*; \tag{30}$$

$$S^{(2)} = \frac{1}{2}\sum_{\substack{n,n'\\k,k'}} \psi_a(\omega_n, x) M_{ab}(\omega_n, k; \omega_n, k')\psi_b(\omega_n'; k'); \tag{31}$$

$$M_{ab} = \frac{(2\pi)^4}{\sqrt{\beta}} S(\vec{k} + \vec{k'}) S(\omega_n + \omega_{n'}) \begin{pmatrix} \varepsilon_k + x_1 & \omega_n \\ -\omega_n & \varepsilon_k + x_2 \end{pmatrix}; \tag{32}$$

$$\varepsilon_k = \vec{k}^2/2m; \tag{33}$$

$$G_a(\omega_2, k) = \frac{1}{\omega_n^2 + E_k^2} \begin{pmatrix} \varepsilon_k + x_2 & \omega_n \\ -\omega_n & \varepsilon_k + x_1 \end{pmatrix} \tag{34}$$

$$E_k = \sqrt{(\varepsilon_k + x_1)(\varepsilon_k + x_2)}; \tag{35}$$

$x_1 = 2gv^2 - \mu, \quad x_2 = gv^2 - \mu$

$$\Omega = V\{-\mu_0 v^2 + g\vartheta^4/2\} + \sum_k E_k + T\sum_k \ln[1 - e^{-\beta E_k}] \tag{36}$$

Conditions

1. $\langle\psi_1\rangle = 0$; $\mathcal{L} = gv^3 \Rightarrow S = 0$
2. $\Sigma_n - \Sigma_m = \mu_1 \Rightarrow \mu_1 = gv^2$; $x_2 = 0$; $E_k \sim ck + 0(k^3)$
3. $\frac{\partial\Omega}{\partial v = 0}$: We can fix $\mu_0$

   No contradictions

## 6. Hartee–Fock–Bogolubov Method in Yukalov–Kleinert Prescription

This prescription can be used not only in path integral functions, but also in operators. The grand Hamiltonians:

$$H = \hat{H} - \mu_0 \hat{N}_0 - \mu_1 \hat{N}_1 - \Lambda; \tag{37}$$

$$\psi \rightarrow v + \tilde{\psi}; \tag{38}$$

$$H = H^0 + H^{(1)} + H^{(2)} + H^{(3)} + H^{(4)} \tag{39}$$

where

$$H^0 = -\mu_0 N_0 + g v^4/2; \tag{40}$$

$$N_0 = \int d\tau v^2 = V \cdot v^2; \tag{41}$$

$$H^{(1)} = 0 \tag{42}$$

with choice of $\mathcal{L}$;

$$H^{(2)} = \int d\tau \{\tilde{\psi^+}(-\frac{\nabla^2}{2m} - \mu_1)\tilde{\psi} + \frac{g v^2}{2}(\tilde{\psi^2} + \tilde{\psi^+}^2 + 4\psi^{\tilde{+}}\tilde{\psi}); \tag{43}$$

$$H^{(3)} = v \int d\tau \{\tilde{\psi^+}\tilde{\psi^2} + \tilde{\psi^+}\tilde{\psi^+}\tilde{\psi}\}; \tag{44}$$

$$H^{(4)} = \frac{g}{2} \int d\tau [\tilde{\psi^+}\tilde{\psi}]^2 \tag{45}$$

Equation of motion

$$i\frac{\partial v}{\partial \tau} = \frac{\delta H}{\delta v} = -\mu_0 v + g[v^3 + \tilde{x}] = 0 \tag{46}$$

$$i\frac{\partial}{\partial \tau}\tilde{\psi} = \frac{\delta H}{\delta \tilde{\psi}^+} = \left(-\frac{\nabla^2}{2m} - \mu_1\right)\tilde{\psi^{\perp}} + g[2v^2\tilde{\psi} + v^2\tilde{\psi_1}^+ + \hat{x}] \tag{47}$$

$$\hat{x} = 2v \mid \tilde{\psi_1} \mid^2 + v\tilde{\psi^2} + \tilde{\psi}^+\tilde{\psi^2} \tag{48}$$

Above equation are exact now we implement Hartree–Fock–Bogolubov (HFB) approximation:

## 7. Second Quantization

$$\tilde{\psi} = \frac{1}{\sqrt{V}} \sum_k a_k e^{i\mathbf{kr}},$$

$$\tilde{\psi^+} = \frac{1}{\sqrt{V}} \sum_k a_k^+ e^{-i\mathbf{kr}}$$

$$\rho_1 = \langle \tilde{\psi}^+ \psi \rangle = \frac{1}{V} \sum_k a_k^+ a_k \equiv \frac{1}{V} \sum_k \rho_k \tag{49}$$

density of uncondensed particle

$$\sigma = \langle \tilde{\psi}\tilde{\psi} \rangle = \frac{1}{V}\sum_k a_k a_{-k} \equiv \frac{1}{V}\sum_k \sigma_k \tag{50}$$

anomaly density. Where

$$\langle a_k^+ a_p \rangle = \delta(\vec{k} - \vec{p})\rho_k; \tag{51}$$

$$\langle a_k a_p \rangle = \delta(\vec{k} + \vec{p})\sigma_k \tag{52}$$

HFB:

$$a_k^+ a_p a_\eta \rightarrow 2\langle a_k^+ a_p \rangle a_\eta + a_k^+ \langle a_p a_\eta \rangle \tag{53}$$

$$\begin{aligned} a_k^+ a_p^+ a_\eta a_m \rightarrow\ & 4a_k^+ a_m \langle a_p^+ a_\eta \rangle \\ & + \langle a_k^+ a_p^+ \rangle a_\eta a_m + a_k^+ a_p^+ \langle a_\eta a_m \rangle - 2N_1^2 - N_\sigma^2 \end{aligned} \tag{54}$$

where $N_1 = \rho_1 \cdot V$; $N_\sigma = \sigma \cdot V$.

Grand Hamiltonian

$$H \rightarrow H_0 + \sum_k a_k^+ a_k \omega_k + \frac{x_1}{4}\sum_k [a_k a_{-k} + a_{-k}^+ a_k^k] \tag{55}$$

$$H^0 = -\mu_0 N_0 + \frac{g v^4}{2} - \frac{Vg}{2}(2\rho_1^2 + \sigma^2) \tag{56}$$

$$\omega_k = \varepsilon_k + 2g\rho_0 + 2g\rho_1 - \mu_1 \tag{57}$$

$$x_1 = 2g[\rho_0 + \sigma] \tag{58}$$

This could be easily diagonalized Bogolyubov theory

$$\begin{cases} a_k = u_k b_k + v_k b_{-k}^+ \\ a_k^+ = u_k b_k^+ + v_k b_{-k} \end{cases}$$

$$[b_k b_p^+] = \delta(\mathbf{k} - \mathbf{p}) \tag{59}$$

with

$$\begin{aligned} u_k &= \sqrt{\frac{\omega_k + E_k}{2E_k}}, \\ v_k &= \sqrt{\frac{\omega_k - E_k}{2E_k}}, \\ E_k &= \sqrt{\omega_k^2 - x_1^2/4}. \end{aligned} \tag{60}$$

Impose conditions: $E_k \sim c(\vec{k}) + O(k^2)$; which $c = \sqrt{\frac{x_1}{2m}}$ – sound velocity. $\Rightarrow \mu_1 = g[\rho + \rho_1 - \sigma] \Rightarrow E_k = \sqrt{\varepsilon_k(\varepsilon_k + x_1)}$.

The diagonalized grand Hamiltonian is:

$$
\begin{aligned}
H &= -\mu_0 N_0 + \frac{Vg\rho_0^2}{2} - \frac{Vg}{2}(2\rho_1^2 + \sigma^2), \\
&\quad + \sum_k b_k^+ b_k E_k + \frac{1}{2}\sum_k (E_k - \varepsilon_k - x_1/2), \qquad (61) \\
x_1 &= 2g[\rho_0 + \sigma], \\
\rho_0 &= \rho - \rho_1, \qquad (62) \\
E_{\text{tot}} &= \langle H \rangle - \mu N = \langle H \rangle + \mu_0 N_0 + \mu_1 N_1. \qquad (63)
\end{aligned}
$$

As to $\mu_0$, it can be found from equation of motion, I brought above:
The result is: $\mu_0 = g[\rho + \rho_1 + \sigma]$.
The densities $\rho_1, \sigma$ and main equation for self energy $x_1$:

$$
\begin{aligned}
\rho_1 &= \frac{1}{V}\langle \sum_k a_k^+ a_k \rangle = \frac{1}{V}\sum_k \langle (\omega_k^2 + v_k^2) n_B + v_k^2 \rangle, \\
&= \frac{1}{2V}\sum_k \left[\frac{\omega_k (2n_B + 1)}{E_k} - 1\right], \qquad (64) \\
n_B &= 1/(e^{\beta E_k} - 1), \\
\omega_k &= \frac{\mathbf{k}^2}{2m + x_1/2}, \qquad (65) \\
\sigma &= \frac{1}{V}\langle \sum_k a_k a_{-k} \rangle = \frac{1}{V}\sum_k \omega_k v_k \,[1 + 2n_B] \\
&= -\frac{x_1}{2V}\sum_k \frac{(n_B + \frac{1}{2})}{E_k}, \qquad (66) \\
x_1 &= 2g\,[\rho - \rho_1 + \sigma] = 2g\left(\rho - \frac{1}{V}\sum_k \left[\frac{(\varepsilon_k + x_1)(n_B + 1/2)}{E_k} - \frac{1}{2}\right]\right), \qquad (67) \\
E_k^2 &= \varepsilon_k(\varepsilon_k + x_1). \qquad (68)
\end{aligned}
$$

$\sum_k \to V \int \frac{d\vec{k}}{(2\pi)^3}$: divergent for $T = 0$
$T = 0$ case

$$
\rho_1 = \frac{1}{2V}\sum_k \frac{\omega_k}{E_k} = \frac{V}{2V}\int \frac{d\vec{k}[\varepsilon_k + x_1/2]}{(2\pi)^3 \sqrt{\varepsilon_k}\sqrt{\varepsilon_k + x_1}} = \frac{1}{8V} I_{\|}(x_1), \qquad (69)
$$

$$
\sigma = -\frac{x_1 V}{4V}\int \frac{d\vec{k}}{(2\pi)^3 E_k} = 3\rho_1, \qquad (70)
$$

anomalous density is times larger than $\rho_1$

$$I_{\alpha\beta}(t) = \sum_k \frac{\varepsilon_k m^{\beta-\alpha}}{E_k}, \tag{71}$$

$$I_n(t) = \frac{V(2+m)^{3/2}}{3\pi^2}, \tag{72}$$

$$x_1 = 2g[\rho - \rho_1 + \sigma]. \tag{73}$$

Using explanation for $I_{\alpha,\beta}$ and $x_1 \equiv 2g\rho z$, $z$-dimensionless

$$Z - 1 = \frac{(2gpzm)^{(3/2)}}{3\pi^2\sqrt{2}\rho}. \tag{74}$$

Let $g = \frac{4\pi a}{m}$, $\gamma = a^3\rho$; $a$–s wave scattering

$$Z^3 - \frac{9\pi}{256\gamma}(Z-1)^2 = 0 \tag{75}$$

cubic equation for $Z$.

This equation could ba solved analytically, as a result

$$x_1 = 2d\rho - \frac{\sqrt{\pi}}{16\sqrt{\gamma}}\{1 + \sqrt{3}\cos(\alpha) - \sin(\alpha)\}, \tag{76}$$

$$\alpha = \frac{1}{3}\arccos\frac{384\gamma}{\pi} - 1 + \frac{\pi}{6}, \tag{77}$$

$$\rho_1 = \frac{(2x_1 m)^{3/2}}{24\pi^2}, \qquad \rho_0 = \rho - \rho_1 \tag{78}$$

$$E_{\text{tot}} = \frac{Vg\rho^2}{2}\left[1 - \frac{896\gamma z^6}{9\pi} + \frac{128\sqrt{\gamma}}{15\sqrt{\pi}}\right], \tag{79}$$

$$P = -\left(\frac{\partial E}{\partial V}\right). \tag{80}$$

## 8. SMALL $\gamma$ $T = 0$

$$n_1 \equiv \frac{\rho_1}{\rho} = \frac{8\sqrt{\gamma}}{3\sqrt{n}} + \frac{64\gamma}{15\pi} + \cdots + O(\gamma^{3/2}), \tag{81}$$

$$E_{tot} = \frac{g\rho^2 m}{2}\left[1 + \frac{128\sqrt{\gamma}}{15\sqrt{\pi}} + \frac{128\gamma}{\pi} + \cdots O(\gamma^{3/2})\right], \tag{82}$$

$$P = \frac{g\rho^2}{2}\left[1 + \frac{64\sqrt{\gamma}}{5\sqrt{\pi}} + \frac{256\gamma}{9\pi} + \cdots O(\gamma^{3/2})\right]. \tag{83}$$

Text book results, based on Bogolyubov–Popov approximation.

Any $\gamma(\gamma \leq 1)$ $T = 0$.

Dependence of $n_1(\gamma)$ is depicted in Fig. 1 it is sees that for $\gamma = 0,016$ BEC disappears.

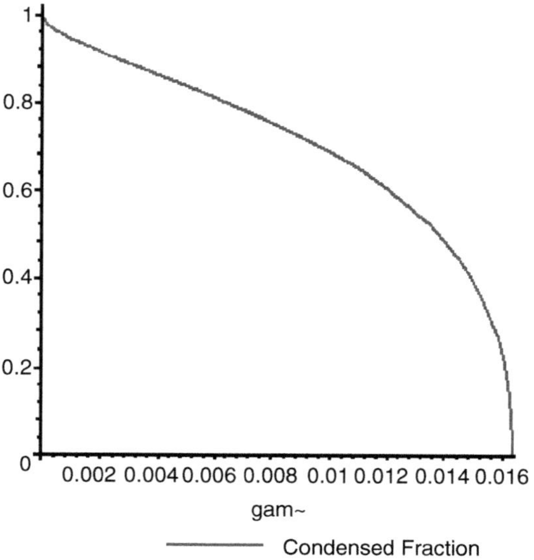


*Figure 1.* Dependence of $n_1(\gamma)$

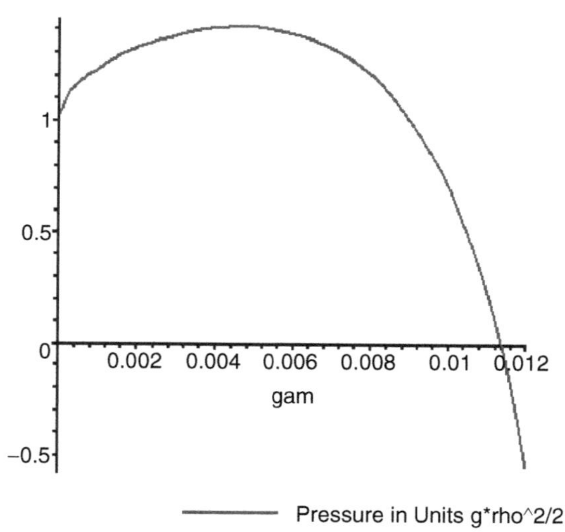


*Figure 2.* Dependence of pressure on $\gamma$

$P(\gamma)$ is presented in Fig. 2. It is seen that at $\gamma = \gamma_{\text{crit}} = 0,012$ $P\langle 0$ (Nozie'res 1995). This signals about the instability of BEC.

## 9. Conclusion

1. It has been shown that H.M. dilemma for a Bose condensate system can ba resolved by introduction of appropriate Lagrange multipliers leading to a representative ensemble. based on the HFB approximation.
2. We show that when the coopling constant of repulsive interaction becomes razer large the condensate collapses. The condensate collapses.

## References

Hohenberg P.C. and Martin P.C.: 1965. *Ann. Phys.*, **34**, 291
Nozie'res P.: 1995. Bose–Einstein Condensation. Griffin A., Snoke D.W. and Stringari S. (eds.): Cambridge University, New York
Yukalov V.I. and Kleinert H.: 2006. *Phys. Rev. A*, **73**, 063612

# PHONONS AND ELECTRON–PHONON INTERACTIONS IN SINGLE-WALLED ACHIRAL CARBON NANOTUBES

B.S. Kandemir
*Department of Physics, Faculty of Sciences, Ankara University, Ankara, Turkey; Bekir.S.Kandemir@science.ankara.edu.tr*

**Abstract.** Exact analytical expressions for the entire phonon spectra in single walled carbon nanotubes with achiral geometries are presented by using an approach, which mainly includes the construction of classical lattice Hamiltonian of single walled carbon nanotubes, then its quantization and finally diagonalization of the resulting second quantized Hamiltonian. Furthermore, within this context, analytical formulas for the relevant electron–phonon interaction coefficients are also obtained for single walled carbon nanotubes having these geometries, by the phonon modulation of the hopping interaction.

**Key words:** Carbon nanotubes; Graphene

The study of phonon dispersion relations in single walled carbon nanotubes (SWCNTs), that is, investigation of wavevector dependence pf phonon frequencies in whole region of the Brilliouin zone, is very important due to their crucial role in understanding electronic and heat transport, infrared and Raman spectra, and most importantly associated electron–phonon interaction (Jishi et al. 1993; Woods and Mahan 2000; Lazzeri et al. 2006; Mahan 2003; Mahan and Jeon 2004; Jeon and Mahan 2005; Jiang et al. 2005; Popov et al. 2005; Popov and Lambin 2006). In this report, I shall concentrate on a new technique we have developed to determine phonon frequencies in achiral SWCNTs, but the methodology described in these works can easily be extended to chiral ones. I shall not go into both theoretical details and the discussion of huge amount of literature concerning the latest developments in this field. The detailed description of theoretical framework together with a survey of the related literature in detail can be found in our previous papers (Kandemir and Altanhan 2008; Kandemir and Keskin 2008). Here, I will rather analyze primarily phonon dispersion relations in achiral SWCNTs, and I shall be concerned slightly with electron–phonon interaction processes. As discussed in Kandemir and Altanhan (2008) and Kandemir and Keskin (2008) in details, the quantization procedure of lattice vibrations of achiral SWCNTs can be formulated precisely in terms of boson operators. Performing afterwards two subsequent unitary transformations for the phonon part of

G. Casati and D. Matrasulov (eds.), *Complex Phenomena in Nanoscale Systems*,
NATO Science for Peace and Security Series B: Physics and Biophysics,

© Springer Science+Business Media B.V. 2009

the resultant Hamiltonian and doing the same things for the interacting part, i.e., for the linear electron–phonon interaction part, the Fröhlich Hamiltonian for SWCNTs then becomes

$$\begin{aligned} H &= \sum_{\mathbf{k}} \left[ E^{(+)}(\mathbf{k})\, C^{\dagger}_{A,\mathbf{k}} C_{A,\mathbf{k}} + E^{(-)}(\mathbf{k})\, C^{\dagger}_{B,\mathbf{k}} C_{B,\mathbf{k}} \right] \\ &\quad + \sum_{\mathbf{q}} \sum_{i} \hbar\widetilde{\omega}_i(\mathbf{q}) \left( a^{\dagger}_{\mathbf{q}i} a_{\mathbf{q}i} + \frac{1}{2} \right) + \sum_{\mathbf{q}} \sum_{\mathbf{k}} \sum_{i} \mathcal{D}_i(\mathbf{k},\mathbf{q}) \left( a^{\dagger}_{\mathbf{q}i} + a_{\mathbf{q}i} \right) \end{aligned} \tag{1}$$

where

$$\begin{aligned} \mathcal{D}_i(\mathbf{k},\mathbf{q}) &= \left[ \mathcal{L}^{(-)}_{i+} C^{\dagger}_{B,\mathbf{k}+\mathbf{q}} C_{B,\mathbf{k}} - \mathcal{L}^{(+)}_{i+} C^{\dagger}_{A,\mathbf{k}+\mathbf{q}} C_{A,\mathbf{k}} \right] \\ &\quad + \left[ \exp(+i\xi)\, \mathcal{L}^{(+)}_{i-} C^{\dagger}_{A,\mathbf{k}+\mathbf{q}} C_{B,\mathbf{k}} + \exp(-i\xi)\, \mathcal{L}^{(-)}_{i-} C^{\dagger}_{B,\mathbf{k}+\mathbf{q}} C_{A,\mathbf{k}} \right] \end{aligned} \tag{2}$$

with

$$\Psi_{\pm} = \{\exp[-i\xi(\mathbf{k})] \pm \exp[+i\xi(\mathbf{k}+\mathbf{q})]\}/2$$

and

$$\begin{aligned} \mathcal{L}^{(-)}_{i\pm} &= \mathcal{M}^{\mathrm{T}(+)}_i \Psi^{*}_{\pm} - \mathcal{M}^{\mathrm{T}(-)}_i \Psi^{*}_{\mp} \\ \mathcal{L}^{(+)}_{i\pm} &= \mathcal{M}^{\mathrm{T}(+)}_i \Psi_{\pm} + \mathcal{M}^{\mathrm{T}(-)}_i \Psi_{\mp}. \end{aligned}$$

To clarify the notation, we use $a_{\mathbf{q}i}\left(a^{\dagger}_{\mathbf{q}i}\right)$ operators as phonon annihilation (creation) operators. Here, the first term in (1) is, after a proper Bogoliobov unitary transformation proposed firstly in Kandemir and Altanhan (2008), the well-known spinless tight-binding Hamiltonian with energies $E^{(\pm)}(\mathbf{k}) = \pm J_0 |\Theta(\mathbf{k})|$, where $J_0$ is the hopping parameter and takes value approximately 2.4–3.1 eV, and $\Theta(\mathbf{k})$ is the sum of phase factors of atom $A$ with its three nearest neighbor $B$ atoms or vice versa. $C^{\dagger}_{A(B),\mathbf{k}}$ and $C_{A(B),\mathbf{k}}$ are the creation and annihilation operators for an electron with $\mathbf{k}$ mode, respectively. In (1) and (2), $i$ is the phonon band index and runs over from 1 to 6. Therefore, the first term in (2) represents the intraband scattering of an electron from the carbon atom $A(B)$ with wavevector $\mathbf{k} \equiv (k, \gamma)$ to the state with wave vector $\mathbf{q} + \mathbf{k} \equiv (q + k, \alpha + \gamma)$ of the carbon atom $A(B)$, the second one describes the interband scattering of an electron from the carbon atom $A(B)$ with wave vector $\mathbf{k} \equiv (k, \gamma)$ to the state with wave vector $\mathbf{q} + \mathbf{k} \equiv (q + k, \alpha + \gamma)$ of the carbon atom $B(A)$.

In obtaining phonon frequencies $\widetilde{\omega}_i(\mathbf{q})$ in (1), i.e., in second unitary transformation stage, we have first chosen a particular frequency, say the component $i$, and then taken into account the correlation of this particular component with the others. In Fig. 1, for a (10, 10) armchair SWCNT, we

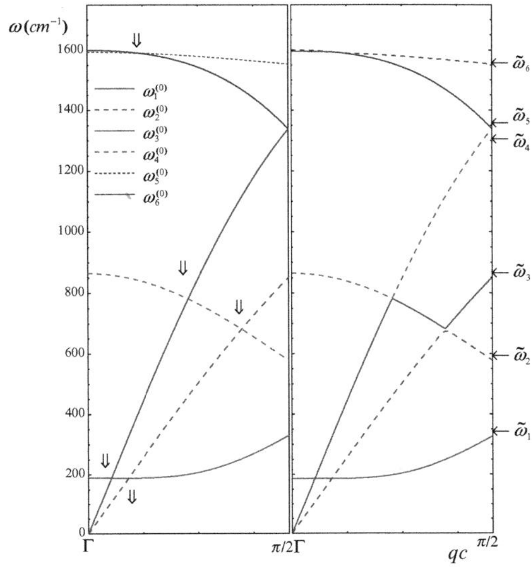


*Figure 1.* For a (10, 10) armchair SWCNT with $\alpha = 0$, $q$-dependence of the phonon spectra according to $\overline{\omega}_i^{(0)}$ (left panel), arising from the first diagonalization. In the right panel, again $q$-dependence of the phonon spectra, but after the second unitary transformation (Kandemir and Altanhan 2008)

compare our analytical phonon modes for $\alpha = 0$ obtained after the first transformation (without phonon correlations) with those obtained after second unitary transformation (with phonon correlations). While the left panel of this figure shows the six non-degenerate modes, of which the two are acoustical and the four are optical, the right panel shows how mode crossings, i.e. degeneracy of phonon branches, are removed by taking into account the correlation effects among the phonons.

Moreover, the overall behavior of calculated results for a (10, 0) zigzag SWCNT are also shown in Fig. 2. In this figure, drawing the phonon dispersion curves so as to span the whole range of $\alpha$ variation, we get a three dimensional picture, i.e., $\omega$ as function of both $qc$ and $\alpha$. We note that $\alpha$ quantum numbers increase as the radius $R$ of the tube increase. This is nothing but, for very large $R$, $\alpha$ becomes continuous so that the shown projectiles create phonon dispersions along the circumferential direction, which corresponds to $\Gamma K'$ direction of the graphene. For tubes with large radius, such a picture helps us to visualize two-dimensional $\mathbf{k}$-vector dependence of energy surfaces of the graphene.

To achieve the effect of electron–phonon interactions on the electronic energy spectrum, the interacting part of the total Hamiltonian can be rewritten as terms linear in $C^{\dagger}_{A(B),\mathbf{k+q}}C_{A(B),\mathbf{k}}$ and in $C^{\dagger}_{A(B),\mathbf{k+q}}C_{B(A),\mathbf{k}}$. Then the complete

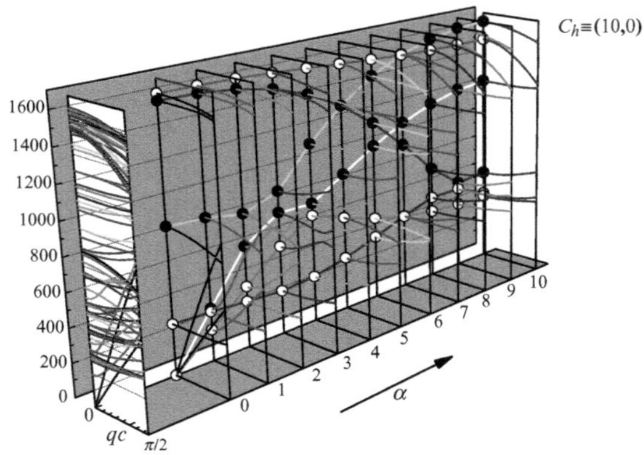


*Figure 2.* Plot of $\widetilde{\omega}_i(\mathbf{q})$ according to our analytical results (Kandemir and Keskin 2008) from $\alpha = 0$ to $\alpha = 10$ for a $(10, 0)$ zigzag SWCNT

Hamiltonian becomes

$$H = \sum_{\mathbf{k}} \sum_{j=1}^{2} E^{(j)}(\mathbf{k}) C^{\dagger}_{j,\mathbf{k}} C_{j,\mathbf{k}} + \sum_{\mathbf{q}} \sum_{i} \hbar \widetilde{\omega}_i(\mathbf{q}) \left( a^{\dagger}_{\mathbf{q}i} a_{\mathbf{q}i} + \frac{1}{2} \right)$$
$$+ \sum_{\mathbf{k},\mathbf{q}} \sum_{i,j} \left[ \mathcal{L}_{ij}(\mathbf{k},\mathbf{q}) a_{\mathbf{q}i} C^{\dagger}_{j,\mathbf{k}+\mathbf{q}} C_{j,\mathbf{k}} + \sum_{\ell \neq j} \mathcal{L}_{ij\ell}(\mathbf{k},\mathbf{q}) a_{\mathbf{q}i} C^{\dagger}_{\ell,\mathbf{k}+\mathbf{q}} C_{j,\mathbf{k}} + \mathrm{H.C.} \right] \quad (3)$$

which can easily be diagonalized by unitary transformation

$$U^{(j)} = \exp \left[ \sum_{\mathbf{q}} \sum_{\mathbf{k}} \left( \lambda^{*}_{\mathbf{k}\mathbf{q}} a^{\dagger}_{\mathbf{q}i} C^{\dagger}_{j,\mathbf{k}} C_{j,\mathbf{k}+\mathbf{q}} - \mathrm{H.C.} \right) \right].$$

The detailed derivation of such a unitary transformation and its direct consequences on the electronic spectrum will be given elsewhere.


## Acknowledgements

It is a great pleasure to thank the organizing committee of the workshop *"Recent Advances in Nonlinear Dynamics and Complex System Physics: From Natural to Social Sciences and Security"* for the very kind invitation.

## References

Jeon G.S. and Mahan G.D.: 2005, Theory of infrared-active phonons in carbon nanotubes. *Phys. Rev. B*, **72**, 155415-1–155415-9

Jiang J., Saito R., Samsonidze Ge.G., Chou S.G., Jorio A., Dresselhaus G. and Dresselhaus M.S.: 2005, Electron-phonon matrix elements in single-wall carbon nanotubes. *Phys. Rev. B*, **72**, 235408-1–235408-11

Jishi R.A., Dresselhaus M.S. and Dresselhaus G.: 1993, Electron–phonon coupling and the electrical conductivity of fullerene nanotubes. *Phys. Rev. B*, **48**, 11385–11389

Kandemir B.S. and Altanhan T.: 2008, Analytical approach to phonons and electron-phonon interactions in single-walled armchair carbon nanotubes. *Phys. Rev. B*, **77**, 045426-1–045426-15

Kandemir B.S. and Keskin M.: 2008, Analytical approach to phonons and electron-phonon interactions in single-walled zigzag carbon nanotubes. *J. Phys.: Condens. Matter*, **20**, 325222-1–325222-7

Lazzeri M., Piscanec S., Mauri F., Ferrari A.C. and Robertson J.: 2006, Phonon linewidths and electron-phonon coupling in graphite and nanotubes. *Phys. Rev. B*, **73**, 155426-1–155426-6

Mahan G.D.: 2003, Electron-optical phonon interaction in carbon nanotubes. *Phys. Rev. B*, **68**, 125409-1–125409-5

Mahan G.D. and Jeon G.S.: 2004, Flexure modes in carbon nanotubes. *Phys. Rev. B*, **70**, 075405-1–075405-11

Popov V.N. and Lambin Ph.: 2006, Intraband electron-phonon scattering in single-walled carbon nanotubes. *Phys. Rev. B*, **74**, 075415-1–075415-13

Popov V.N., Henrard L., and Lambin Ph.: 2005, Electron-phonon and electron-photon interactions and resonant Raman scattering from the radial-breathing mode of single-walled carbon nanotubes. *Phys. Rev. B*, **72**, 035436-1–035436-10

Woods L.M. and Mahan G.D.: 2000, Electron-phonon effects in graphene and armchair (10,10) single-wall carbon nanotubes. *Phys. Rev. B*, **61**, 10651–10663

# FAST-FORWARD PROBLEM IN MICROSCOPIC AND MACROSCOPIC QUANTUM MECHANICS

Sh. Masuda*
*Department of Applied Mathematics and Physics, Graduate School of Informatics, Kyoto University, Kyoto, Japan; syunpei710@yahoo.co.jp*

K. Nakamura
*Heat Physics Department of the Uzbek Academy of Sciences, 28 Katartal Street, 100135 Tashkent, Uzbekistan; nakamura@a-phys.eng.osaka-cu.ac.jp*

**Abstract.** We showed the way to speed up the time-evolution of a wave function (WF), i.e. to fast-forward the WF in microscopic and macroscopic quantum mechanics, by controlling the driving potential with resultant regulation of the additional phase of the WF, so that a target state is obtained in a shorter time (Phys. Rev. A 78:062108, 2008). We first presented a general framework of the fast-forwarding of WF in quantum mechanics and provided an example of the fast-forward of WF in two-dimensional (2-d) free space. Then the framework of the fast-forward was extended to macroscopic quantum mechanics described by the non-linear Schrödinger equation. We showed the fast-forward of (1) transport of Bose Einstein condensates trapped by a moving 2-d harmonic potential and (2) propagation of a soliton both in free space and through a potential barrier (macroscopic quantum tunnelling).

**Key words:** Time-evolution; Fast-forwarding; Quantum transport

## 1. Theoretical Framework of Fast-Forward in Quantum Mechanics

Suppose that $\Psi_0(\mathbf{x}, t)$ is a known function of space ($\mathbf{x}$) and time ($t$) and is called a standard state. Schrödinger equation is given as

$$i\hbar\frac{d}{dt} \mid \Psi_0(t)\rangle = \hat{H}_0 \mid \Psi_0(t)\rangle, \tag{1}$$

with

$$\hat{H}_0 = \frac{\hat{\mathbf{p}}^2}{2m_0} + \hat{V}_0(\hat{\mathbf{x}}, t). \tag{2}$$

Let $\Psi_\alpha(\mathbf{x}, t)$ be a fast-forwarded state of $\Psi_0(\mathbf{x}, t)$ by $\alpha$ times, i.e.

$$\mid \Psi_\alpha(t)\rangle = \mid \Psi_0(\alpha t)\rangle, \tag{3}$$

where $\alpha$ is a time-independent magnification factor of the fast-forward. The time-evolution of the WF is speeded up for $\alpha > 1$ and slowed down for

G. Casati and D. Matrasulov (eds.), *Complex Phenomena in Nanoscale Systems*,
NATO Science for Peace and Security Series B: Physics and Biophysics,

© Springer Science+Business Media B.V. 2009

$0 < \alpha < 1$ like a slow-motion. A rewind can occur for $\alpha < 0$, and the WF pauses when $\alpha = 0$.

In general, the magnification factor can be time-dependent $\alpha = \alpha(t)$. The time-evolution of a WF is accelerated and decelerated when $\alpha(t)$ is increasing and decreasing, respectively. In this case, the fast-forwarded state is defined as,

$$| \Psi_\alpha(t)\rangle = | \Psi_0(\Lambda(t))\rangle, \tag{4}$$

where

$$\Lambda(t) = \int_0^t \alpha(t')dt'. \tag{5}$$

### 1.1. FAST-FORWARD OF AMPLITUDE OF WAVE FUNCTIONS

It is not possible to make $| \Psi_\alpha\rangle$ itself by controlling the potential because the potential to generate $\Psi_\alpha$ in (4) is complex in general. The problem is resolved by introducing the additional phase factor. We first explored the fast-forward of the amplitude of a WF. Let $| \Psi_{FF}\langle$ be a realistic fast-forwarded state, which is defined with use of $\Psi_\alpha$ multiplied by the additional phase factor $e^{if(\mathbf{x},t)}$ as,

$$\langle \mathbf{x} | \Psi_{FF}(t)\rangle = \exp[if(\mathbf{x}, t)]\langle \mathbf{x} | \Psi_\alpha(t)\rangle. \tag{6}$$

$\Psi_{FF}$ is not the exactly fast-forwarded state, but its amplitude is certainly a fast-forwarded one of $\Psi_0$.

The Hamiltonians operating on $| \Psi_\alpha\rangle$ and $| \Psi_{FF}\rangle$ are respectively given as,

$$\hat{H}_\alpha = \frac{\hat{\mathbf{p}}^2}{2m_\alpha} + \hat{V}_\alpha(\hat{\mathbf{x}}, t), \tag{7}$$

$$\hat{H}_{FF} = \frac{\hat{\mathbf{p}}^2}{2m_0} + \hat{V}_{FF}(\hat{\mathbf{x}}, t), \tag{8}$$

where $m_\alpha$ and $\hat{V}_\alpha$, which correspond to $\Psi_\alpha$, are defined with use of $\alpha(t)$ as,

$$m_\alpha = m_0/\alpha(t), \tag{9}$$

$$\hat{V}_\alpha(\hat{\mathbf{x}}, t) = \alpha(t)\hat{V}_0(\hat{\mathbf{x}}, \Lambda(t)). \tag{10}$$

$\hat{V}_{FF}$ is the driving potential to generate $| \Psi_{FF}\rangle$. Note that the mass in $\hat{H}_{FF}$ in (8) is the same as that in $\hat{H}_0$.

From Schrödinger equations corresponding Hamiltonians in (7) and (8), we have

$$\hbar\frac{\partial f}{\partial t}\Psi_\alpha = w\nabla^2\Psi_\alpha + (V_\alpha - V_{FF})\Psi_\alpha + \eta[i(\nabla^2 f)\Psi_\alpha + 2i\nabla f \cdot \nabla\Psi_\alpha - (\nabla f)^2\Psi_\alpha], \tag{11}$$

with $w = -\frac{\hbar^2}{2m_0}(\alpha(t) - 1)$ and $\eta = \frac{\hbar^2}{2m_0}$. Multiplying $\Psi_\alpha$ on the complex conjugate of (11) and decomposing it into its real and imaginary parts, we have

$$|\Psi_\alpha|^2\nabla^2 f + 2\mathrm{Re}[\Psi_\alpha\nabla\Psi_\alpha^*]\cdot\nabla f + (\alpha(t) - 1)\mathrm{Im}[\Psi_\alpha\nabla^2\Psi_\alpha^*] = 0, \quad (12)$$

$$V_{FF} = V_\alpha - \hbar\frac{\partial f}{\partial t} - \eta(\nabla f)^2 + \mathrm{Re}[w\nabla^2\Psi_\alpha/\Psi_\alpha + 2i\eta\nabla f\cdot\nabla\Psi_\alpha/\Psi_\alpha]. \quad (13)$$

Equation (12) should be satisfied by $f(\mathbf{x}, t)$. The driving potential $V_{FF}$ for the fast-forward is given by (13), which is real and available in terms of $f(\mathbf{x}, t)$ that is a solution of (12).

While it looks difficult to obtain $\nabla f$ from (12), one can alternatively reach $\nabla f$ by having recourse to the continuity equation,

$$\frac{\partial}{\partial t}(\Psi^*\Psi) = -\frac{\hbar}{m}\nabla\cdot\mathrm{Im}[\Psi^*\nabla\Psi], \quad (14)$$

which leads to

$$\nabla f(\mathbf{x}, t) = (\alpha(t) - 1)\frac{\mathrm{Im}[\Psi_\alpha^*\nabla\Psi_\alpha]}{|\Psi_\alpha|^2}(\mathbf{x}, t) = (\alpha(t) - 1)\frac{\mathrm{Im}[\Psi_0^*\nabla\Psi_0]}{|\Psi_0 mid^2}(\mathbf{x}, \Lambda(t)), \quad (15)$$

Since (15) is a solution of (12), we can obtain the driving potential $V_{FF}(\mathbf{x}, t)$ without solving (12). $f(\mathbf{x}, t)$ in (15) vanishes except for arbitrary purely time-dependent terms when $\alpha(t)$ becomes 1 due to the factor $\alpha(t) - 1$. On the other hand, $V_{FF}(\mathbf{x}, t)$ in (13) becomes $V_0(\mathbf{x}, t)$ when $\frac{\partial\alpha}{\partial t}(\mathbf{x}, t) = 0$ as well as $\alpha = 1$, because $V_{FF}(\mathbf{x}, t)$ includes $\frac{\partial f}{\partial t}(\mathbf{x}, t)$.

### 1.2. FAST-FORWARD OF EXACT TARGET STATES

As mentioned above, the additional phase $f(\mathbf{x}, t)$ in (15) can vanish when $\alpha = 1$. Therefore, if we choose $\alpha(t)$ such as increasing from 1, keeping a desired fixed value larger than 1 for a while and then decreasing back to 1, and apply the corresponding driving potential $V_{FF}$, $\Psi_{FF}$ can reach the exact target state at the goal without the extra phase in a desired short time, though it has a nonvanishing additional phase on the way of fast-forwarding. There is no additional phase also at the initial time, that is, $\Psi_{FF}$ can now recover the phase of the exact target state as well as its amplitude at the final time. $V_{FF}$ can smoothly merges to $V_0$ at the initial and final time when $\alpha = 1$ and $\partial\alpha/\partial t = 0$ [see (10), (13) and (15)]. In fact, it is convenient that the extra potential $V_{FF} - V_0$ for the fast-forward is gradually applied, and smoothly removed after the fast-forwarding. Below we shall give an example of the fast-forward of a WF in quantum mechanics.

In numerical calculation, we chose a magnification factor $\alpha(t)$ as

$$\alpha(t) = (\bar{\alpha} - 1)\cos\left(\frac{2\pi}{T/\bar{\alpha}}t + \pi\right) + \bar{\alpha} \tag{16}$$

for $0 \leq t \leq T/\bar{\alpha}$, which starts from 1, increases up to $2\bar{\alpha} - 1$ (acceleration) and decreases gradually back to 1 (deceleration). $T$ (standard final time) is the time that the standard state spends to reach the target state. Because $\bar{\alpha}$ is equal to the time average of $\alpha(t)$, we can expect that $\Psi_{FF}$ reaches the target state in a $1/\bar{\alpha}$ amount of $T$. We call the time

$$T_{FF} = T/\bar{\alpha}, \tag{17}$$

as the final time of the fast-forward.

## 2. Fast-Forward in Macroscopic Quantum Mechanics

The theory of the fast-forward was extended to the macroscopic quantum mechanics. Let us consider the fast-forward of a WF for the Bose-Einstein condensates (BEC) governed by the nonlinear Schrödinger equation:

$$i\hbar\frac{\partial\Psi_0}{\partial t} = -\frac{\hbar^2}{2m_0}\nabla^2\Psi_0 + V_0\Psi_0 - c_0|\Psi_0|^2\Psi_0, \tag{18}$$

where $c_0$ is a nonlinearity constant. In this case, the driving potential is represented as

$$\begin{aligned} V_{FF} &= V_\alpha - (\alpha(t) - 1)c_0|\Psi_\alpha|^2 - \hbar\frac{\partial f}{\partial t} - \eta(\nabla f)^2 \\ &\quad + \mathrm{Re}[w\nabla^2\Psi_\alpha/\Psi_\alpha + 2i\eta\nabla f \cdot \nabla\Psi_\alpha/\Psi_\alpha], \end{aligned} \tag{19}$$

while the additional phase is given as (15).

### 2.1. EXAMPLES OF FAST-FORWARD

Using the driving potential evaluated with use of (15) and (19) together with a numerically obtained standard state $\Psi_0$, we showed the fast-forward of the transport of BEC (Leggett 2001; Ketterle 2002) in a moving harmonic potential in 2-d. Suppose an initial Gaussian WP to be static and located at the center of the harmonic potential. If we move the trapping potential slowly enough, the WP keeps staying around the center of the potential during and after the transport. But if we try to transport the WP fast by moving the trapping potential rapidly, it moves away from the the potential minimum during the transport, and keeps oscillating in a wide range in the trapping region even after the movement of the potential is ceased. Such large deviation from the center of the trapping potential is serious because, in practical experiments on BEC, the confinement of WP is not complete.[1] However if we apply the driving potential $V_{FF}$ based on the present theory of the fast-forward, we can transport the WP with neither giving the large unfavorable energy transfer from the potential nor serious disturbance of the WP, that is, we can obtain the target state of the WP in any desired short time, as if we move the harmonic potential slowly. To evaluate the accuracy of the fast-forward, we calculated the fidelity defined as

$$F = \mid \langle \Psi_{FF}(t) \mid \Psi_0(\Lambda(t)) \rangle \mid, \tag{20}$$

i.e. the overlap between the fast-forwarded state $\Psi_{FF}(t)$ and the corresponding standard one $\Psi_0(\Lambda(t))$. It is unity when $\Psi_{FF}$ agrees with the target state, and is now around 0.9995 at the final time of the fast-forward $t = T_{FF}$, while it is less than unity during the fast-forward (see Fig. 1). The profiles of $\mid \Psi_{FF} \mid^2$ at the final time $T_{FF}$ and the target state $\mid \Psi_0(T) \mid^2$ are shown in the inset in Fig. 1. They are overlapping completely. We have also confirmed the coincidence of the phase of the WP (which is not shown here). In this way we checked that the WP carried by the potential moving slowly, is accessible in a half amount of time with use of $V_{FF}$.

We also obtained the driving potential for the fast-forward of soliton propagation and barrier penetrations, and confirmed that we could generate target states in any desired short time.

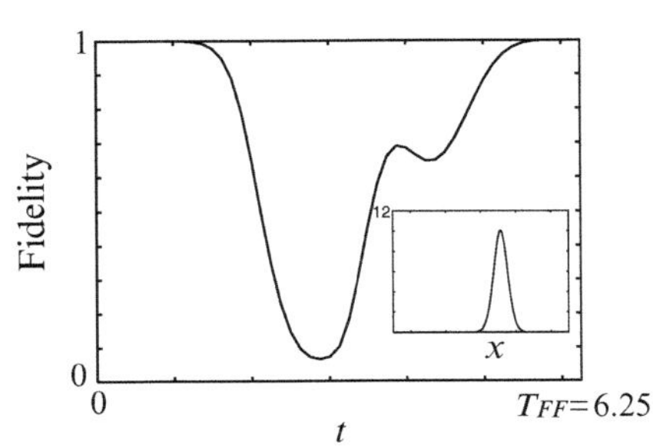


*Figure 1.* Time-dependence of the fidelity in (20). The inset represents $x$-dependence of the amplitude of $\Psi_{FF}$ and the target state at the final time. They are overlapping precisely

---

[1] On the other hand, a simply deepened trapping potential together with its fast movement could transport the WP rapidly keeping it near the potential minimum, but the WP again acquires the unfavorable transferred energy.

## References

Ketterle W.: 2002. *Rev. Mod. Phys.*, **74**, 1131

Leggett A.J.: 2001. *Rev. Mod. Phys.*, **73**, 307

Masuda S. and Nakamura K.: 2008. *Phys. Rev. A*, **78**, 062108

# COLLECTIVE OSCILLATIONS OF A QUASI ONE DIMENSIONAL BOSE CONDENSATE UNDER DAMPING

F.Kh. Abdullaev, R.M. Galimzyanov, and Kh.N. Ismatullaev*
*Physical-Technical Institute of the Academy of Sciences, G. Mavlyanov 2-b, Tashkent 100084, Uzbekistan; hayoti@yahoo.com*

**Abstract.** Affect of the damping on collective oscillations of a quasi one-dimensional trapped repulsive Bose gas has been studied. Based on the phenomenological damping approach (Zh. Eksp. Teor. Fiz. 35:408, 1958a; Sov. Phys. JETP 35:282, 1958b) developed by Pitaevskii variational equations for the parameters of the condensate wave function have been derived. Analytical expressions for the condensate parameters in the steady-state have been obtained. Combined effect of the *resonant periodical variation of the trap strength* and *the damping* has been shown to change drastically asymptotical behavior of the driven norm oscillations. Bistability in nonlinear oscillations of the condensate under periodic variations of the trap potential is predicted.

**Key words:** Bose–Einstein condensate; Damping; Nonlinear oscillations

## 1. Introduction

The dynamics of a one-dimensional trapped ultra-cold Bose gas has attracted a great attention for last years (Pitaevskii and Stringari 1958; Brazhnyi et al. 2003). Recently 1D regime has been realized experimentally in Moritz et al. (2003). Measurements of the collective oscillations of such a system should give a lot of information about the BEC dynamics. In particular this is important for the analysis of the condensate dynamics in a magnetic waveguide, being a fundamental atom optical element (Ott et al. 2003).

A trapped 1D repulsive Bose gas is known (Moritz et al. 2003) to be characterized by a single parameter $\zeta = mg_{1D}/(\hbar^2 n_{1D})$ which is the ratio between interaction energy and the kinetic energy of the ground state, $m$, $g_{1D}$ and $n_{1D}$ being atomic mass, the strength of interaction and 1D density, respectively. Different regimes in one dimensional geometry are possible depending on the density of gas. In the high density regime ($\zeta \ll 1$) the dynamics at low temperatures is described by a one-dimensional Gross–Pitaevskii equation with cubic mean field nonlinearity. The low density regime ($\zeta \gg 1$, Tonks–Girardeau (TG) regime) is characterized by the strong quantum correlations

G. Casati and D. Matrasulov (eds.), *Complex Phenomena in Nanoscale Systems*,
NATO Science for Peace and Security Series B: Physics and Biophysics,
© Springer Science+Business Media B.V. 2009

and a fermionic behavior of the system (Lieb and Liniger 1963; Lieb 1963; Girardeau 1960). Modern experiments cover both of these limiting cases. Computations of the collective excitations frequencies of a trapped 1D repulsive Bose gas for different 1D configurations varying from the mean field regime to the TG regime were performed in Menotti and Stringari (2002). In the present work we will concentrate our attention on the description of the quasi-one-dimensional dynamics of a repulsive BEC in the mean field regime.

Most of the theoretical descriptions have mainly dealt with conservative systems (e.g. see Ott et al. 2003; Abdullaev and Garnier 2004), where collective oscillations of a quasi 1D Bose–Einstein condensate (BEC) in the low- and high-density regimes were investigated. However, the dissipation inheres in real systems. So, theoretical study of the effect of damping on collective oscillations of a one-dimensional trapped repulsive Bose gas is of importance. The damping of the radial BEC oscillations in a cylindric trap connected with the parametric resonance and leading to the energy transfer from collective oscillations to longitudinal sound waves has been studied in Kagan and Maksimov (2001).

We consider here the problem of using phenomenological approach developed by Pitaevskii (1958a, b) and employed later in Choi et al. (1998). Recently this approach has been successfully applied to a series of problems. For example, conditions for the parametric driving of dark solitons in repulsive quasi-one-dimensional BEC were found in Proukakis et al. (2004), the analysis of the existence of stable 3D droplets in attractive BEC with nonlinearity management was carried out in Saito and Ueda (2004), Faradey patterns in 2D BEC with damping were studied in Staliunas et al. (2002).

## 2. The Model

To take into account the damping caused by the interaction of the condensate with the thermal cloud atoms we employ the phenomenological damping approach developed by Pitaevskii (1958a, b). The dynamics of a trapped one-dimensional repulsive Bose gas with the damping is described in the framework of the modified 1D Gross–Pitaevskii equation

$$i\hbar\phi_t = (1 + i\gamma)(-\frac{\hbar^2}{2m}\phi_{xx} + V(x,t)\phi + g_{1D}|\phi|^2\phi - \mu\phi), \tag{1}$$

with the total number of atoms $N = \int |\phi|^2 dx$. The constant $\gamma$ is the damping constant introduced phenomenologically to describe evolution toward equilibrium between the thermal cloud atoms and the condensate

(Choi et al. 1998). Approximate estimate obtained from the collision integral is given as $\gamma \sim 4Cm(akT)^2/(\pi\hbar^3), C \approx 3$ (Penckwitt et al. 2002). The dissipation constant can vary with the changes in temperature and atomic scattering length.

Equation (1) is obtained for the case of a highly anisotropic external potential under the assumption that the transversal trapping potential is harmonic: $V(y,z) = m\omega_\perp^2(y^2+z^2)/2$ and $\omega_\perp \gg \omega_x$. Under such conditions we can consider the solution of 3D equation to have the form $U(x,y,z;t) = R(y,z)\phi(x,t)$ where $R_0^2 = m\omega_\perp \exp(-m\omega_\perp\rho^2/\hbar)/(\pi\hbar)$. Averaging the condensate wave function in radial direction we come to (1) describing the dynamics of the gas in longitudinal direction. The condition of 1D approximation is $\omega_\perp \gg \omega_x, \mu \ll \hbar\omega_\perp$, where $\mu$ is the chemical potential.

The potential $V(x,t)$ is assumed to be $V(x,t) = m\omega_x^2 x^2 F(t)$, where $F(t)$ describes the time dependence of the potential. The effective one dimensional mean field nonlinearity coefficient is given as $g_{1D} = 2\hbar a_s\omega_\perp$, where $a_s$ is the atomic scattering length, and $a_s > 0$ corresponds to the Bose gas with a repulsive interaction between atoms while $a_s < 0$ to an attractive interaction. In this work we will study the case of repulsive condensate. Exact expression for this coefficient is given (Olshanii 1998) by $g_{1D} = 2\hbar a_s\omega_\perp/(1+1.03a_s/l)$.

It is convenient to work with the dimensionless form of (1)

$$i\psi_t + \frac{1}{2}\psi_{xx} - \frac{x^2}{2}F(t)\psi - g|\psi|^2\psi + \mu\psi =$$

$$= i\gamma(-\frac{1}{2}\psi_{xx} + \frac{x^2}{2}F(t)\psi + g|\psi|^2\psi - \mu\psi) = R(\psi,\psi^*), \tag{2}$$

by setting: $t = \omega_x t$, $l = \sqrt{\hbar/(m\omega_x)}$, $x = x/l$, $\psi = \sqrt{2|a_s|\omega_\perp/\omega_x}\phi$, with $g = 1$ for the repulsive two-body interaction.

Any damping process eventually leads to an equilibrium state. Corresponding stationary solution of the GPE can be found from the equation:

$$-\frac{1}{2}\psi_{xx} + \frac{x^2}{2}F\psi + g|\psi|^2\psi - \mu\psi = 0, \tag{3}$$

which, naturally, does not depend on the dissipative constant.

## 3. Moments Method

In this section we derive the moment equations for the dissipative GP equation. They are exact equations and can be useful both for deriving the equations for the wave packet parameters and for the control of the stability of numerical simulations.

Let us introduce the following integral quantities:

$$N = \int |\psi|^2 dx, P = i \int \psi\psi_x^* dx,$$
$$I_1 = \int x|\psi|^2 dx, I_2 = \int (x - x_0)^2 |\psi|^2 dx,$$
$$I_3 = i \int (x - x_0)(\psi^*\psi_x - \psi\psi_x^*) dx, \quad (4)$$

where $x_0 = I_1/N$. Taking their time-derivative and substituting $\psi_t$ from (2) the following evolution equations for the generalized moments can be derived:

$$\frac{dN}{dt} = \gamma \int |\psi_x|^2 dx + \gamma F \int x^2 |\psi|^2 dx + 2\gamma g \int |\psi|^4 dx - 2\gamma\mu N, \quad (5)$$

$$\frac{dP}{dt} = -Fx_0 N - i\gamma \int \psi_{xx}\psi_x^* + i\gamma F \int \frac{x^2}{2}(\psi\psi_x^* - \psi^*\psi_x) dx + $$
$$+ i\gamma g \int |\psi|^2 (\psi\psi_x^* - \psi^*\psi_x) dx - 2\gamma\mu P, \quad (6)$$

$$\frac{dI_1}{dt} = P + \gamma \int x|\psi_x|^2 dx + \gamma F \int x^3 |\psi|^2 dx +$$
$$+ 2\gamma g \int x|\psi|^4 dx - 2\gamma\mu x_0 N, \quad (7)$$

$$\frac{dI_2}{dt} = -iI_3 - \gamma N + \gamma \int (x - x_0)^2 |\psi_x|^2 dx + \gamma F \int (x - x_0)^2 x^2 |\psi|^2 dx +$$
$$+ 2\gamma g \int (x - x_0)^2 |\psi|^4 dx - 2\gamma\mu I_2, \quad (8)$$

$$\frac{dI_3}{dt} = 2x_{0t}P + 2FI_2 - 2 \int |\psi_x|^2 dx - g \int |\psi|^4 dx -$$
$$- \int (x - x_0)(\psi_x R^* + \psi_x^* R) dx. \quad (9)$$

The choice of the trial function is crucial for obtaining solutions with the desired properties. In a harmonic trap potential taking the ansatz in the Gaussian form is a good approximation for the condensate wave function. In Abdullaev and Garnier (2004) the Gaussian ansatz describes well the dynamics of one-dimensional BEC. Therefore, we take the Gaussian ansatz:

$$\psi(x,t) = A(t) exp\left(-\frac{(x - x_0)^2}{2a^2(t)} + \frac{ib(t)(x - x_0)^2}{2} + ik(x - x_0) + i\varphi(t)\right), \quad (10)$$

where A, a, b, k, $x_0$ and $\varphi$ are the amplitude, width, chirp, velocity, center of mass and linear phase, respectively.

Let us express the moments in terms of the variable parameters of the ansatz. Substitution (10) into the moment equations gives the following expressions:

$$a_{tt} = \frac{1}{a^3} - aF + \frac{gN}{\sqrt{2\pi}a^2} - \gamma\left(\frac{3}{a^2} + a_t^2 + a^2F + \frac{gN}{2\sqrt{2\pi}a^2}\right)a_t + \frac{\gamma a^3 F_t}{2} = 0,$$
$$N_t = -\gamma N\left(\frac{1}{2a^2} + \frac{a_t^2}{2} + \frac{a^2F}{2} + \frac{2gN}{\sqrt{2\pi}a^2} - 2\mu\right). \quad (11)$$

It should be noted that here the trap strength $F$ may depend on time.

Taking into consideration that in an equilibrium state the derivatives $a_t = 0$, $N_t = 0$ one can obtain the following stationary solutions:

$$a_s = \sqrt{\frac{4\mu + \sqrt{16\mu^2 + 60F}}{10F}}, \qquad N_s = \frac{8\sqrt{\pi}}{5g}\left(\mu a_s - \frac{1}{2a_s}\right) \quad (12)$$

## 3.1. RESONANT SUPPRESSION OF THE NORM

Let us consider a periodical variation of the parabolic trap strength in time

$$F(t) = F_0(1 + h\sin\omega t), \quad (13)$$

where $F_0$ is constant part of the trap strength F, h is the relative amplitude of oscillations which is supposed to be small, $\omega$ is the frequency of driven oscillations. To describe evolution of the width $a$ and norm $N$ under periodical variation of the trap strength we expand them near the stationary points $a(t) = a_s + a_1(t), N(t) = N_s + N_1(t)$. Corresponding stationary values $a_s$ and $N_s$ are determined by expressions (12). Substituting the expansion for $a$ into the first equation of (11) and keeping only the first-order terms, $a_1$, $h$ and $\gamma$ we come to the following equation

$$a_{1tt} + \lambda(N)a_{1t} + \omega_0^2 a_1 = -ha_sF_0\sin\omega t, \quad (14)$$

where

$$\omega_0^2 = \frac{1}{a_s^4} + 3F_0, \quad \lambda(N_s) = \gamma\left(\frac{3}{a_s^2} + a_s^2F_0 + \frac{gN_s}{2\sqrt{2\pi}a_s}\right) \quad (15)$$

are the eigenfrequency and effective damping coefficient, respectively. It should be noted that $\lambda > 0$, for $\gamma > 0$.

As readily seen, this equation describes main resonance in the width oscillations if $\omega \approx \omega_0$. At $\omega = \omega_0$ the amplitude of the width oscillations is maximal and determined by formula (Landau and Lifshitz 1973)

$$a_{1max} = \frac{h a_s F_0}{\lambda(N_s)\omega_0}. \tag{16}$$

Substituting obtained resonant solution for $a$ into the second equation of set (11) and averaging it over the period $T = 2\pi/\omega$, for the steady-state norm we get

$$\tilde{N}_s = N_s - \frac{\sqrt{2\pi}a_s}{2g}\left(\mu + \frac{1}{2a_s^2} + \frac{a_s^2\omega_0^2}{4}\right)\left(\frac{hF_0}{\lambda(N_s)\omega_0}\right)^2. \tag{17}$$

One can see that the *resonant* variation of the trap strength causes *decreasing* in the steady-state value of the norm, $N$ which is in inverse proportion to *the damping*, $\gamma$.

## 4. Numerical Simulations

We have carried out a series of time dependent simulations of the system based on the variational approach using (11) as well as exact numerical calculations using (2). In our numerical calculations we discretize the problem in a standard way, with the time step $dt$, and spatial step $dx$, so $\psi_j^k$ approximates $\psi(jdx, kdt)$. More specifically we approximate the governing equation (2) with the semi-implicit Crank–Nickolson scheme using split-step method (Adhikari and Muruganandam 2002). The results of numerical simulations of both PDE and ODE models are presented below.

Figure 1 presents the behavior of the norm when *the trap strength is periodically varied in time* as in (13). An interesting behavior of the norm is observed here. If an external periodical perturbation of the resonant frequency is applied to a trapped BEC which is already in the equilibrium state then the norm of the condensate starts to decrease going to new steady state. In the figure the frequency $\omega$ of the periodical trap perturbation is taken to be equal to the eigenfrequency $\omega_0$ of the system determined from (15). Numerical simulations have been carried out for different values of $\gamma$ at the same amplitude of the trap oscillations, $h = 0.06$. One can see that combined effect of *the damping* and *resonant periodical variation of the trap strength* causes suppression of the averaged steady state value of the driven norm, smaller values of the damping constant $\gamma$ leading to more strong suppression of the norm. The effect is explained by that at smaller values of $\gamma$ the amplitude of the width oscillations becomes greater and in accordance with the second equation of set (11) averaged value of the steady state norm decreases.

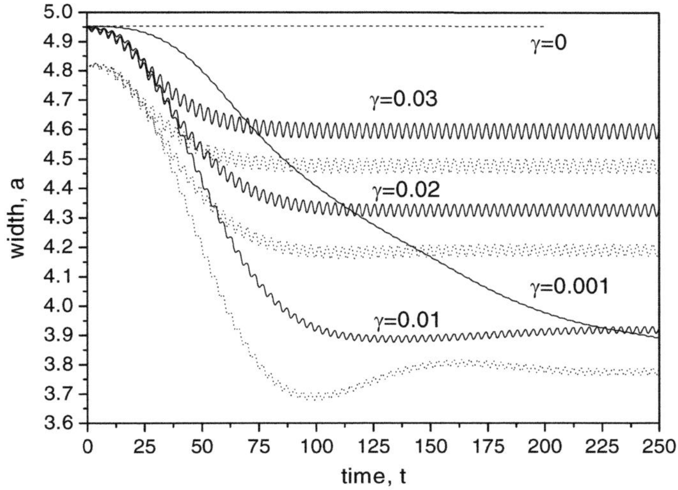


*Figure 1.* Behavior of the norm of a trapped BEC when the trap starts oscillating with the amplitude $h = 0.06$. Initially the BEC is in the equilibrium state. The solid lines stand for full numerical simulations of the PDE, while the dotted lines represent the ODE results

## 5. Conclusions

In this paper we have studied collective oscillations of a quasi-one-dimensional Bose gas in the presence of dissipative effects. The modified Gross–Pitaevskii equation in the framework of the phenomenological approach (Pitaevskii 1958a, b; Choi et al. 1998) has been employed. To describe evolution of oscillations we use the moments approach taking into account the dissipation. The results obtained from computation of the system of equations for the wave function parameters are confirmed direct numerical simulations of the full GP equation.

The expressions for the width and the norm of a condensate in an equilibrium state have been derived analytically.

Main resonance in the condensate oscillations has been studied. We found that periodical resonant modulation of the trap potential in the modified GP equation (1) drastically change asymptotical behavior of the norm. As known a ground state of the governing equation (1) does not depend on the damping constant $\gamma$ *by definition* and corresponding asymptotical value of the solution norm depends only on the chemical potential $\mu$. In the case of time dependent modulation of the trap we have shown the combined effect of *resonant periodical modulation of the trap strength* and *the damping* to change drastically asymptotical behavior of the driven norm oscillations and cause suppression of the averaged steady state value of the driven norm, smaller values of the damping constant $\gamma$ leading to more strong suppression of the norm.

## References

Abdullaev F.Kh. and Garnier J.: 2004. *Phys. Rev. A*, **70**, 053604
Adhikari S.K. and Muruganandam P.: 2002. *J. Phys. B*, **35**, 2831
Brazhnyi V.A., Kamchatnov A.M. and Konotop V.V.: 2003. *Phys. Rev. A*, **68**, 035603
Choi S., Morgan S.A. and Burnett K.: 1998. *Phys. Rev. A*, **57**, 4057
Girardeau M.D.: 1960. *J. Math. Phys.*, **1**, 516
Kagan Yu. and Maksimov L.A.: 2001. cond-mat/0112292
Landau L.D. and Lifshitz E.M.: 1973. *Mechanics*. Pergamon Press, Oxford
Lieb E.H.: 1963. *Phys. Rev.*, **130**, 1616
Lieb E.H. and Liniger W.: 1963. *Phys. Rev.*, **130**, 1605
Menotti C. and Stringari S.: 2002. *Phys. Rev. A*, **66**, 043610
Moritz H., Stoferle T., Kohl M. and Esslinger T.: 2003. *Phys. Rev. Lett.*, **91**, 250402
Olshanii M.: 1998. *Phys. Rev. Lett.*, **81**, 938
Ott H., Fortagh J., Kraft S., Gunther A., Komma D. and Zimmermann C.: 2003. *Phys. Rev. Lett.*, **91**, 040402
Penckwitt A., Ballagh R. and Gardner C.W.: 2002. *Phys. Rev. Lett.*, **89**, 260402
Pitaevskii L.P.: 1958a. *Zh. Eksp. Teor. Fiz.*, **35**, 408
Pitaevskii L.P.: 1958b. *Sov. Phys. JETP*, **35**, 282
Pitaevskii L.P. and Stringari S.: 2003. Bose–Einstein condensation. OUP, Oxford
Proukakis N.P., Parker N.G., Barendi C. and Adams C.C.: 2004. *Phys. Rev. Lett.*, **93**, 130408
Saito H. and Ueda M.: 2004. *Phys. Rev. A*, **70**, 053610
Staliunas K., Longhi S. and de Valcarcel G.: 2002. *Phys. Rev. Lett.*, **89**, 210406

# NONLINEAR DYNAMICS OF THE KICKED SQUARE BILLIARD

U.R. Salomov*, D. Matrasulov, and N.E. Iskandarov
*Heat Physics Department of the Uzbek Academy of Sciences, 28 Katartal Street, 100135 Tashkent, Uzbekistan; salukrah@yahoo.com, dmatrasu@phys.ualberta.ca*

**Abstract.** Nonlinear dynamics of a kicked particle in a square billiard with the central symmetric kick source is studied. Time-dependence of the average energy is explored and compared with that for a kicked free particle. Dependence of the acceleration on different localizations of the kick source is considered.

**Key words:** Nonlinear dynamics; Delta-kicked systems; Billiards

Kicked particle dynamics is one of the simplest systems which provides comprehensive study of nonlinear dynamics and quantum chaos.

Extensive studies of a paradigm of nonlinear dynamics, kicked rotor show that the dynamics of the kicked system can be regular, mixed and chaotic depending on the kick parameters (Chirikov 1979; Escande 1985; Lieberman and Lichtenberg 1972; Izrailev 1990). In particular, in the chaotic regime the average energy of the kicked rotor is a monotonically growing linear function of time. However, corresponding quantum system quantum system exhibits localization of classical chaos (Izrailev 1990; Casati et al. 1979).

In this work we investigate an extension of the previous studies of kicked system dynamics to the case of confined systems. Namely, we explore nonlinear dynamics of a kicked particle whose motion is confined by rectangular billiard boundaries, by calculating time dependence of the energy both for a single trajectory and ensemble of the trajectories. Also, we treat momentum transfer distribution for this system and compare it with that of kicked rotor.

The obtained results show that depending on the type of the kick potential, localization and strength of the perturbation force the dynamics can be different. It is found that the average energy of the kicked particle grows diffusively as a function of time, as does the kicked rotor average energy. However, this growth is more rapid than that of kicked rotor.

In this work, we treat a system consisting of a particle moving inside a two-dimensional square billiard with an additional kicking source located at the center of the billiard. The kicking potential is given by

$$V(x,y,t) = \begin{cases} \alpha\cos(\frac{8\pi\rho}{a})\sum_n \delta(t-nT), & \rho \le a/4 \\ 0, & \rho > a/4 \end{cases} \quad (1)$$

G. Casati and D. Matrasulov (eds.), *Complex Phenomena in Nanoscale Systems*,
NATO Science for Peace and Security Series B: Physics and Biophysics,

© Springer Science+Business Media B.V. 2009

with

$$\rho = \sqrt{\left(x - \frac{a}{2}\right)^2 + \left(y - \frac{a}{2}\right)^2},$$

where $\alpha$, $T$ and $a$ are the coupling constant, the kicking period and the side length of the square, respectively. The Hamiltonian of the system can be written as

$$H = H_0 + V(\rho, t), \tag{2}$$

with $H_0$ being the Hamiltonian of the particle moving in the square billiard without any kicking. Since particles move balistically in between collision with the billiard boundary as well as in between kicks, a discrete mapping is used to solve the corresponding equations of motion. In Fig. 1, the energy $E(t)$ as a function of time for a typical trajectory is shown. The curve can be decomposed in characteristic segments, each corresponding to a representative dynamics of the particle: Firstly, parts with rapid oscillations of the energy can be seen, see the inset of Fig. 2. During such phases, the particle is trapped for certain time inside the kicking area and experiences successive kicks, leading to the typical oscillations of the energy shown in the inset of Fig. 1. Secondly, single vertical lines in the curve of Fig. 1 correspond

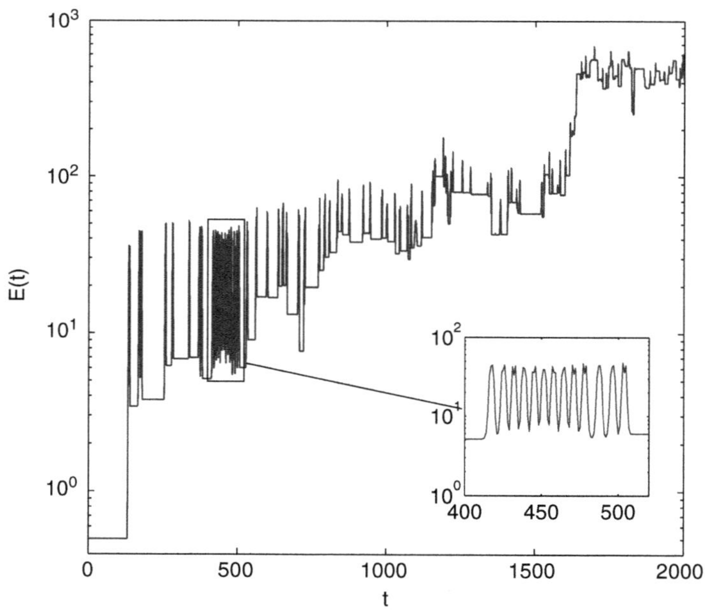


*Figure 1.* Time-dependence of the energy for a typical particle in the kicked billiard ($\alpha = 0.2$, $T = 0.01$). Inset: rapid oscillations of the energy, the particle is trapped for a certain time inside the kicking area

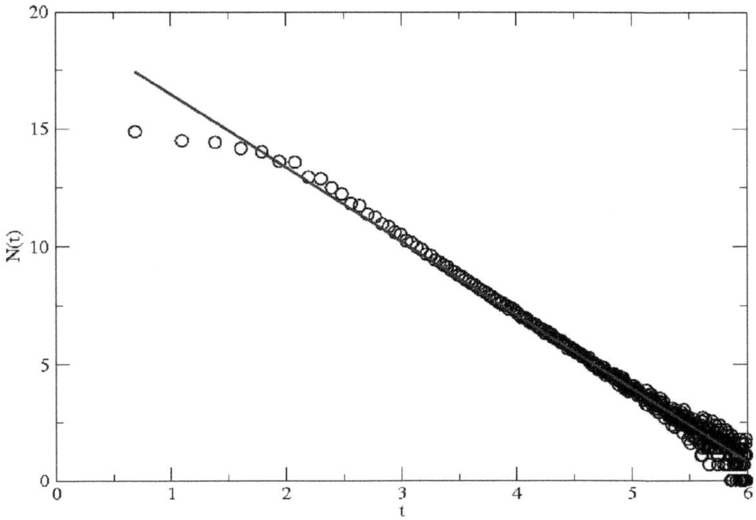


*Figure 2.* Distribution of constant energy time intervals. Circles are the results of direct numerical calculations, solid line is the linear regression

to single kicks, the energy of the particle is either increased or decreased, depending on where the kick happens. Finally, parts of the curve where the energy remains constant. During such times, the particle moves outside the kicking area (quasiperiodic orbits) or crosses the kicking area without getting a kick, the latter means it is in a way not synchronized with the kicking period.

To understand the kicked billiard dynamics more deeply we need to explore distribution, $N(t)$ of the time intervals during which the energy remains constant, i.e. constant-energy-time intervals. Figure 2 presents the plot of such distribution in double logarithmic scale obtained using direct numerical computation. In addition, this figure compares also $N(t)$ with the curve $N(t)$ obtained from the linear regression. Such power law (with the exponent equal to $-3$) behavior of $N(t)$ can be explained as follows: appearing constant energy intervals is caused by two factors. One of them corresponds to the situation when billiard particle moves along the quasiperiodic orbits which initially doesn't cross the kicking area. For this case one can obtain the estimate $N(t) \sim t^{-1}$. However, we found that the probability for appearing such orbits in our system is quite small. Therefore such regime of motion doesn't make contribution to the above distribution of constant-energy-time intervals. Second type of motion corresponds to the above mentioned synchronized motion. It is easy to show that the distribution for the constant energy time intervals for this case behaves as $N(t) \sim t^{-3}$.

When considering not only a single trajectory, but rather an ensemble of particles, a diffusive growth of the ensemble averaged energy $\langle E_b(t)\rangle$

(averaged of 1,000 trajectories) can be observed, see Fig. 2. More precisely, the energy growth linearly (normal diffusion) with the time $t$. The proportionality constant is naturally just the diffusion coefficient $D_b(\alpha, T)$, so that $\langle E_b(t)\rangle = D_b(\alpha, T) \cdot t$. As indicated, $D_b$ depends on the coupling constant $\alpha$ and the kicking period $T$. $D_b$ increases monotonically with increasing $\alpha$ and decreases monotonically with increasing $T$.

It is reasonable to compare the diffusive growth of the ensemble averaged energy $\langle E_b(t)\rangle = D_b(\alpha, T) \cdot t$ of the kicked billiard with the evolution of energy $\langle E_r(t)\rangle$ of the kicked rotor (again with $\alpha$ being the coupling constant and $T$ being the period between two successive kicks). From Fig. 3, it can be seen that $\langle E_r(t)\rangle = D_r(\alpha, T) \cdot t$, but with $D_r(\alpha, T) \ll D_b(\alpha, T)$, so the energy growths much faster in the case of the kicked billiard. We note that in the case of the kicked rotor the dynamics is effectively governed by a single parameter $K_r = \alpha T$ only, see e.g. Izrailev (1990). The energy of the kicked rotor as a function of the dimensionless time $n = t/T$ can than for $K_r \gtrsim 5$ be written as $E_r(n) = K_r^2/4 \cdot n$, so

$$D_r(\alpha, T) = D_r(K_r) = K_r^2/4. \tag{3}$$

In the kicked billiard the situation is different, the dynamics depends on $\alpha$ and $T$ individually, thus there is no simple representation for $D_b(\alpha, T)$ as in (3)

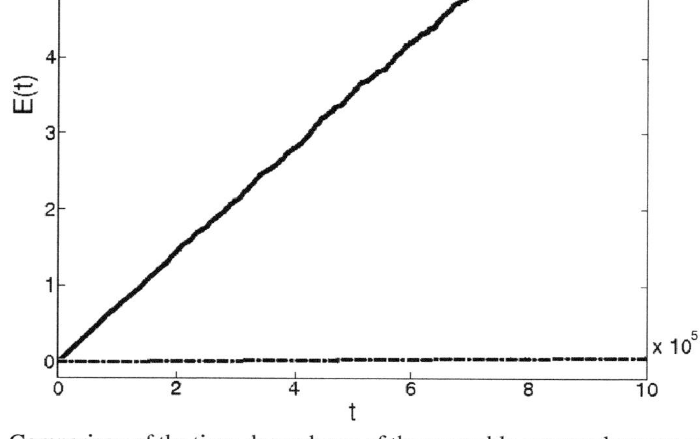


*Figure 3.* Comparison of the time-dependence of the ensemble averaged energy of the kicked billiard and kicked rotor for $\alpha = 5.0$, $T = 1.0$

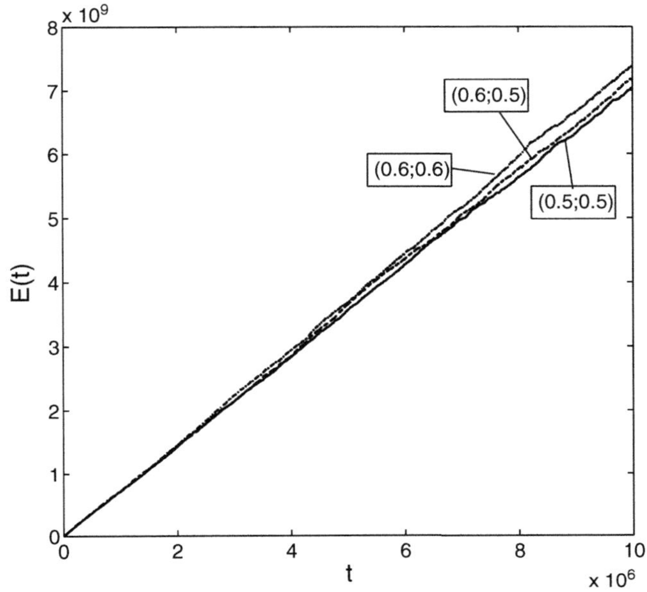


*Figure 4.* Time-dependence of the ensemble averaged energy for different localizations of the center of the kicking source in the billiard ($\alpha = 5$, $T = 1$): The central localization is compared to the case, when kicking source is shifted to the left (position of the center at (0.6; 0.5)) and when it is shifted along the diagonal of billiard (center at (0.6; 0.6))

possible. The large deviation between $D_b$ and $D_r$ becomes immediately clear when considering the maximum momentum transfer $\Delta P_{max}$ at a single kick. In the case of the kicked billiard, $\Delta P_{max} = 8\pi\alpha/a$, whereas for the kicked rotor, $\Delta P_{max} = \alpha$. In Fig. 4, the time-dependence of the average energy $\langle E(t)\rangle$ is plotted for different localizations of the kicking source and compared to the one when kicking source is located at the center of the billiard. The shift of the kick source leads to minor changes in $\langle E(t)\rangle$ only, so the dynamics is rather robust against this shift. We tested this for other delocalizations as well, with similar results, thus there are not shown here.

Summarizing, in this work we studied nonlinear dynamics of a kicked particle whose motion is confined in a square billiard with a kick source localized inside the billiard with central symmetric spatial distribution. We showed that for this type of kick potential the average energy of the particle as a function of time grows diffusively.

The above studied model is relevant to the particle transport in quantum dots, and other confined systems. Time-dependent external field can be used in these systems as an additional tool for manipulating by particle dynamics and transport.

## Acknowledgements

This work is supported in part by a grant of Volkswagen Foundation and INTAS Young Scientist Fellowship (Nr 06-100023-6008).

## References

Casati G, Chirikov B.V., Ford J. and Izrailev F.M.: 1979. *Lecture Notes in Physics*, Springer, **57**, 334
Chirikov B.V.: 1979. *Phys. Rep.*, **52**, 263
Escande D.F.: 1985. *Phys. Rep.*, **121**, 165
Izrailev G.M.: 1990. *Phys. Rep.*, **196**, 299
Lieberman M.A. and Lichtenberg A.J.: 1972. *Phys. Rev. A*, **5**, 1852

# CHAOTIC INSTANTONS AND EXPONENTIAL WIDENING OF THE GROUND QUASIENERGY DOUBLET IN KICKED DOUBLE WELL POTENTIAL

V.I. Kuvshinov, A.V. Kuzmin, and V.A. Piatrou*
*Joint Institute for Power and Nuclear Research, Krasina Streeet 99, Minsk 220109, Belarus; PiatrouVadzim@tut.by*

**Abstract.** Kicked double-well system is investigated both analytically and numerically. Formula for ground quasienergy splitting is obtained using resonances' overlap criterion in the framework of chaotic instanton approach. Results of numerical calculations for the ground quasienergy splitting dependence on both the perturbation strength and frequency are in good agreement with the derived analytical formula. Thus providing a support for the further development of chaotic instanton approach to tunneling in quantum systems with mixed classical dynamics.

**Key words:** Instantons; Kicked well; Tunneling

## 1. Introduction

The connection between the semiclassical properties of perturbed nonlinear systems and purely quantum processes such as tunneling is a reach rapidly developing field of research nowadays. Our insight in some novel phenomena in this field was extended during the last decades. The most intriguing among them are the chaos assisted tunneling (CAT) and the closely related coherent destruction of tunneling (CDT). The former in particular is an enhancement of tunneling in the perturbed low-dimensional systems at relatively high external field strengths and high driving frequencies (in order the singlet-doublet crossing to occur) (Lin and Ballentine 1990). The later is a suppression of tunneling when values of amplitude and frequency of driving force belong to some one-dimensional manifold in the perturbation parameters' space (Grossmann et al. 1991).

CAT phenomenon as well as CDT were experimentally observed in a number of real physical systems: CAT for ultracold atoms (Steck et al. 2001; Hensinger et al. 2001), both phenomena in two coupled optical waveguides (Vorobeichik et al. 2003; Valle et al. 2007). The most common methods

G. Casati and D. Matrasulov (eds.), *Complex Phenomena in Nanoscale Systems*,
NATO Science for Peace and Security Series B: Physics and Biophysics,

© Springer Science+Business Media B.V. 2009

which are used to investigate the CAT are numerical methods based on Floquet theory (Shirley 1965). Among other approaches to CAT we would like to mention the scattering approach for billiard systems (Frischat and Doron 1998), quantum mechanical amplitudes in complex configuration space (Shudo and Kensuke 1998) and resonance assisted tunneling (Brodier et al. 2002).

In this paper we will consider the original approach based on instanton technique, which was proposed in Kuvshinov et al. (2002, 2003) and independently used in Igarashi and Yamada (2006). Alternative approach based on quantum instantons which are defined using an introduced notion of quantum action was suggested in Jirari et al. (2001). The purpose of the present study is to prove the ability of proposed chaotic instanton approach to give quantitative analytical description of tunneling well agreed with independent numerical calculations based on Floquet theory. It will give additional support and pulse for the further development of analytical methods to investigate tunneling phenomenon in quantum systems with mixed classical dynamics.

## 2. Chaotic Instantons and Ground Quasienergy Splitting

Hamiltonian of the particle in the double-well potential with kick-type perturbation can be written in the following form:

$$H = H_0 + V = \frac{p^2}{2m} + a_0\, x^4 - a_2\, x^2 + \epsilon\, x^2 \sum_{n=-\infty}^{+\infty} \delta(t - nT), \tag{1}$$

where $H_0$ – nonperturbed part of the Hamiltonian, $V$ – perturbation, $m$ – mass of the particle, $a_0, a_2$ – parameters of the potential, $\epsilon$ – perturbation strength, $T$ – perturbation period, $t$ – time.

Euclidean equations of motion of the particle in the nonperturbed double-well potential ($\epsilon = 0$) have a well known solution – instanton. In the phase space it lies on the separatrix. Perturbation destroys the separatrix forming stochastic layer. In this layer a number of chaotic instantons appears. Chaotic instanton solution is a solution of the Euclidean equations of motion and it can be written in the following form:

$$x_{chaos} = x_{inst} + \epsilon\, \Delta x_{chaos},$$

where $x_{chaos}$ and $x_{inst}$ – chaotic and nonperturbed instanton solutions, respectively, $\Delta x_{chaos}$ – stochastic correction assumed to be small for small $\epsilon$. These solutions are used in chaotic instanton approach (Kuvshinov et al. 2002, 2003) for description of the dynamical tunneling in the system. We use this approach in our paper.

In this section we will use Euclidean space-time only. Thus we introduce an Euclidean time ($t \to -i\tau$) and use the following assumptions: small value of the perturbation strength ($\epsilon < 0.1$), uniform stochastic layer, Euclidean chaotic instanton action is put equal to nonperturbed instanton action corresponding to some nonmaximal energy. It can be approximated by the following linear form (see Kuvshinov et al. 2003)

$$S[x_{chaos}(\tau,\xi)] = S[x_{inst}(\tau,0)] - \alpha\sqrt{\frac{m}{a_2}}\,\xi, \tag{2}$$

where $S[x_{inst}(\tau,0)] = 2\sqrt{m}\,a_2^{3/2}/(3\,a_0)$ – nonperturbed instanton action, $\alpha = (1+18\ln 2)/6$ – numerical coefficient, $E$ – energy, $E_{sep}$ – energy on separatrix and $\xi = E_{sep} - E$.

In the framework of the chaotic instanton approach the width of the stochastic layer determines the contribution of these chaotic solutions of Euclidean equations of motion to the tunneling amplitude (Kuvshinov et al. 2003). The stochastic layer width is estimated using Chirikov's resonances overlap criterion and it can be write down in the following form

$$\Delta\mathcal{H}_s = E_{sep} - E_{bor} \approx \tilde{k}\,\epsilon\,\nu, \tag{3}$$

where $E_{bor}$ is the estimated energy on the "border" between stochastic and regular regions, $\tilde{k}$ – some numerical parameter which cannot be rigorously obtained in the framework of the criterion used.

In the framework of the chaotic instanton approach the tunneling amplitude for the perturbed system is assumed to be a sum of the amplitude in the nonperturbed case and the contribution of chaotic instantons. The borders of the integration are lie on the separatrix and on the "border" between stochastic and regular regions. Using expression (2) this integral can be transformed to the integral over the energy difference from zero up to the width of the stochastic layer (3):

$$A_{chaos} = \alpha\sqrt{\frac{m}{a_2}}\,\tilde{N}\int_0^{\Delta\mathcal{H}_s} d\xi\int_{-\infty}^{+\infty} dc_0\,\sqrt{S[x_{chaos}(\tau,\xi)]}\,e^{-S[x_{chaos}(\tau,\xi)]},$$

where $\tilde{N}$ is a normalization factor. In order to estimate the contribution of chaotic instantons we use the approximate expression for the chaotic instanton action (2). Integration over $c_0$ gives the contribution of zero modes (Vainshtein et al. 1982). As the result we get the following expression for the amplitude:

$$A = A_{inst} + A_{chaos} \approx \tilde{N}\,\sqrt{S^{inst}}\,e^{-S^{inst}}\,\Gamma\,exp\left(\alpha\sqrt{\frac{m}{a_2}}\,\Delta\mathcal{H}_s\right), \tag{4}$$

where $A_{inst}$ is well known tunneling amplitude in the nonperturbed system, $\Gamma$ – is a time of the tunneling which is put to infinity at the end of the calculations. The last exponential factor in expression (4) is responsible for the tunneling enhancement in the perturbed system when the width of the stochastic layer ($\Delta\mathcal{H}_s$), i.e. perturbation strength, is small. In the nonperturbed case the width of the stochastic layer is equal to zero and expression (4) coincides with the well known expression describing the ordinary tunneling amplitude.

Using expressions (3) and (4) we can write down analytical formula for the quasienergy splitting

$$\Delta\eta(\epsilon, \nu) = 2\sqrt{\frac{6}{\pi}}\,\sqrt{S^{inst}}\,e^{-S^{inst}}\,e^{k\,\epsilon\,\nu}, \tag{5}$$

where $k = \alpha\sqrt{m/a_2}\,\tilde{k}$.

We fix this parameter value $k$ using the results of numerical calculations. For this purpose we perform the linear fitting of the numerical data for the dependencies of ground quasienergy splitting on the perturbation strength and take average value of the parameter over these dependencies. As the result we have the *single* numerical parameter $k$ for our formula (5) explaining the dependencies of ground quasienergy splitting on both the perturbation strength and frequency.

## 3. Numerical Calculations

For the computational purposes it is convenient to choose the eigenvectors of harmonic oscillator as basis vectors. In this representation matrix elements of the parts ($H_0$ and $V$) of the full Hamiltonian (1) are real and symmetric. They have the following forms ($n \geq m$):

$$\begin{aligned} H^0_{m\,n} &= \delta_{m\,n}\left[\hbar\omega\left(n+\frac{1}{2}\right)+\frac{g}{2}\left(\frac{3}{2}\,g\,a_0\,(2m^2+2m+1)-a_2'(2m+1)\right)\right] \\ &\quad +\delta_{m+2\,n}\,\frac{g}{2}\left(g\,a_0(2m+3)-a_2'\right)\sqrt{(m+1)(m+2)} \\ &\quad +\delta_{m+4\,n}\frac{a_0 g^2}{4}\sqrt{(m+1)(m+2)(m+3)(m+4)}, \\ V_{m\,n} &= \epsilon\,\frac{g}{2}\left(\delta_{m+2\,n}\,\sqrt{(m+1)(m+2)}+\delta_{m\,n}(2m+1)\right), \end{aligned}$$

where $g = \hbar/m\omega$ and $a_2' = a_2 + m\,\omega^2/2$, $\hbar$ is Planck constant which we put equal to 1, $\omega$ – frequency of harmonic oscillator which is arbitrary, and so may be adjusted to optimize the computation. We use the value $\omega = 0.2$ with

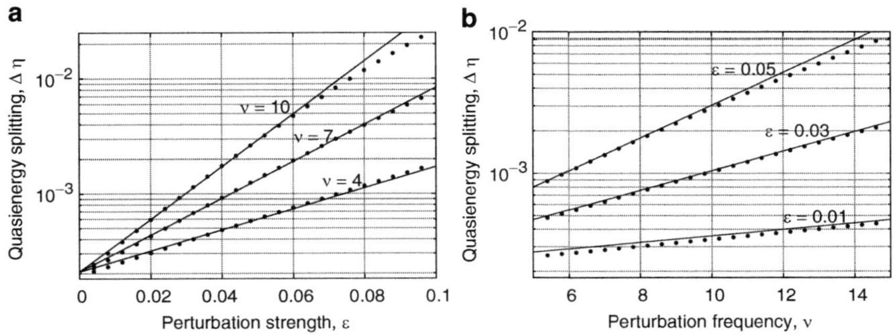


*Figure 1.* Quasienergy splitting as a function of the strength (**a**) and frequency (**b**) of the perturbation. Lines – analytical formula, points – numerical results. The model parameters are $m = 1$, $a_0 = 1/128$, $a_2 = 1/4$

parameters $m = 1$, $a_0 = 1/128$, $a_2 = 1/4$. The matrix size is chosen to be equal to $200 \times 200$. Calculations with larger matrices give the same results. System of computer algebra Mathematica was used for numerical calculations.

We calculate eigenvalues of the evolution operator on the one-period $e^{-iHT}e^{-iV}$ and obtain quasienergy levels $(\eta_k)$ which are related with the evolution operator eigenvalues $(\lambda_k)$ through the expression $\eta_k = i \ln \lambda_k/T$. Then we get ten levels with the lowest one-period average energy which is calculated using the formula $\langle v_i| H_0 + V/T |v_i\rangle$ ($|v_i\rangle$ are the eigenvectors of the one-period evolution operator).

Performed numerical calculations give the dependence of the ground quasienergy splitting both on the strength (Fig. 1a) and the frequency (Fig. 1b) of the perturbation. These dependencies are exponential as it is predicted by chaotic instanton approach and obtained analytical formula (5). Using least square technique we calculate the *single* numerical parameter $k$ which describes *all* numerical dependencies demonstrated in Fig. 1a, b. Relative error in determining of the parameter $k$ from numerical results is less than 2%. Analytical results are plotted in Fig. 1a, b by straight solid lines. Numerical points lie close to these lines. The agreement between numerical simulations and analytical expression is good in the parametric region considered.

## 4. Conclusions

Double-well system is investigated in presence of external kick-type perturbation. Analytic chaotic instanton approach is extended and applied for this system in order to obtain the analytical formula for the ground quasienergy splitting dependence on both the perturbation strength and frequency. This formula predicts exponential dependence of the ground quasienergy splitting on these parameters for small perturbation strength values. It has the single numerical parameter which is determined from the numerical calculations. Numerical results for the quasienergy splitting as a function of the perturbation frequency and strength demonstrate exponential dependence as well. They are in a good agreement with formula (5).

## References


Brodier, O., Schlagheck, P., and Ullmo, D. (2002) Resonance-assisted tunneling, *Ann. Phys.* **300**, 88–136

Frischat, S. D. and Doron, E. (1998) Dynamical tunneling in mixed systems, *Phys. Rev. E* **57**, 1421–1443

Grossmann, F., Dittrich, T., Jung, P., and Hänggi, P. (1991) Coherent destruction of tunneling, *Phys. Rev. Lett.* **67**, 516–519

Hensinger, W. K., Häaffner, H., Browaeys, A., Heckenberg, N. R., Helmerson, K., McKenzie, C., Milburn, G. J., Phillips, W. D., Rolston, S. L., Rubinsztein-Dunlop, H., and Upcroft, B. (2001) Dynamical tunnelling of ultracold atoms, *Nature* **412**, 52–55

Igarashi, A. and Yamada, H. S. (2006) Quantum dynamics and delocalization in coherently driven one-dimensional double-well system, *Physica D: Nonlin. Phenom.* **221**, 146–156

Jirari, H., Kroger, H., Luo, X. Q., Moriarty, K. J. M., and Rubin, S. G. (2001) Quantum instantons and quantum chaos, *Phys. Lett. A* **281**, 1

Kuvshinov, V. I., Kuzmin, A. V., and Shulyakovsky, R. G. (2002) Chaos enforced instanton tunnelling in one-dimensional model with periodic potential, *Acta Phys. Polon.* **B33**, 1721–1728

Kuvshinov, V. I., Kuzmin, A. V., and Shulyakovsky, R. G. (2003) Chaos assisted instanton tunneling in one dimensional perturbed periodic potential, *Phys. Rev. E* **67**, 015201

Lin, W. A. and Ballentine, L. E. (1990) Quantum tunneling and chaos in a driven anharmonic oscillator, *Phys. Rev. Lett.* **65**, 2927–2930

Shirley, J. H. (1965) Solution of the Schrödinger equation with a Hamiltonian periodic in time, *Phys. Rev.* **138**, B979–B987

Shudo, A. and Kensuke, S. (1998) Chaotic tunneling: A remarkable manifestation of complex classical dynamics in non-integrable quantum phenomena, *Physica D: Nonlin. Phenom.* **115**, 234–292

Steck, D. A., Oskay, W. H., and Raizen, M. G. (2001) Observation of chaos-assisted tunneling between islands of stability, *Science* **293**, 274–278

Vainshtein, A. I., Zakharov, V. I., Novikov, V. A., and Shifman, M. A. (1982) ABC of instantons, *Sov. Phys. Usp.* **25**, 195

Valle, G. D., Ornigotti, M., Cianci, E., Foglietti, V., Laporta, P., and Longhi, S. (2007) Visualization of coherent destruction of tunneling in an optical double well system, *Phys. Rev. Lett.* **98**, 263601

Vorobeichik, I., Narevicius, E., Rosenblum, G., Orenstein, M., and Moiseyev, N. (2003) Electromagnetic realization of orders-of-magnitude tunneling enhancement in a double well system, *Phys. Rev. Lett.* **90**, 176806

# A FRESH VIEW ON FERMI ACCELERATION IN DRIVEN TWO-DIMENSIONAL BILLIARDS

F. Lenz*, C. Petri, F.N.R. Koch, and P. Schmelcher
*Physikalisches Institut, Universität Heidelberg, Philosophenweg 12, 69120 Heidelberg, Germany; lenz@physi.uni-heidelberg.de*

**Abstract.** We briefly describe the state-of-the-art concerning Fermi acceleration in two-dimensional driven billiards by reviewing the corresponding literature with a focus on our recent work on the time-dependent ellipse and some new results on the breathing mode of a certain oval billiard. In a way, these two systems represent the "missing links" in the studies of two-dimensional driven billiards and help to clarify the conditions under which Fermi acceleration will arise.

**Key words:** Driven billiards; Fermi acceleration; Nonlinear dynamics

## 1. Introduction

Fermi acceleration (FA) is the unbounded energy gain of an ensemble of particles exposed to time-dependent external forces. It was first proposed by Fermi (1949) to explain the high energies of cosmic ray particles interacting with a time-dependent magnetic field. The simplest system which allows the investigation of FA is the one dimensional Fermi-Ulam model (FUM) (Lieberman and Lichtenberg 1972; Lichtenberg and Lieberman 1992), where non-interacting particles move between an oscillating and a fixed wall. Note that the corresponding static system, particles bouncing between two static walls, is integrable. The FUM and its variants have been the subject of extensive theoretical (see Lieberman and Lichtenberg 1972 and references therein) and experimental (Kowalik et al. 1988; Carvalho and Zimmerman 1987) studies. It has been proven (Lichtenberg and Lieberman 1992) that as long as the driving law of the oscillating wall is sufficiently smooth, invariant spanning curves in phase space prohibit the unlimited energy growth of particles. In particular, a stochastic driving law, which is obviously non-smooth, will lead to FA. This means that in one dimensional billiards the existence of FA exclusively depends on the driving law, since the corresponding static system is always integrable.

G. Casati and D. Matrasulov (eds.), *Complex Phenomena in Nanoscale Systems*,
NATO Science for Peace and Security Series B: Physics and Biophysics,

© Springer Science+Business Media B.V. 2009

In two dimensions, the phase space structure of static billiard systems is already much richer than for one dimensional billiards and can range from integrable over mixed to fully chaotic (Berry 1981). This leads to the question whether already a smooth, especially harmonic, driving law is sufficient in two dimensional billiards to obtain FA. In Loskutov and Ryabov (2002), the existence of FA was shown for a harmonically oscillating stadium-like billiard and the authors conjectured that a sufficient condition for the occurrence of FA in a $2D$ smoothly driven billiard is the existence of a chaotic part in the phase space of the corresponding static system. This so-called "LRA conjecture" is supported by the absence (presence) of FA in the oscillating circular billiard (Kamphorst and Carvalho 1999) (eccentric annular billiard, Carvalho et al. 2006), where the corresponding static system is integrable (has a mixed phase space).

## 2. Billiard Geometries and Driving Laws

We consider six different static billiards. The circle, the ellipse and the concentric annular (Fig. 1a–c) are integrable, the oval and the eccentric annular (Fig. 1d, e) possess a mixed phase space, whereas the stadium (Fig. 1f) is fully chaotic. For details on these static systems, see e.g. Berry (1981) for the circle, the ellipse and the stadium, Carvalho et al. (2006) for the concentric and eccentric annulus and Kamphorst et al. (2007) for the oval. The energy is conserved in all six billiards (due to elastic reflections upon collisions with the billiard boundary). Additionally, in the circle and concentric annulus the angular momentum is conserved, whereas in the static ellipse the product of

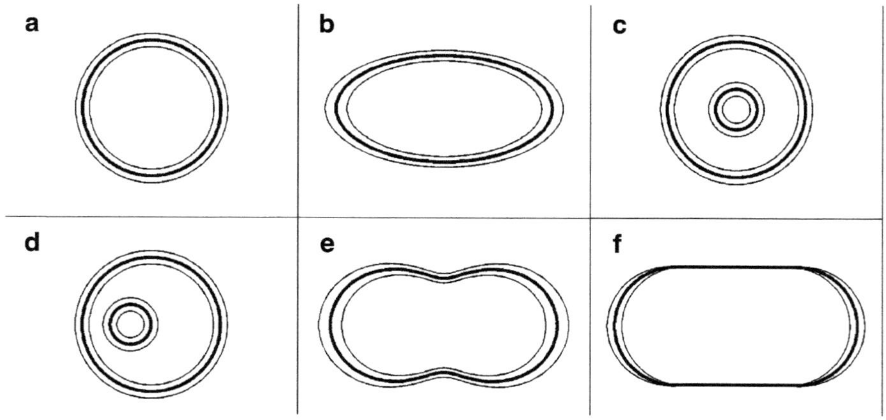


*Figure 1.* Billiard geometries, thick lines represent the equilibrium position, thin lines are the minimal and maximal extension of the driven billiards, respectively: circle (**a**), ellipse (**b**), concentric (**c**) and eccentric annulus (**d**), oval (**e**), stadium (**f**)

the angular momenta around the two foci is a constant of motion. Due to ballistic motion in between collisions, the dynamics can be described by an, in general, implicit $2D$ map using e.g. the variables $(\phi_n, \alpha_n)$, where $\phi_n$ is a $2\pi$-periodic parameter specifying the position on the boundary (e.g. the polar angle) and $\alpha_n$ is the angle between $\boldsymbol{v}_n$ and the tangent on the boundary at the $n$-th collision, $\boldsymbol{v}_n$ is the velocity of the particle.

When driving these billiards, the resulting mapping is four dimensional and a suitable choice for the variables is $(\xi_n, \phi_n, \alpha_n, v_n)$, where $v_n = |\boldsymbol{v}_n|$ is the modulus of the particles velocity and $\xi_n = \omega t \bmod 2\pi$ is the phase of the boundary oscillation. The circle and the annuli are driven by letting the radius, the radii, respectively, evolve harmonically, i.e. $r(t) = r_0 + c\sin(\omega t)$, where $c$ is the driving amplitude. In the stadium, the two parallel line segments change their length harmonically. The boundary of the time-dependent ellipse is given by

$$\begin{pmatrix} x(t,\phi) \\ y(t,\phi) \end{pmatrix} = \begin{pmatrix} a(t)\cos\phi \\ b(t)\sin\phi \end{pmatrix} = \begin{pmatrix} (a_0 + c\sin\omega t)\cos\phi \\ (b_0 + c\sin\omega t)\sin\phi \end{pmatrix}, \tag{1}$$

and the one of the driven oval by

$$\begin{pmatrix} x(t,\phi) \\ y(t,\phi) \end{pmatrix} = \begin{pmatrix} \cos\phi\,[1 + \eta_2\cos t + \epsilon(1 + \eta_1\cos t)\cos 2\phi] \\ \sin\phi\,[1 + \eta_2\cos t + \epsilon(1 + \eta_1\cos t)\cos 2\phi] \end{pmatrix}. \tag{2}$$

For $\eta_1 = \eta_2 = 0$ the static oval is recovered and the geometry can be described in polar coordinates by $r(\phi) = 1 + \epsilon\cos 2\phi$. The parameter $\epsilon \in [0, 1)$ controls the amount of chaos in phase space (for $\epsilon = 0$ the integrable circle is obtained). In the driven case, $\eta_1 = \eta_2$ corresponds to a breathing billiard, since (2) reduces to $r(t,\phi) = [1 + \eta_1\cos t](1 + \epsilon\cos 2\phi)$, whereas for $\eta_1 \neq \eta_2$ the billiard changes its shape. The different billiards together with their minimal and maximal extension are shown in Fig. 1.

## 3. Fermi Acceleration

When studying FA, an ensemble of particles is propagated for a certain number of collisions and the evolution of the ensemble averaged energy is the main quantity of interest. For the initial conditions, usually a fixed modulus of the velocity $v_0$ is chosen, whereas the remaining variables $\xi_0, \phi_0, \alpha_0$ are distributed uniformly randomly. In Kamphorst and Carvalho (1999), the breathing circle is investigated. The authors rigorously proof that there is no FA, i.e. any admissible initial condition will have a bounded velocity. The key ingredient of the proof is the fact that the angular momentum, which is a constant of motion of the static circle, is also in the driven case preserved. This effectively reduces the dimensions of the phase space to two and allows the construction of invariant spanning curves that limit the energy growth. The driven concentric annular billiard is considered in Carvalho et al. (2006). The ensemble averaged modulus of the velocity $\langle v \rangle(n)$ as a function of the number of collisions $n$ saturates for all considered parameters, so there is no FA. Like in the breathing circle, invariant spanning curves in phase space prohibit unlimited energy growth. In contrast, the eccentric annulus (Carvalho et al. 2006) shows FA, with $\langle v \rangle(n) \sim n^{\delta}$. The exponent $\delta$ depends on the geometry parameters of the billiard, but not on $v_0$. A similar behavior (FA with $\langle v \rangle(n) \sim n^{\delta}$) is obtained in Loskutov and Ryabov (2002) and Loskutov et al. (2000) for the oscillating stadium billiard and in Kamphorst et al. (2007) for the driven oval in the case $\eta_1 \neq \eta_2$. Surprisingly, Kamphorst et al. did not observe FA in the breathing mode ($\eta_1 = \eta_2$), which contradicts the LRA-conjecture (see Sect. 1).

According to the LRA-conjecture, all driven billiards with a static counterpart that has a mixed phase space will show FA, whereas this it not clear for integrable (in the static case) billiards when they are driven. Besides the static circle, the ellipse is also integrable and it is worthwhile testing whether there will be FA when driving the system. The ensemble averaged modulus of the velocity $\langle v \rangle(n)$ as a function of the number of collisions $n$ for two different driving amplitudes ($c = 0.1$ and $0.5$) is shown in Fig. 2a ($v_0 = 0.1$). Clearly, the system exhibits FA, after an initial transient $\langle v \rangle(n) \sim n^{\delta}$ (Lenz et al. 2008). The system shows a crossover after approximately $10^8$ collisions from amplitude dependent subdiffusion to universal normal diffusion. This means that the diffusion exponent $\delta_c < 0.5$ depends on the driving amplitude $c$ for $n < n_{cr} \approx 10^8$, whereas for $n > n_{cr}$, $\delta = 0.5$ independent of the amplitude.

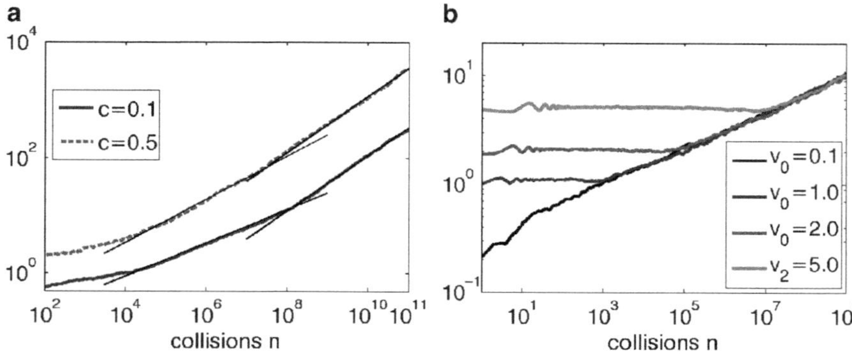


*Figure 2.* Ensemble averaged velocity $\langle v \rangle(n)$ as a function of $n$ (collisions) in the driven ellipse (**a**) and oval (**b**). Two different driving amplitudes $c$ are shown in (**a**), different initial velocities $v_0$ in (**b**)

The crossover is caused by a change of the phase space composition with increasing velocity. The phase space geometry evolves from a large chaotic sea with rich stickiness structures due to regular islands (low $v$), to large dominating regular domains with thin chaotic acceleration channels (high $v$). The underlying mechanism is in detailed described in Lenz et al. (2009).

Since already the ellipse shows FA, it is rather unlikely that the oval, which has a static counterpart with mixed phase space, does not show FA in the breathing mode. We thus rechecked the results of Kamphorst et al. (2007) by numerically simulating an ensemble of particles in the driven oval, with the same parameters as used by Kamphorst et al. While we are able to reproduce the results in the $\eta_1 \neq \eta_2$ case, we did find FA even in the breathing case ($\eta_1 = \eta_2$), see Fig. 2b. Possibly, the long transient of the order of $10^7$ collisions ($v_0 = 5$) has been misinterpreted by Kamphorst et al. as the absence of FA. As can be seen from Fig. 2b, the ensemble averaged velocity will eventually, after a transient, follow a power law $\langle v \rangle(n) = dn^{\delta}$, where the exponent $\delta$ is independent of $v_0$ ($d$ is the diffusion constant). The length $n_t$ of the transient depends on $v_0$ and scales as $n_t \sim (v_0/d)^{1/\delta}$ (Petri et al. 2009).

## 4. Conclusion and Outlook

Whether a time-dependent two-dimensional billiard will show Fermi acceleration (FA) or not depends on the specific driving law and on the phase space of the underlying static system. When restricting the driving laws to harmonic ones, solely the composition of phase space of the corresponding static system is decisive for the occurrence of FA. In that sense, the LRA-conjecture (Loskutov and Ryabov 2002) states that a chaotic part in the phase space of the underlying static billiard is sufficient for the presence of FA in the driven counterpart. The absence of FA in the breathing mode of the oval (Kamphorst et al. 2007) seemed to disprove the LRA-conjecture. While resolving this contradiction by showing that the breathing oval does exhibit FA, our discovery of FA in the driven ellipse demonstrates that the LRA-conjecture cannot be the full story, since there are clearly no chaotic parts in the phase space of the static ellipse. Unlike in the circle, where the angular momentum is conserved in the static and in the driven case, in the ellipse the product of the angular momenta around the foci is conserved in the static billiard only but not in the driven one. This difference explains why there is FA in the time-dependent ellipse but not in the circle, although the static counterparts are both integrable. The phase space of the static ellipse possesses two hyperbolic fixed points, whereas there are no hyperbolic fixed points in the phase space of the static circle. Note that if the phase space of a static billiard is mixed or chaotic, it contains always hyperbolic fixed points.

We thus conjecture that a necessary and sufficient condition for the presence of FA in a driven billiard is the existence of at least one hyperbolic fixed point in the phase space of the corresponding static counterpart.

## References

Berry M.V.: 1981, Regularity and chaos in classical mechanics, illustrated by three deformations of a circular billiard. *European Journal of Physics*, **2**, 91

de Carvalho R.E., Caetano Souza F. and Leonel E.D.: 2006, Fermi acceleration on the annular billiard. *Physical Review E*, **73**, 066229

Celaschi S. and Zimmerman R.L.: 1987, Evolution of a two-parameter chaotic dynamics from universal attractors. *Physics Letters A*, **120**, 447

Fermi E.: 1949, On the origin of the cosmic radiation. *Physical Review*, **75**, 1169

Gelfreich V. and Turaev D.: 2008, Fermi acceleration in non-autonomous billiards. *Journal of Physics A*, **41**, 212003

Kamphorst S.O. and de Carvalho S.P.: 1999, Bounded gain of energy on the breathing circle billiard. *Nonlinearity*, **12**, 1363

Kamphorst S.O., Leonel E.D. and da Silva J.K.L.: 2007, The presence and lack of Fermi acceleration in nonintegrable billiards. *Journal of Physics A*, **40**, F887

Kowalik Z.J., Franaszek M. and Pierański P.: 1988, Self-reanimating chaos in the bouncing-ball system. *Physical Review A*, **37**, 4016

Lenz F., Diakonos F.K. and Schmelcher P.: 2008, Tunable fermi acceleration in the driven elliptical billiard. *Physical Review Letters*, **100**, 014103

Lenz F., Petri C., Koch F.R.N., Diakonos F.K. and Schmelcher P.: 2009, Evolutionary phase space in driven elliptical billiards. in press

Lichtenberg A.J. and Lieberman M.A.: 1992, Regular and Chaotic Dynamics. Springer, New York, 38

Lieberman M.A. and Lichtenberg A.J.: 1972, Stochastic and adiabatic behavior of particles accelerated by periodic forces. *Physical Review A*, **5**, 1852

Loskutov A. and Ryabov A.: 2002, Particle dynamics in time-dependent stadium-like billiards. *Journal of Statistical Physics*, **108**, 995

Loskutov A., Ryabov A.B. and Akinshin L.G.: 2000, Properties of some chaotic billiards with time-dependent boundaries. *Journal of Physics A*, **33**, 7973

Petri C., Lenz F., Koch F.R.N., Diakonos F.K. and Schmelcher P.: 2009. in preparation

Warr S., Cooke W., Ball R.C. and Huntley J.M.: 1996, Probability distribution functions for a single-particle vibrating in one dimension: Experimental study and theoretical analysis. *Physica A*, **231**, 551

# TIME DEPENDENT NEUTRINO BILLIARDS

Z.A. Sobirov*, D. Matrasulov, Sh. Ataev, and H. Yusupov
*Heat Physics Department of the Uzbek Academy of Sciences, 28 Katartal Street, 100135 Tashkent, Uzbekistan; sobirovzar@yahoo.com, dmatrasu@phys.ualberta.ca*

**Abstract.** A one dimensional massless Dirac equation with time-dependent boundary condition is treated. An exact analytical wave functions and eigenvalues are obtained for the case of linear time-dependence of the boundary position.

**Key words:** Neutrino billiard; Dirac equation; Moving boundaries

Neutrino billiard was discussed first by Berry and Mondragon in the context of time-reversal symmetry breaking and quantum chaos in relativistic systems (Berry and Mondragon 1987). Recently it become extensively developing topic because of its relevance to so-called graphene billiard. Graphene is a planar monolayer of carbon atoms tightly packed into a two-dimensional honeycomb lattice. Electron dynamics in graphene is described by massless Dirac equation, i.e. quasielectrons in graphene can be treated as relativistic massless spinors (Katsnelson and Novoselov 2007; Brey and Fertig 2006). In this work we address the problem of neutrino billiard with moving boundaries. Besides its relevance to graphene billiard such system is of importance in the context of relativistic quantum Fermi acceleration, too. Time-dependent quantum billiard problem requires solution of the two-dimensional Schrödinger or Dirac equation with time-dependent boundary conditions. Quantum mechanical wave equations with time-dependent boundary conditions have been subject of extensive study during past three decades in the context of nonrelativistic quantum mechanics (Doescher and Rice 1969; Scheiniger and Kleber 1991). The early treatment of the one-dimensional Schrödinger equation with non-stationary boundary conditions dates back to Doescher and Rice (1969) who explored quantum dynamics of a particle in square well with moving walls. Later Munier et al. (1981) and Pinder (1990) studied this problem for some special cases. More comprehensive treatment of the Schrödinger equation with time-dependent boundary conditions has been done in a series of papers by Makowski and Dembinski (1991) and Makowski and Peptowski (1992). Seba has studied this problem in the context of quantum Fermi acceleration (Seba 1990). The case of time-periodic boundary condition is treated by Scheiniger and

G. Casati and D. Matrasulov (eds.), *Complex Phenomena in Nanoscale Systems*,
NATO Science for Peace and Security Series B: Physics and Biophysics,
© Springer Science+Business Media B.V. 2009

Kleber (1991). Three-dimensional case was treated by Yuce for the case of spherical boundaries (Yuce 2004). Despite the progress made in the study of Schrödinger equation with time-dependent boundary conditions, conditions, relativistic extension of this problem for the case of Dirac equation with moving boundary conditions has not yet been considered. We note that Klein–Gordon equation with time-dependent boundary conditions has been extensively studied in the literature in the context of dynamical Casimir effect and related problems (Moore 1970; Cole and Schieve 1995). Here we address the Dirac equation for massles particle with time-dependent boundary conditions. To solve the Dirac equation with time-dependent boundary condition we use the same prescription as that used by Makowski et al. in the case corresponding Schrödinger equation (Makowski and Dembinski 1991). For the case of special time-dependence of the boudnary conditions we obtain analytically eigenfunctions and eigenvalues of the Dirac equation for massless particle.

## 1. One Dimensional Box

Thus we intended to solve the following Dirac system ($\hbar = c = 1$)

$$i\frac{\partial \Psi}{\partial t} = (\alpha p + m\beta)\Psi, \tag{1}$$

with $\Psi(t,x) = (\Psi_1(t,x), \Psi_2(t,x))^T$ being two-component spinors, in the domain

$$D = \{(t,x),\ 0 < t < T,\ \ 0 < x < L(t)\},$$

where the right boundary $L(t)$, is time-dependent and

$$p = -i\frac{\partial}{\partial x}$$

and

$$\alpha = \begin{pmatrix} 0 & -i \\ i & 0 \end{pmatrix}, \qquad \beta = \begin{pmatrix} 1 & 0 \\ 0 & -1 \end{pmatrix},$$

are the Dirac matrices. For which the boundary conditions are given as

$$\Psi_1(t,0) = 0, \quad \Psi_1(t, L(t)) = 0, \quad 0 \le t \le T \tag{2}$$

The Dirac equation for (fixed) boundary conditions, given in the one-dimensional box, ($L(t)$ = const) was treated in Alonso et al. (1997) and Alonso and de Vincenzoz (1999) where the solution of the Dirac equation

$$(\alpha p + mc^2\beta)\Psi = E\Psi$$

is obtained as

$$\Psi_n(x) = A_n \begin{pmatrix} \sin(k_n x) \\ -\frac{ck_n}{E_n+mc^2}\cos(k_n x) \end{pmatrix},$$

with $A_n$ being the normalization constant and $k_n = n\pi/L,\ n = 1, 2, ..., E_n = [k_n^2 + (mc^2)^2]^{1/2}$.

To solve (1) we will restrict ourselves by considering the massless case, i.e. we assume that $m = 0$. Equation (1) with the boundary conditions (2) cannot be solved until time-dependent boundary conditions are not replaced with fixed ones. To do this we use in (1) the following substitution: $y = \frac{x}{L(t)}$ that reduced the domain $D$ to $D_1 = \{(t, x), 0 < t < T, 0 < y < 1\}$.

In this case (1) can be rewritten as

$$\begin{cases} i\frac{\partial\Psi_1}{\partial t} = L^{-1}\frac{\partial\Psi_2}{\partial y} + i\dot{L}L^{-1}y\frac{\partial\Psi_1}{\partial y}, \\ i\frac{\partial\Psi_2}{\partial t} = -L^{-1}\frac{\partial\Psi_1}{\partial y} + i\dot{L}L^{-1}y\frac{\partial\Psi_2}{\partial y}, \end{cases} \tag{3}$$

where $\dot{L} = \frac{dL}{dt}$, and the boundary conditions are given as

$$\Psi_1(t, 0) = 0, \quad \Psi_1(t, 1)) = 0, \tag{4}$$

Time and coordinate variables in (3) cannot be separated for arbitrary time-dependence of $L(t)$. The only case for which variables can be separated is $L(t) = at + b$.

Introducing new time variable

$$\tau = \int_0^t \frac{ds}{L(s)} = \frac{1}{a}\ln\left(\frac{at+b}{b}\right),$$

and using the substitutions $\Psi_1(\tau, y) = e^{-i\lambda\tau}f(y)$, $\Psi_1(\tau, y) = e^{-i\lambda\tau}g(y)$ we get from (3)

$$\begin{cases} \frac{dg}{dy} + iay\frac{df}{dy} = \lambda f, \\ -\frac{df}{dy} + iay\frac{dg}{dy} = \lambda g. \end{cases} \tag{5}$$

This system can be reduced to the second-order equation as

$$\left\{(1 - a^2y^2)\frac{d^2}{dy^2} - 2ay(i\lambda + a)\frac{d}{dy} - ia\lambda\right\} f = -\lambda^2 f. \tag{6}$$

Using the substitution

$$z = \frac{1 - ay}{2},$$

Equation (6) for $|a|<1$ can be reduced to the following equation

$$z(1-z)f'' + (c-(a+b+1)z)f' - abf = 0$$

whose solutions are hypergeometric functions.

In our case the hypergeomtric series can be written as

$$F_1(z) = F(\alpha_1, \beta_1, \gamma_1, z) = \sum_{k=0}^{+\infty} \frac{(\beta_1)_k}{k!} z^k = (1-z)^{\beta_1}.$$

Second solution can be found as

$$F_2(z) = z^{\beta_1}.$$

Then the general solution of (6) can be written as

$$f(y) = A\left(\frac{1-ay}{2}\right)^{-i\lambda/a} + B\left(\frac{1+ay}{2}\right)^{-i\lambda/a}, \tag{7}$$

where $a^i = \exp(i\ln(a))$.

Then from the boundary conditions given by (4) we get

$$A = -B, \quad \left(\frac{1-a}{2}\right)^{-i\lambda/a} - \left(\frac{1+a}{2}\right)^{-i\lambda/a} = 0. \tag{8}$$

Solving this equation we obtain the eigenvalues as

$$\lambda_n = 2a\pi n \left|\ln|1-a| - \ln|1+a|\right|^{-1}. \tag{9}$$

Corresponding eigenfunctions are

$$\begin{aligned} f_n(y) &= M\left((1-ay)^{-i\lambda_n/a} - (1+ay)^{-i\lambda_n/a}\right), \\ g_n(y) &= \tfrac{M}{i\lambda_n - a}\left((1-ay)^{-i\lambda_n/a+1} + (1+ay)^{-i\lambda_n/a+1}\right) - \\ &\quad iayM\left((1-ay)^{-i\lambda_n/a} - (1+ay)^{-i\lambda_n/a}\right) \end{aligned} \tag{10}$$

We note that for $a \to 0$ the eigenvalues coincide with $\lambda_n \to \pi n$, and eigenfunctions become

$$\begin{aligned} f_n(y) &\to A\sin(\pi n y), \\ g_n(y) &\to -\tfrac{A}{\pi n}\cos(\pi n y). \end{aligned}$$

Then the complete set of solutions of (1) for linearly moving boundaries can be written as

$$\begin{aligned} \Psi_{1n}(t,x) &= M\exp\left[-\tfrac{i\lambda_n}{a}\ln\left(\tfrac{at+b}{b}\right)\right]\left(\left(1-\tfrac{ax}{at+b}\right)^{-i\lambda_n/a} - \left(1+\tfrac{ax}{at+b}\right)^{-i\lambda_n/a}\right), \\ \Psi_{2n}(t,x) &= M\exp\left[-\tfrac{i\lambda_n}{a}\ln\left(\tfrac{at+b}{b}\right)\right]\cdot \\ &\cdot\left\{\tfrac{1}{i\lambda_n - a}\left(\left(1-\tfrac{ax}{at+b}\right)^{-i\lambda_n/a+1} + \left(1+\tfrac{ax}{at+b}\right)^{-i\lambda_n/a+1}\right) - \right. \\ &\left. -iay\left(\left(1-\tfrac{ax}{at+b}\right)^{-i\lambda_n/a} - \left(1+\tfrac{ax}{at+b}\right)^{-i\lambda_n/a}\right)\right\}. \end{aligned} \tag{11}$$

Thus we have treated the Dirac equation for massless particle in one-dimensional infinite square well with a time-dependent wall.

## 2. Time-Dependent Circular Billiard

Consider the dirac equation with the boundary conditions given at the circle with time-dependent radius $x^2 + y^2 < r_0^2(t)$. Radial equation can be written as

$$\begin{aligned} i\frac{\partial P(r,t)}{\partial t} &= \frac{\partial Q(r,t)}{\partial r} - \frac{k}{r}Q(r,t), \\ i\frac{\partial Q(r,t)}{\partial t} &= -\frac{\partial P(r,t)}{\partial r} - \frac{k}{r}P(r,t), \end{aligned} \tag{12}$$

where $k$ is an integer number. Boundary condition is given by

$$P(r_0(t), t) = 0. \tag{13}$$

Using substitution

$$y = \frac{r}{r_0(t)}, \quad \tau = \int_0^t \frac{ds}{r_0(s)},$$

Equations (12) and (13) can be rewritten as

$$\begin{aligned} i\frac{\partial P(y,\tau)}{\partial \tau} &= i\dot{r}_0 y\frac{\partial P(y,\tau)}{\partial y} + \frac{\partial Q(y,\tau)}{\partial y} - \frac{k}{y}Q(y,\tau), \\ i\frac{\partial Q(y,\tau)}{\partial \tau} &= i\dot{r}_0 y\frac{\partial Q(y,\tau)}{\partial y} - \frac{\partial P(y,\tau)}{\partial y} - \frac{k}{y}P(y,\tau), \end{aligned} \tag{14}$$

where the boundary condition is time-independent now and given as

$$P(1, t) = 0. \tag{15}$$

Separating time and coordinate variables in (14) for $\ddot{r}_0(t) = 0$ and $r_0(t) = at+b$ and reducing the obtained first-order system into second order equation we have

$$y^2(a^2y^2 - 1)\frac{d^2 f}{dy^2} + 2a(a + i\lambda)y^3\frac{df}{dy} + [(ia\lambda - \lambda^2)y^2 + k(k+1)]f = 0 \tag{16}$$

Solution of this equation can be written in terms of (regular at $y = 0$) hypergeometric function as

$$f_\lambda(y) = y^\gamma F\left(\alpha, \alpha + \frac{1}{2}, \gamma, a^2y^2\right),$$

where

$$\alpha = \frac{1}{2}\left|k + \frac{1}{2}\right| + \frac{i\lambda}{2m} + \frac{1}{4} \quad \gamma = \left|k + \frac{1}{2}\right| + 1.$$

For second component we have

$$g_\lambda(y) = (\lambda - i(1-k)a)y^k \int_0^y \frac{f_\lambda(s)ds}{s^k} - iayf_\lambda(y).$$

The eigenvalues $\lambda_n$ can be found from the boundary condition:

$$F\left(\frac{1}{2}\left|k+\frac{1}{2}\right| + \frac{i\lambda}{2m} + \frac{1}{4}, \frac{1}{2}\left|k+\frac{1}{2}\right| + \frac{i\lambda}{2m} + \frac{3}{4}, \gamma, a^2\right) = 0. \tag{17}$$

For $k = 0$ the solution of (17) can be found analytically:

$$\lambda_n = 2\pi n\, a / \ln\left(\frac{1+a}{1-a}\right)$$

The solutions of (12) and (13) for $k = 0$ can be written as

$$P_n(r,t) = N \exp\left(-\frac{i\lambda_n}{2a}\ln\left(\frac{at+b}{b}\right)\right) f_{\lambda_n}\left(\frac{r}{r_0(t)}\right),$$

$$Q_n(r,t) = N \exp\left(-\frac{i\lambda_n}{2a}\ln\left(\frac{at+b}{b}\right)\right) g_{\lambda_n}\left(\frac{r}{r_0(t)}\right),$$

where $N$ is the normalization constant. Thus we have obtained analytically the solution of the massless Dirac equation for time-dependent circular billiard. The circle is considered as monotonically expanding (contracting) with constant velocity. For general case solution can be obtained numerically. The above system is of importance due to two reasons. The first one is related to the fact that it can be used for describing time-dependent graphene billiard, while another important point is caused by its relevance to relativistic quantum Fermi acceleration. In this context the extension of the above problem to the case of non-integrable billiard geometries and oscillating boundaries is of importance. Currently such studies are in progress.

## References

Abramowitz M.A. and Stegun I.A.: 1964, Handbook of Mathematical Functions. *Nat. Bur. Stand*, Washington, DC

Alonso V. and De Vincenzoz S.: 1999. *J. Phys. A*, **32**, 5277

Alonso V., De Vincenzoz S. and Mondinox L.: 1997. *Eur. J. Phys.*, **18**, 315

Berry M.V. and Mondragon R.J.: 1987. *Proc. Roy. Soc. A.*, **412**, 53

Brey L. and Fertig H.A.: 2006. *Phys. Rev. B*, **73**, 235411

Cole C.K. and Schieve W.C.: 1995. *Phys. Rev. A*, **52**, 4405

Doescher S.W. and Rice H.H.: 1969. *Am. J. Phys.*, **37**, 1246

Katsnelson M.I. and Novoselov K.S.: 2007. *Solid. State Commun.*, **143**, 3
Makowski A.J. and Dembinski S.T.: 1991. *Phys. Lett. A*, **154**, 217
Makowski A.J. and Peptowski P.: 1992. *Phys. Lett. A*, **163**, 142
Moore G.T.: 1970. *J. Math. Phys.*, **11**, 2679
Munier A., Burgan J.R., Feix M. and Fijalkow E.: 1981. *J. Math. Phys.*, **22**, 1219
Pinder D.N.: 1990. *Am. J. Phys.*, **58**, 54
Seba P.: 1990. *Phys. Rev. A*, **41**, 2306
Scheiniger C. and Kleber M.: 1991. *Physica D*, **50**, 391
Yuce C.: 2004. *Phys. Lett. A*, **321**, 291

# STATISTIC THEORY OF MULTIPLE EXCITON GENERATION IN QUANTUM DOT BASED SOLAR CELLS

B.L. Oksengendler and N.N. Turaeva*
*Institute of Polymer Chemistry and Physics AS of Uzbekistan, 7B Kadiriy Street, 100128 Tashkent, Uzbekistan; nturaeva@sarkor.com*

A. Zakhidov
*UTD NanoTech Institute, BE26, Richardson TX 75083-0688*

**Abstract.** In this work the statistical theory of multiple exciton generation in quantum dots is presented based on the Fermi approach to the problem of multiple generation of elementary particles at nucleon–nucleon collisions. Our calculations show that the quantum efficiencies of multiple exciton generation in various quantum dots at absorption of single photon are in a good agreement with the experimental data.

**Key words:** Solar cells; Quantum dots; Exciton generation

Carrier multiplication (CM) effect in quantum dot(QD)-based solar cells is one of the conceptual ways to improve dramatically the efficiency of third solar cells generation (Lewis and Crabtree 2005) beyond the Schottky–Quiser limit. This CM phenomenon was predicted by Nozik of NREL in 2002 (Nozik 2002). In 2004, Klimov, et al. reported that PbSe nanocrystals could respond to absorption of a single photon by producing two or more electron-hole pairs with 200% efficiency (the carrier multiplication (CM) phenomenon, also known as multiple exciton generation (MEG) (Schaller and Klimov 2004). Later, the effect of MEG has been confirmed also by NREL team of Nozik (Elingson et al. 2005) in PbSe and PbS QDs of three various sizes and quantum yield was increased up to 300%. Most recently it was found by Schaller et al. (2006) that MEG is a real phenomenon with generation up to seven excitons in PbSe QDs ($E_g = 0.3$ eV, $R = 20$ nm) in absorption of a single photon (with the energy $h\nu = 7.8\,E_g$). Using the carrier multiplication Schaller and Klimov (2006) has demonstrated "exotic" non-Poissonian distributions of carrier populations. Recently an activation threshold near the two-energy-gap $2E_g$ limit (as defined by the energy conservation) has been also observed (Schaller et al. 2006). Following all these initial spectroscopic observations, several groups have reported the formation of multiple excitons in device structures (Schaller et al. 2005a, 2006; Allan 2007; Jiang et al. 2007), and several attempts have been made to construct solar cells, based on Polymer/QD

G. Casati and D. Matrasulov (eds.), *Complex Phenomena in Nanoscale Systems*,
NATO Science for Peace and Security Series B: Physics and Biophysics,
© Springer Science+Business Media B.V. 2009

mixtures for separation of multiple excitons into several electrons and holes (Schaller et al. 2005a, 2006, 2007; Allan 2007; Jiang et al. 2007).

The high efficiency of such solar cells which have QDs as absorbers can be determined not only by single e–h pair, but by the MEG effect in QDs in single photon absorption, too. Therefore the power efficiency of solar cells can exceed the Schottky–Quiser limit that is a great promise, which needs to be demonstrated yet.

The theoretical analysis of MEG effect related to two- or three- excitons generation was made by Schaller et al. (2005b) and by Elingson et al. (2005). The problem of theoretical explanation of MEG of higher multiplicities in QDs remains as open question, since the application of the perturbation theory to the charge multiplication effect (e.g. as used in Schaller et al. 2005b) is complicated for the higher multiplicity processes (Lewis et al. 1948). Recently some other approaches to theoretical understanding the mechanisms of MEG have been considered. In particular, the spectral densities of multi-exciton states have been calculated and the possibility for direct and instantaneous photo-generation of multi-excitons has been explored.

They confirm the importance of the multi-exciton spectral densities in the MEG problem caused by their (multiexciton densities) rapid variation (several orders of magnitude) as a function of the energy. Also, recently Allan (2007) showed that the high MEG efficiencies in PbSe and Si nanocrystals (up to seven excitons per photon), would imply a very efficient relaxation in multi-exciton states into other lower energy states, whereas they are characterized by a negligible density. So, the overall conclusion is that the mechanisms of MEG generation and relaxation are not fully understood yet, and other mechanisms must be considered to explain the highest MEG efficiencies.

In the present paper as one of possible mechanisms of MEG we propose a statistical theory of MEG in QDs based on Fermi approach (Fermi 1950) originally coming from the theory of multiple elementary particle(such as nucleons and mesons) generation, in nucleon–nucleon collisions. Unlike to perturbation theory used in Schaller et al. (2005b) this approach is based on the strong interaction of correlated electrons with electromagnetic field in QD. We argue that the Fermi theory seems more realistic at high energy range when the number of possible exciton states with given energy is large, and this factor sharply increases the probability for establishing of the statistic equilibrium.

Based on the analogy with the Fermi approach we suppose that in the case of multiple excitons generation a high energetic photon is absorbed by QD and according to statistic laws the absorbed energy is rapidly distributed among the various degrees of freedom which are presented in QD volume $\Omega$, so that $n/2$ excitons are generated ($n$ is the total number of produced electrons and holes). In this case we can calculate the probability of $n$ particles generation with a given energy distribution in such a small volume on the basis of the Fermi formula for statistic weights. In our case of MEG the total kinetic energy is defined by the difference of photon energy and energy gap multiplied by the number of excitons $\left(T = h\nu - \frac{n}{2}E_g\right)$. Finally, we can show that the probability of $n$ particles generation in the volume $\Omega$ is proportional to the statistic weight:

$$S(n) = \frac{m^{3n/2}\Omega^n\left(h\nu - \frac{n}{2}\tilde{E}_g\right)^{\frac{3n}{2}-1}}{2^{3n/2}\pi^{3n/2}\left(\frac{3n}{2}-1\right)}.$$

Here $m$ is the electron mass, $\tilde{E}_g = E_g - \frac{1.78\times e^2}{\varepsilon R}$ (Suzdalev 2006). The second term in the last expression describes the electron interaction energy with hole in exciton. The number of particles, $n$ should be even.

$$n = 2, 4, 6, 8, 10, 12, 14.$$

The relative probability of $n$ particles generation is given by

$$W(n) = \frac{S(n)}{\sum\limits_n S(n)}.$$

The quantum efficiency(QE) of multiple excitons production by a single photon can be defined by the expression QE $= 100\% \times \langle N_{exc}\rangle$, where the average number of excitons in QD $\langle N_{exc}\rangle$ is calculated using the following equation:

$$\bar{n} = \frac{\sum\limits_n nS(n)}{\sum\limits_n S(n)}$$

Let us consider some cases of MEG in various QDs using the above statistical approach.

1. We can first calculate the statistic weights for small numbers of excitons, i.e. for $\langle N_{exc}\rangle = 1, 2, 3$ excitons generation in PbSe QDs in the absorption of single photon ($h\nu = 3.63\,E_g$) (Table 1) with the experimentally measured parameters ($E_g = 0.64$ eV and $R = 3.9$ nm) (Schaller et al. 2005b). To estimate the quantum efficiency we choose the following values of $m = m_e$, $E_{exc} = 0.14$ eV.
   In this case the quantum efficiency can be easily obtained using the relation QE $= 100\% \times \langle N_{exc}\rangle = 210\%$ which is in a good agreement with the experimental data (QE = 200%) from Schaller et al. (2005b).

Table 1. Statistical weights (in $J^{-1}$) of MEG in PbSe QDs

| $n$ | $S(n)$ | $W(n)$ | $\langle N_{exc}\rangle$ (theory) | $\langle N_{exc}\rangle$ (exper.) |
|---|---|---|---|---|
| 2 | $1.06 \times 10^{22}$ | 0.04 | | |
| 4 | $2.198 \times 10^{23}$ | 0.829 | 2.10 | 2.00 |
| 6 | $3.469 \times 10^{22}$ | 0.131 | | |

$E_g = 0.64$ eV, $R = 3.9$ nm, $h\nu = 3.63\,E_g$

Table 2. Statistical weights (in $J^{-1}$) of MEG in PbSe QDs

| $n$ | $S(n)$ | $W(n)$ | $\langle N_{exc}\rangle$ (theory) | $\langle N_{exc}\rangle$ (exper.) |
|---|---|---|---|---|
| 2 | $2.215 \times 10^{22}$ | $1.916 \times 10^{-3}$ | | |
| 4 | $2.409 \times 10^{24}$ | 0.208 | | |
| 6 | $8.442 \times 10^{24}$ | 0.73 | 2.85 | 3.25 |
| 8 | $6.891 \times 10^{23}$ | 0.06 | | |
| 10 | $1.402 \times 10^{20}$ | $1.213 \times 10^{-5}$ | | |

$E_g = 0.64$ eV, $R = 3.9$ nm, $h\nu = 4.9\,E_g$

2. Statistical weights of 1-, 2-, 3-, 4- and 5-excitons generation (with the parameters, $E_g = 0.64$ eV and $R = 3.9$ nm of PbSe QDs taken from the experiment, Schaller et al. 2005b) in the absorption of single photon ($h\nu = 4.9\,E_g$) are presented in Table 2. The following values of $m = m_e$, $E_{exc} = 0.14$ eV are chosen to estimate QE of MEG.
   The quantum efficiency calculated on the basis of statistical weights is QE $= 100\% \times \langle N_{exc}\rangle = 285\%$ that is slightly different than that of experimental one (QE=325%) (Schaller et al. 2005b).
3. In Table 3 the statistic weights of 1-, 2-, 3-, 4-, 5-, 6-, 7-, 8-excitons generation in PbSe QDs calculated using the following experimentally measured parameters (Schaller et al. 2005b): ($E_g = 0.3$ eV and $R =$ 20 nm and $h\nu = 7.8\,E_g$) are presented. For the estimation of QE we choose the following values of the parameters: $m = m_e$, $E_{exc} = 0.063$ eV.

   The efficiency of seven excitons generation is estimated as QE $= 100\% \times \langle N_{exc}\rangle = 684\%$. This estimate is in good agreement with the experimental data QE=700% from Schaller et al. (2005b).
4. For the Si NCs with the parameters $E_g = 1.2$ eV, $R = 9.5$ nm, $h\nu = 3.4\,E_g$ we have calculated the probabilities for 1-, 2-, 3-excitons generations.

Table 3. Statistical weights of (in $J^{-1}$) MEG in PbSe QDs

| $n$ | $S(n)$ | $W(n)$ | $\langle N_{exc}\rangle$ (theory) | $\langle N_{exc}\rangle$ (exper.) |
|---|---|---|---|---|
| 2 | $2.339 \times 10^{26}$ | 0 | | |
| 4 | $3.37 \times 10^{32}$ | 0 | | |
| 6 | $3.706 \times 10^{37}$ | $7.011 \times 10^{-11}$ | | |
| 8 | $4.82 \times 10^{41}$ | $9.119 \times 10^{-7}$ | 6.84 | 7.00 |
| 10 | $7.284 \times 10^{44}$ | $1.378 \times 10^{-3}$ | | |
| 12 | $9.155 \times 10^{46}$ | 0.173 | | |
| 14 | $4.26 \times 10^{47}$ | 0.806 | | |
| 16 | $1.03 \times 10^{46}$ | 0.019 | | |

$E_g = 0.3$ eV, $R = 2.0$ nm, $h\nu = 7.8\,E_g$

Table 4. Statistical weights of (in $J^{-1}$) MEG in Sin Cs QDs

| $n$ | $S(n)$ | $W(n)$ | $\langle N_{exc}\rangle$ (theory) | $\langle N_{exc}\rangle$ (exper.) |
|---|---|---|---|---|
| 2 | $4.256 \times 10^{24}$ | $6 \times 10^{-5}$ | | |
| 4 | $2.667 \times 10^{28}$ | 0.379 | 2.62 | $2.6 \pm 0.2$ |
| 6 | $4.371 \times 10^{28}$ | 0.621 | | |

$E_g$=1.2 eV, $R = 9.5$ nm, $h\nu = 3.4\,E_g$

The results are presented in Table 4. The parameters are chosen as $m = m_e/10$, $E_{exc} = 0.067$ eV.
For the quantum efficiency we have an estimate QE $= 100\% \times \langle N_{exc}\rangle =$ 262% and this that is in a good agreement with the experimental data QE = (260 ± 20)% (Beard et al. 2007).

## Conclusion

Thus we have developed a statistical approach to MEG in quantum dots effect based on the simple assumption that the probability for $n$ particles generation in the volume QDs is defined by its statistic weight $S(n)$, which depends on such parameters as the size of QD, photon energy, gap energy, exciton bind energy, electron and hole effective masses. By analyzing the above results we can conclude that in statistic approach the probability of multiple exciton generation is very sensitive to the changes of the exciton binding energy and electron (hole) effective mass.

## References

Allan G., Delerue C.: 2008. In: *Book of Abstracts*, CECAM Workshop on *"Critical materials issues in photovoltaics"*, 2-4th June, http://www.cecam.org/workshop-2-202.html

Beard M. et al: 2007. *Nano Lett.*, **7**, 2506–2612

Elingson et al: 2005. *Nano Lett.*, **5**, 865

Fermi E.: 1950. *Prog. Theor. Phys.*, **5**, 570

Jiang X., Schaller R.D., Lee S.B., Pietryga J.M., Klimov V.I. and Zakhidov A.A.: 2007. *J. Mater. Res.*, **22**, 2204–2210

Lewis N.S. and Crabtree G.W.: 2005, Basic research needs for Solar Energy Utilization. In: *Report of the Basic Energy Sciences Workshop on Solar Energy Utilization*, 18021. Second Printing: U.S. Department of Energy (DOE), Office of Basic Energy Science, Washington, DC

Lewis H.W., Oppenheimer J.R. and Wouthuysen S.A.: 1948. *Phys. Rev.*, **73**, 127

Nozik A.: 2002. *J. Physica E*, **14**, 115

Schaller R.D. and Klimov V.I.: 2004. *Phys. Rev. Lett.*, **92**, 186601

Schaller R.D. and Klimov V.I.: 2006. *Phys. Rev. Lett.*, **96**, 097402-1-4

Schaller R.D., Petruska M.A. and Klimov V.I.: 2005a. *Appl. Phys. Lett.*, **87**, 253102

Schaller R.D., Agranovich V.M. and Klimov V.I.: 2005b. *Nat. Phys.*, **1**, 189

Schaller R.D., Sykora M., Pietryga J.M. and Klimov V.I.: 2006. *Nano Lett.*, **6**, 424

Schaller R.D., Pietryga J.M. and Klimov V.I.: 2007. *Nano Lett.*, **7**, 3469–3476

Suzdalev I.P.: 2006. *Nanotechnology: Nanoclasters, Nanostructures and Nanomaterials Physico-Chemistry*, Moscow, DomKniga

# MODELLING OF QUANTUM WIRES IN THE INTERFACE LAYER OF THE SEMICONDUCTOR-OXIDE STRUCTURES WITH CHARGE BUILT IN OXIDE

A.E. Atamuratov
*Urganch State University, Urganch, Uzbekistan; atabek@ursu.uzpak.uz*

**Abstract.** In this we study formation of a nanoscale potential well with nanowires near the oxide-semiconductor interface by special a charge distribution built-in in oxide. The $SiO_2$-Si structure with a cylindrical substrate covered by a coaxial oxide layer is considered. The dependence of the potential well parameters on the geometry, size and built-in charge density is analyzed. The case of two charged rings also is considered. Parallel rings is also considered

**Key words:** Quantum wires; Semiconductor nanostructures; Charge density

## 1. Introduction

Study of quantum wire formation in solid state structures is of importance because of their relevance to the nanoscale device technologies. Traditionally, nanoscale structures in semiconductors are created using multilayer and local epitaxy technologies (Alferov 1998), high precision lithography (Kawamato 1999; Thompson 1994), precision doping (Broers 1984; Kazor et al. 1994) and others (Ledentsov et al. 1998; Emeleus et al. 1998; Deng and Krishnamurthy 1998; Shi et al. 1999). Also, it is important in constructing of nanostructures controlling properties (e.g. geometry, charge density, size, etc.) of the structure.

Recently principally different (than that above mentioned) method for constructing stable and transformable semiconducting nano-scale structures, such as quantum dost, wells, wires and superlattices has been developed (Goldman et al. 2001) on the basis of regular and spontaneous variation of the charge distribution near the surface of semiconductor. Achieving of the surface density of the built in charge $3 \times 10^{13}\,\mathrm{s\,m^{-2}}$ which is possible using the space-charge ionic polarization (Verwey et al. 1990; Nicolian and Brews 1982) or tunnel (or avalanche) injection into the oxide layer in not destroying electrical fields less than $5 \times 10^6\,\mathrm{V/s\,m}$ (Sah 1990). Scanning tunnel microscope also have the same possibility (Di Maria and Stasiak 1989).

G. Casati and D. Matrasulov (eds.), *Complex Phenomena in Nanoscale Systems*,
NATO Science for Peace and Security Series B: Physics and Biophysics,

© Springer Science+Business Media B.V. 2009

The local charge injection into dielectric structures is also possible in tip-dielectric-semiconductor system in a strong electric field at the maintenance of positioning accuracy of a metal tip on the surface. Such method provides selecting and positioning of the charged area with the size of few tens nanometers (Binning et al. 1982; Guntherodt and Wiesendanger 1991) and creation of the potential wells with needed parameters. By using opposite polarity voltage to the tip-dielectric-semiconductor system one can remove built-in charge.

In this work we discuss one of the ways for simulation of the above methods on the oxide-silicon system.

Namely, we treat formation of nanosized potential well under built-in oxide charge by using analytical and numerical methods.

For simplification of computations the charge in oxide layer is supposed to be distributed at one and two finite thickness rings. The dependence of the parameters of quantum wires created by a potential well on the distribution and density of the built-in charge is analyzed.

## 2. Theoretical Analysis

To simplify our calculations we consider a structure consisting of a semiconductor substrate of the cylindrical form covered by a coaxial oxide layer with inner and external radiuses, $R_1$ and $R_2$, respectively (Fig. 1). The charge with the density $\rho$ is built-in on the ring shaped oxide layer with the longitudinal thickness $d = Z_b - Z_a$. The potential distribution of a resulting field of these charges in the semiconductor can be found from the Poisson equation written in cylindrical coordinates:

$$\frac{\partial^2 \varphi}{dr^2} + \frac{1}{r}\frac{\partial \varphi}{\partial r} + \frac{\partial^2 \varphi}{\partial z^2} = -4\pi\rho \tag{1}$$

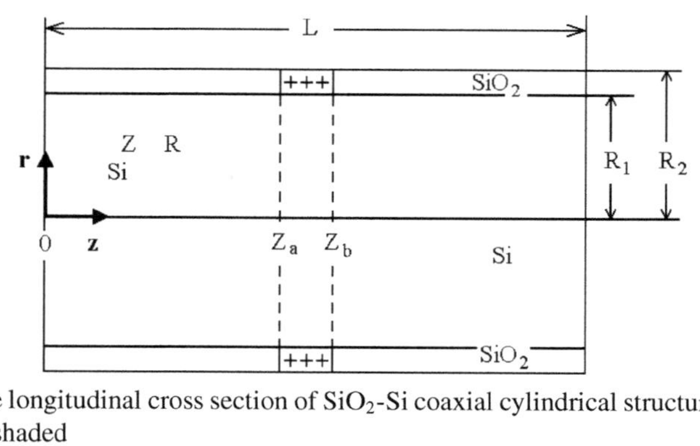


*Figure 1.* The longitudinal cross section of $SiO_2$-Si coaxial cylindrical structure. The charged ring areas are shaded

The equation can be solved numerically using the finite difference scheme suggested in Huges (1971). Consider $(r, z)$-plane covered by a uniform mesh with a constant spacing $\Delta r\ \Delta z$ along the $r$ and $z$ directions

$$z = \left(i + \frac{1}{2}\right) \cdot \Delta z, \qquad i = 0, 1, \ldots N_{Z-1},$$

$$r = j \cdot \Delta r, \qquad j = 0, 1, \ldots N_R,$$

where $\Delta z = \frac{L}{N_Z}$, $\Delta r = \frac{R_2}{N_R}$ and $L$ is the length of the structure along $Z$ axis, $R_2$ is the external radius of the structure (see Fig. 1). Furthermore, we will use the following expansion for the potential:

$$\varphi(r, z) = \frac{2}{N_Z} \sum_{K=0}^{N_Z-1} \tilde{\varphi}_k(r) cos \frac{\pi k z}{\Delta z N_z}. \tag{2}$$

Substituting of (2) into Poisson equation (1), after straightforward simplification we obtain equation for $\tilde{\varphi}_{k,j}$-factors

$$\frac{\partial^2 \tilde{\varphi}_k}{\partial r^2} + \frac{1}{r} \frac{\tilde{\varphi}_k}{\partial r} - \frac{1}{\Delta z^2} \left(\frac{\pi k}{N_z}\right)^2 \tilde{\varphi}_k = -4\pi \tilde{\rho}_k \tag{3}$$

where $\rho$ has been similarly transformed.

Applying the above mentioned finite-difference scheme to (3) we get the following algebraic system

$$\tilde{\varphi}_{k,j} \cdot \left[2 + \frac{\Delta r^2}{\Delta z^2} \left(\frac{\pi k}{N_z}\right)^2\right] - \tilde{\varphi}_{k,j-1} \left(1 - \frac{1}{2 \cdot j}\right) - \tilde{\varphi}_{k,j+1} \left(1 + \frac{1}{2 \cdot j}\right) = 4\pi \Delta r^2 \tilde{\rho}_{k,j}. \tag{4}$$

In this system there are $N_R - 1$ equations, while number of unknown function is $N_R + 1$. Therefore to solve this system one should know the value of the potential on the wall of the cylinder ($j = N_R$). However, in most of the cases it is not possible to find such value and we will introduce an additional equation following from the continuity condition with respect to normal part of the field on oxide-semiconductor border. This condition is given as

$$\varepsilon_1 \frac{\partial \varphi_1}{\partial n} = \varepsilon_2 \frac{\partial \varphi_2}{\partial n} \tag{5}$$

where $\varepsilon_1$, $\varepsilon_2$ are the dielectric constants for semiconductor and oxide, $\varphi_1$, and $\varphi_2$ are the potentials near the borders of semiconductor and oxide, respectively. In the finite-difference form (5) can be written as

$$\tilde{\varphi}_{N+2} = (\varepsilon + 1)\tilde{\varphi}_N - \varepsilon \tilde{,} \varphi_{N-2} \tag{6}$$

where we assumed that the semiconductor-oxide border correspond to the point $j = N$ and $\varepsilon = \varepsilon_1/\varepsilon_2$. Equation (6) can be solved together with the system (4).

Solving (4) and (6) by giving the value of $\varphi(r,z)$ on the cylinder's axis and using inverse Fourier transformation we can find the distribution of the potential on the plane $\varphi(r,z)$.

The potentials on the axis can be found analytically by integrating the field strength of the charged annulus. The field on the axis can be found from the superposition principle, $dq = \rho\, dV$ ($dV = dy\, dr\, dl$, $dr = R_2 - R_1$) for the fields of the elementary space charges with account of the cylindrical symmetry of the system (see Fig. 2).

Thus the resulting field can be found by integrating charge distribution over the ring: for the interval between $Z_a$ and $Z_b$ (Fig. 1) we have

$$E_I(z) = \frac{dr\rho R_0}{2\varepsilon_1\varepsilon_0}\left(\frac{1}{\sqrt{(Z-Z_a)^2+R_0^2}} - \frac{1}{\sqrt{(Z_b-Z)^2+R_0^2}}\right) \tag{7a}$$

and for the left side of the point $Z_a$ we have

$$E_{II}(z) = \frac{dr\rho R_0}{2\varepsilon_1\varepsilon_0}\left(\frac{1}{\sqrt{(Z_a-Z)^2+R_0^2}} - \frac{1}{\sqrt{(Z_b-Z)^2+R_0^2}}\right) \tag{7b}$$

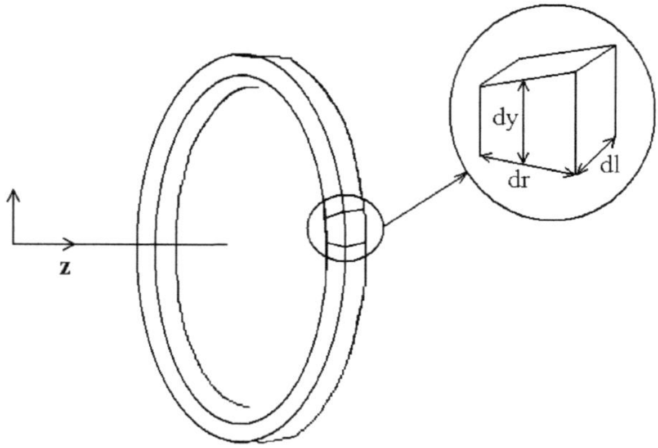


*Figure 2.* The charged ring in the oxide layer and elementary charged volume in this layer. The selected area can be considered elementary space charge by choosing the width, *dr* of a ring along radius smaller than that of ring radius

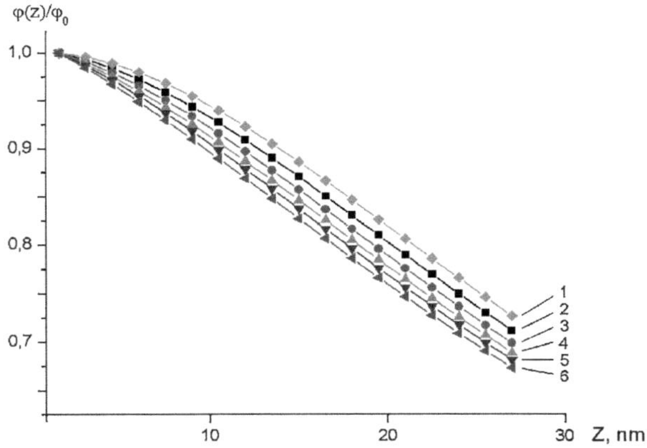


*Figure 3.* The normalized potential distributions along $z$-axis for charged ideally thin ring (curve 1) and for the charged rings with the (same) thickness (along $z$-axis), 3 nm (curve 2), 6 nm (curve 3), 9 nm (curve 4), 12 nm (curve 5), 16 nm (curve 6)

Where $R_0 = (R_1 + R_2)/2$ and $\varepsilon_1\varepsilon_0$ is the dielectric constant of the semiconductor. For corresponding potentials we have

$$\varphi_I(z) = \frac{dr\rho R_0}{2\varepsilon_1\varepsilon_0}\left(Arsh\frac{Z - Z_a}{R_0} - Arsh\frac{Z_b - Z}{R_0}\right) \tag{8a}$$

and

$$\varphi_{II}(z) = \frac{dr\rho R_0}{2\varepsilon_1\varepsilon_0}\left(Arsh\frac{Z_a - Z}{R_0} - Arsh\frac{Z_b - Z}{R_0}\right) \tag{8b}$$

In Fig. 3 the normalized potential distributions along $Z$ axis for the charged rings with limited different thickness calculated using (8) are plotted for different thicknesses of the ring and for the circle (for which the thickness is zero). It is clear from this plots that for small thicknesses our results becomes closer to the potential of the circle which implies the high accuracy of (8).

## 3. Numerical Results and Discussion

Using (8) we have numerically computed the potential distribution on the cylinder axis for different various thickness of the charged rings and charge densities. In Figs. 4 and 5 the results of such calculations are plotted. It is clear from these figures that for considered charge distributions the one-dimensional potential well consisting of a system of quantum wires localized under the charged surface is formed.

The potential distribution on volume and on a surface of the semiconductor near oxide-semiconductor interface is numerically calculated using the obtained potential distributions as the boundary conditions for (4) (Fig. 6).

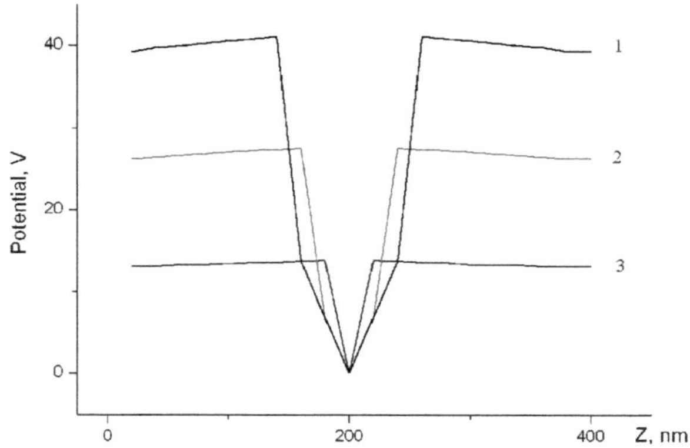


*Figure 4.* The potential distribution on the cylinder axis for different thicknesses of the charged ring: 120 nm (curve 1), 80 nm (curve 2), 40 nm (curve 3)

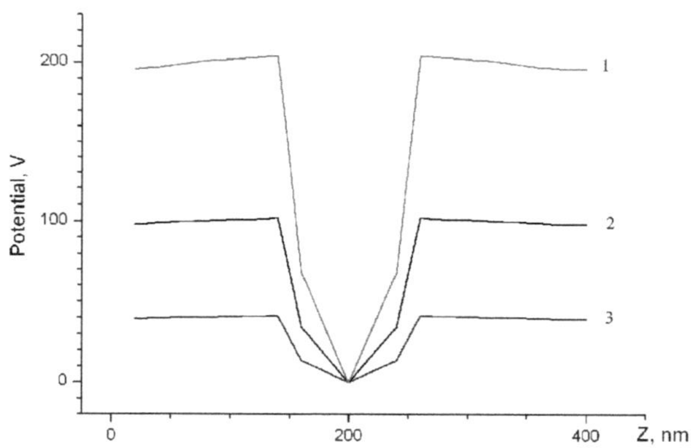


*Figure 5.* The potential distribution on an axis of the cylinder for different charge densities (the thickness of the ring is 120 nm): $5 \times 10^7 Q/m^3$ (curve 1), $2.5 \times 10^7 Q/m^3$ (curve 2), $10^7 \times Q/m^3$ (curve 3)

As is seen from this plot nanowires are formed near the Si-$SiO_2$ interface, on distance approximately 0.05 $R_1$ from the interface. Nanowires are located on a plane which is parallel to the basis of the cylinder on circles with the centre on an axis. Width and depth of the potential of quantum wires depend on the width of charged ring (Fig. 7). Increasing of the distance from the oxide-semiconductor interface leads to disappearing if quantum wires.

Also, we have considered the case of two parallel (with the same sizes) two parallel charged rings located along the axis of cylinder. In this case the number of nanowires in the well increases near the surface of semiconductor. The density of quantum wires is increased by increasing of the distance between the rings, while their width is decreased.

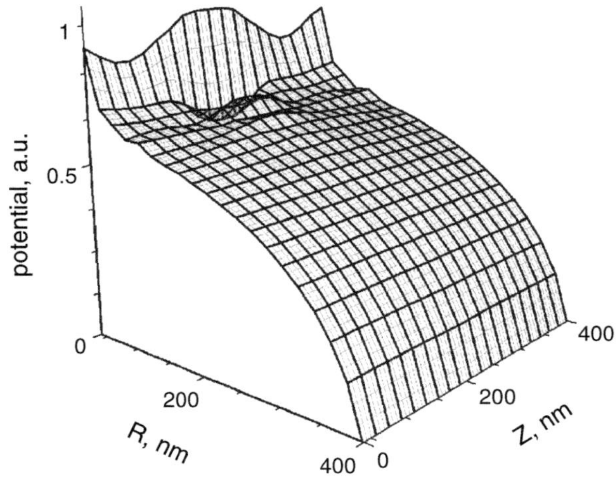


*Figure 6.* The two-dimensional distribution of the potential along the radius $R$ and axis of the cylindrical structure for surface charge density $9.5 \times 10^{12}$ s m$^{-2}$

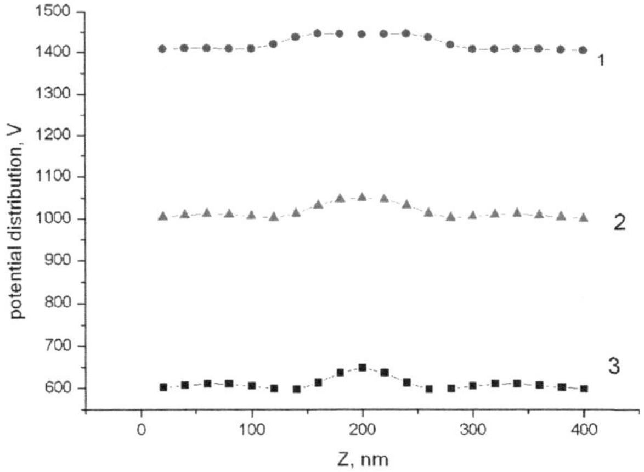


*Figure 7.* The potential distribution along $z$-axis in semiconductor at the distance 20 nm from the oxide-semiconductor interface for the thicknesses of the ring: 120 nm (curve 1), 80 nm (curve 2), 40 nm (curve 3)

## 4. Conclusion

Thus in this work we have studied quantum wire formation near the semiconductor surface in oxide-semiconductor structures. It is shown that potential wells consisting of the system of quantum wires can appear in such structures for certain charge distributions in the oxide layer.

Formation of the quantum wells is possible by creating of the nanosized charged areas. Such wells can be reconstructed by changing the position, size and the charge density of the structure. The depth of the well is basically defined by the charge density, while its width depends on width of the charged area. In the case of two charged parallel rings it depends also on the distance between the rings.

## References


Alferov Zh.I.: 1998. *Semiconductors*, **32**, 1

Binning G., Rohrer H., Gerher Ch. and Weide E.: 1982. *Phys. Rev. Lett.*, **49**, 57

Broers A.N.: 1984. *Materials for Microlithography.* Washington, DC, American Chemical Society, p. 11

Deng X. and Krishnamurthy M.: 1998. *Appl. Phys. Lett.*, **81**, 1473

Di Maria D.I. and Stasiak J.W.: 1989. *J. Appl. Phys.*, **65**, 2342

Emeleus C.J., Milton B., Long A.R. et. al: 1998. *Appl. Phys. Lett.*, **73**, 1412

Goldman E.I., Gulyaev Yu.V., Jdan A.G. and Chucheva G.V.: 2001. *Russian Microelectron.*, **30**, 312

Guntherodt W.J. and Wiesendanger R. eds. 1991. *Scanning Tunnel Microscopy.* Berlin, Springer

Huges M.H.: 1971. *Comput. Phys. Commun.*, **2**, 157

Kawamato E., Kimura K. and Nakazato J.: 1999. *Hitachi Rev.*, **48**, 334

Kazor A., Gwilliam R.M., Jeynes C. et al: 1994. *Ion Implantation Technology – 94*, Amsterdam, Elsevier, p. 994

Ledentsov N.N., Ustinov V.M., Shchukin V.A., Kop'ev P.S., Alferov Zh.I. and Bimberg D.: 1998 *Semiconductors*, **32**, 343

Nicolian E.H. and Brews J.R.: 1982. *MOS (Metal Oxide Semiconductor) Physics and Technology*, New York, Wiley, p. 906

Sah S.T.: 1982. *Sol. St. Electr.*, **33**, 147

Shi Y., Saito K., Ishikuro H. and Hiramoto T.: 1999. *Jpn. J. Appl. Phys.*, **38**, 2453

Thompson L.F.: 1994. *Introduction in Microlithography.* Washington, DC, ASC, p. 85

Verwey J.F., Amerasekera E.A. and Bisschop J.: 1990. *Rep. Prog. Phys.*, **53**, 1297

# NONLINEAR RESPONSES IN HARD DISK SYSTEMS

T. Miyaguchi
*Department of Applied Physics, Osaka City University, Japan;*
*tomo@a-phys.eng.osaka-cu.ac.jp*

**Abstract.** Periodically driven shear flow of hard disk systems is numerically studied using the fluctuation theorem (FT). Relations between the cumulants of integrated momentum flux, which can be derived from FT, are used in the numerical analysis. It is found that the system exhibits nonlinear responses (NRs). Moreover, non-dissipative components of the momentum flux as well as the dissipative component (entropy production) are found to obey a relation between NRs and nonequilibrium fluctuations even for far from equilibrium states.

**Key words:** Hard disks; Shear flow; Dissipation

Understanding macroscopic responses to perturbations, such as temperature gradients and electric fields, in terms of microscopic dynamics is one of the most important tasks of statistical mechanics. For near equilibrium systems (linear response regimes), the fluctuation-dissipation theorem (FDT) tells us that linear responses are related to correlation functions of equilibrium states (Kubo et al. 1991). By contrast, for systems far from equilibrium (NR regimes), FT has been considered to be important, because it is a generalization of the linear response theory (LRT). In fact, FDT (Gallavotti 1996; Hayashi and Sasa 2006) and a NR formula (Andrieux and Gaspard 2007) have been explicitly derived from FT.

The steady state FT (SSFT) was first observed for a thermostatted steady shear flow (Evans et al. 1993), and then derived analytically for Anosov systems (Gallavotti and Cohen 1995a, b). In addition to the results for these deterministic systems, SSFT was proved for Markovian stochastic models, such as Langevan systems and jump processes (Kurchan 1998; Lebowitz and Spohn 1999; Maes 1999). A generalization to time-dependent perturbations has been established in Crooks (1999), in which a connection to the Jarzynski equality (Jarzynski 1997) has also been clarified. Furthermore, FT has been verified numerically for various systems (Bonetto et al. 1997, 1998; Bonetto and Lebowitz 2001; Lepri et al. 1998; Sano 2000; Ayton et al. 2001), and there are also some experimental studies (Goldburg et al. 2001; Carberry et al. 2004).

G. Casati and D. Matrasulov (eds.), *Complex Phenomena in Nanoscale Systems*,
NATO Science for Peace and Security Series B: Physics and Biophysics,
© Springer Science+Business Media B.V. 2009

In this paper NRs for the shear flow of hard disk systems are investigated numerically with emphasis on their relation to nonequilibrium fluctuations (correlation functions). For our settings of the parameters, the kinetic theory is not applicable, because the system size is too small and therefore the boundary effects are not negligible. The main purpose of the present paper is as follows. Many of the numerical studies concerning FT so far devoted to steady nonequilibrium systems (Evans et al. 1993; Ayton et al. 2001; Bonetto et al. 1997, 1998; Bonetto and Lebowitz 2001; Lepri et al. 1998; Sano 2000) (see, Mittag and Evans 2003; Crooks 1999 for exceptions). In the present paper, however, a system with time periodic perturbations is investigated. Moreover, we study non-dissipative Fourier components of the flux as well as the dissipative component (entropy production). It should be noted that FT has information only of the dissipative component of the flux.

Now, the statement of FT is briefly explained. Let $\sigma_t$ be the entropy production rate of the system at time $t$. We also define the integrated entropy production $q_\tau$ as $q_\tau \equiv \int_{-\tau/2}^{\tau/2} \sigma_s ds$, where $\tau$ is the observation time. We assume that the perturbation, $\gamma(t)$, is symmetric, $\gamma(-t) = \gamma(t)$. Then, FT states that $\pi_\tau(q)/\pi_\tau(-q) \approx e^q$, where $\pi_\tau(q)$ is the probability density of $q_\tau$ (Crooks 1999). For non-symmetric perturbations, we should take into account time-reversed paths (Crooks 1999). From FT, we can derive the following relations between the cumulants ($k = 1, 3, ...$):

$$\langle q_\tau^k \rangle_c = \langle q_\tau^{k+1} \rangle_c/2 - \langle q_\tau^{k+2} \rangle_c/12 + \langle q_\tau^{k+4} \rangle_c/720 + ..., \tag{1}$$

where $\langle \cdot \rangle_c$ is the cumulant. This linear equation system is formally equivalent to FT. From the cumulant (1), we have the following equation:

$$\begin{aligned} \langle j(t) \rangle &= \int_0^\infty ds \Phi^{\gamma,1}(t,s)\gamma(t-s) \\ &+ \int_0^\infty ds_1 \int_{s_1}^\infty ds_2 \Phi^{\gamma,2}(t,s_1,s_2)\gamma(t-s_1)\gamma(t-s_2) + f^\gamma(t), \end{aligned} \tag{2}$$

where $j(t)$ is the flux associated with the perturbation $\gamma(t)$, that is, $q_\tau = \int_{-\tau/2}^{\tau/2} dt \gamma(t) j(t)$, and the functions $\Phi^{\gamma,1}(t,s)$ and $\Phi^{\gamma,2}(t,s_1,s_2)$ are defined as

$$\Phi^{\gamma,1}(t,s) = \langle j(t) j(t-s) \rangle_c \tag{3}$$

$$\Phi^{\gamma,2}(t,s_1,s_2) = \langle j(t) j(t-s_1) j(t-s_2) \rangle_c/2. \tag{4}$$

The function $f^\gamma(t)$ is a non-dissipative part of the flux and satisfies $\int_{-T_p/2}^{T_p/2} dt\, f^\gamma(t)\gamma(t) = 0$, where $T_p$ is the period of the periodic perturbation $\gamma(t)$. Without further assumptions, we can not determine the function $f^\gamma(t)$ explicitly.

Therefore, the above relation has been numerically checked for the time periodic shear flow. Note that the functions $\Phi^{\gamma,1}(t,s)$ and $\Phi^{\gamma,2}(t,s_1,s_2)$ depend on the perturbation $\gamma(t)$, and periodic with respect to $t$. In the derivation of (2), the fifth and higher order terms were neglected, and we have assumed that the correlation functions $\langle j(t)j(t-s)\rangle_c$ and $\langle j(t)j(t-s_1)j(t-s_2)\rangle_c$ decrease as $s\to\infty$ and $s_1, s_2\to\infty$, respectively.

Next, let us consider the following Fourier representations: $\gamma(t) \equiv \sum_{n=-\infty}^{\infty}\gamma_n e^{i\omega nt}$, $j(t) \equiv \sum_{n=-\infty}^{\infty} j_n e^{i\omega nt}$, and $f^{\gamma}(t) \equiv \sum_{n=-\infty}^{\infty} f_n e^{i\omega nt}$, where $\omega = 2\pi/T_p$. Using these representations for (2), we have

$$\langle j_k\rangle = \sum_{k'=-\infty}^{\infty}\gamma_{-k'}\hat{\Phi}^{\gamma,1}_{k',k+k'} + \sum_{k',k''=-\infty}^{\infty}\gamma_{-k'}\gamma_{-k''}\hat{\Phi}^{\gamma,2}_{k',k'',k+k'+k''} + f_k, \qquad (5)$$

where $\Phi^{\gamma,1}_{k',k}$ and $\Phi^{\gamma,2}_{k',k'',k}$ are defined using the Fourier–Laplace transforms of the correlation functions $\Phi^{\gamma,1}(t,s)$ and $\Phi^{\gamma,2}(t,s_1,s_2)$.

In what follows, we consider the case that the only single mode ($\gamma_{\pm1}$) is applied as the perturbation. In this case, the above equation is rewritten as

$$\mathrm{Re}\langle j_k\rangle = \gamma_1\sum_{k'=\pm1}\mathrm{Re}\hat{\Phi}^{\gamma_1,1}_{k',k+k'} + \gamma_1^2\sum_{k',k''=\pm1}\mathrm{Re}\hat{\Phi}^{\gamma_1,2}_{k',k'',k+k'+k''} + \mathrm{Re}f_k, \qquad (6)$$

$$\mathrm{Im}\langle j_k\rangle = \gamma_1\sum_{k'=\pm1}\mathrm{Im}\hat{\Phi}^{\gamma_1,1}_{k',k+k'} + \gamma_1^2\sum_{k',k''=\pm1}\mathrm{Im}\hat{\Phi}^{\gamma_1,2}_{k',k'',k+k'+k''} + \mathrm{Im}f_k, \qquad (7)$$

where $\gamma_1$ and $\gamma_{-1}$ are real ($\gamma_1=\gamma_{-1}$), i.e. $\gamma(t)=\gamma_1\cos(\omega t)$. Note that (6) with $k=1$ is equivalent to the cumulant equation (1) with $k=1$, that is, the mode $\mathrm{Re}\langle j_1\rangle$ is just the total entropy production. On the other hand, the other modes do not contribute to the entropy production, because they are orthogonal to the perturbation $\gamma(t)$.

Now, let us consider the hard disk system in a two-dimensional square box $[-l,l]^2$. The disks obey the Hamiltonian dynamics inside the box with hard core interactions. The left and right walls ($x=\pm l$) are identified by the periodic boundary condition, while stochastic boundary conditions are employed at the top and bottom walls ($y=\pm l$). Namely, if a hard disk collides with the top or bottom wall, it is reflected with a random velocity according to the Maxwell-Boltzmann distribution at the wall (Lebowitz and Spohn 1978; Goldstein et al. 1985; Chernov and Lebowitz 1997; Bonetto and Lebowitz 2001), $f(v_x,v_y) = (2\pi T^3)^{-1/2}|v_y|\exp(-\{(v_x-v_0(t))^2+v_y^2\}/2T)$, where $T$ is the temperature of the heat reservoirs. The time dependent function $v_0(t)$ represents the effect of shear stress induced by the moving walls; we consider the situation that the two walls always move in the opposite directions to one another with the same velocity.

In simulations, system parameters are set as follows: the hard disks are assumed to be identical, and the diameter $r$ and the mass $m$ of each disk are fixed as $r = m = 1$. The volume fraction $\rho$, the total number of particles $N$, and the temperature of the walls $T$ are fixed as $\rho = 0.3$, $N = 20$ and $T = 1.0$ in all the following numerical simulations. The amplitude of the function $v_0(t)$ is changed as a perturbation parameter; the perturbation $\gamma(t)$ for the time periodic shear flow is defined as $\gamma(t) = 2v_0(t)/T$. The associated current $j(t)$ is given by the momentum flux $j_m(t)$. More precisely, the momentum flux is defined as $j_m(t) = (j_T(t) + j_B(t))/2$, where $j_T(t)$ $(j_B(t))$ is the momentum transfer to (from) the top (bottom) wall from (to) the system. In numerical simulations, the frequency $\omega$ was fixed as $\omega = 0.05$. Integrations with respect to $s$, $s_1$ and $s_2$ were performed over the range $[0, T_p/10]$. It was confirmed that the correlation functions (3) and (4) sufficiently decay in this range.

In Fig. 1a, the real part of the first Fourier mode of the momentum flux $\mathrm{Re}\langle j_1\rangle$ is shown as a function of the perturbation intensity $\gamma_1$ (solid curve). The dashed line is the prediction of LRT, and a deviation from LRT, i.e. NRs, can be observed. This is clearer in Fig. 1b, in which the deviation from LRT is plotted. This deviation is mainly caused by the third order NR, because there is no second order response due to a symmetry of the system. In the same figure, the first term of RHS of (6) is shown by circles. These values are consistent with a solid curve [$\mathrm{Re}\langle j_1\rangle$] for small values of $\gamma_1$, but deviates for larger values. The sum of the first and second terms of RHS of (6) is also shown in Fig. 1a by squares, which are consistent with $\mathrm{Re}\langle j_1\rangle$ (solid curve) in a wider range. The imaginary part $\mathrm{Im}\langle j_1\rangle$ is also shown in Fig. 1c (solid curve). In this case, LRT (dashed line) is consistent with the solid curve, and NR does not seem to exist. However, the first and second terms of RHS of (7) with $k = 1$ have the third and higher order dependence on $\gamma_1$, because, as shown in Fig. 1c by circles, the first term deviates from the solid curve. But the sum of the first and second terms (squares) coincides well with this curve, and thus $\mathrm{Im} f_1 \equiv 0$.

Similarly, the real and imaginary part of the second Fourier component $\langle j_2\rangle$ is displayed (solid curves) in Fig. 1d, e, respectively. In these figures, the values are almost vanishing. The linear response theory predicts that there are no higher harmonics of the linear order, but NR can not be also observed. Each term of the RHSs of (6) and (7) with $k = 2$ also vanishes. Thus, (6) and (7) are also valid in this case, in particular $f_2 \equiv 0$. Finally, the real part of the third Fourier component $\mathrm{Re}\langle j_3\rangle$ is shown in Fig. 1f (solid curve). There are no first and second order responses as explained above, and the third order NR appears. The first term of RHS of (6) with $k = 3$ (circles) deviates from the solid curve even for small values of $\gamma_1$. This is because the first term is only the second order approximation. On the other hand, the sum of the

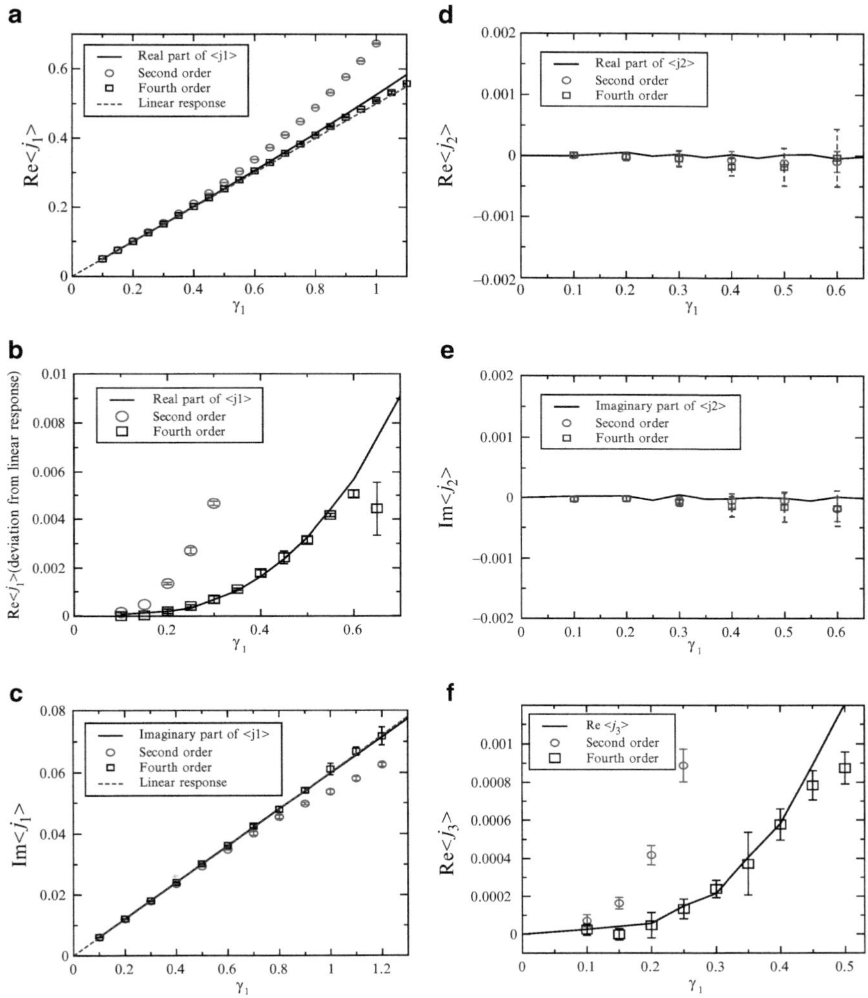


*Figure 1.* (**a**) The real part of the first Fourier coefficient $\mathrm{Re}\langle j_1 \rangle$ vs perturbation intensity $\gamma_1$ (solid curve). The dashed line is the prediction of LRT. The symbols represent the values of RHS of (6): the circles are the values of the first term (second order approximation), and the squares the sum of the first and second terms (fourth order approximation). (**b**) The same as (**a**), but deviations from LRT are plotted. (**c**) The same as (**a**), but for the imaginary part $\mathrm{Im}\langle j_1 \rangle$. (**d**) $\mathrm{Re}\langle j_2 \rangle$ (solid curve). The meanings of the symbols are the same as those in (**a**). (**e**) The same as (**d**), but for the imaginary part $\mathrm{Im}\langle j_2 \rangle$. (**f**) $\mathrm{Re}\langle j_3 \rangle$ (solid curve). The meanings of the symbols are the same as those in (**a**)

first and second terms (squares), which is the fourth order approximation, is consistent with the solid curve in more wider region, in which the relation $\mathrm{Re} f_3 \equiv 0$ holds.

In conclusion, nonequilibrium properties of the hard disk systems driven by the time periodic shear was investigated in this paper. In particular, properties of NRs were studied for the first three Fourier modes, and it was found that NRs exist in the first and third modes. Moreover, (2) was numerically checked and it is found that $f^{\gamma}(t) \equiv 0$. But there is no theoretical explanation of this relation – exception is the linear response regime, in which this relation is obviously valid from LRT. It is also found that the second Fourier component vanishes as shown in Fig. 1d, e, but this is easy to understand from a symmetry of the system.

## References

Andrieux D. and Gaspard P.: 2007, A fluctuation theorem for currents and non-linear response coefficients. *J. Stat. Mech.*, P02006

Ayton G., Evans D.J. and Searles D.J.: 2001, A local fluctuation theorem. *J. Chem. Phys.*, **115**, 2033–2037

Bonetto F. and Lebowitz J.L.: 2001, Thermodynamic entropy production fluctuation in a two-dimensional shear flow model. *Phys. Rev. E*, **64**, 056129

Bonetto F., Gallavotti G. and Garrido P.L.: 1997, Chaotic principle: An experimental test. *Physica D*, **105**, 226–252

Bonetto F., Chernov N.I. and Lebowitz J.L.: 1998, (Global and local) fluctuations of phase space contraction in deterministic stationary nonequilibrium. *Chaos*, **8**, 823–833

Carberry D.M., Reid J.C., Wang G.M., Sevick E.M., Searles D.J. and Evans D.J.: 2004, Fluctuations and irreversibility: An experimental demonstration of a second-law-like theorem using a colloidal particle held in an optical trap. *Phys. Rev. Lett.*, **92**, 140601

Chernov N.I. and Lebowitz J.L.: 1997, Stationary nonequilibrium states in boundary-driven Hamiltonian systems: Shear flow. *J. Stat. Phys.*, **86**, 953–990

Crooks G.E.: 1999, Entropy production fluctuation theorem and the nonequilibrium work relation for free energy differences. *Phys. Rev. E*, **60**, 2721–2726

Evans D.J., Cohen E.G.D. and Morriss G.P.: 1993, Probability of second law violations in shearing steady states. *Phys. Rev. Lett.*, **71**, 2401–2404

Gallavotti G.: 1996, Extension of Onsager's Reciprocity to large fields and the chaotic hypothesis. *Phys. Rev. Lett.*, **77**, 4334–4337

Gallavotti G. and Cohen E.G.D.: 1995a, Dynamical ensembles in nonequilibrium statistical-mechanics. *Phys. Rev. Lett.*, **74**, 2694–2697

Gallavotti G. and Cohen E.G.D.: 1995b, Dynamical ensembles in stationary states. *J. Stat. Phys.*, **80**, 931–970

Goldburg W.I., Goldschmidt Y.Y. and Kellay H.: 2001, Fluctuation and dissipation in liquid-crystal electroconvection. *Phys. Rev. Lett.*, **87**, 245502

Goldstein S., Kipnis C. and Ianiro N.: 1985, Stationary states for a mechanical system with stochastic boundary-conditions. *J. Stat. Phys.*, **41**, 915–939

Hayashi K. and Sasa S.: 2006, Linear response theory in stochastic many-body systems revisited. *Physica A*, **370**, 407–429

Jarzynski C.: 1997, Nonequilibrium equality for free energy differences. *Phys. Rev. Lett.*, **78**, 2690–2693

Kubo R., Toda M. and Hashitume N.: 1991. *Statistical Physics II. Nonequilibrium Statistical Mechanics*, Berlin, Springer

Kurchan J.: 1998, Fluctuation theorem for stochastic dynamics. *J. Phys. A*, **31**, 3719–3729

Lebowitz J.L. and Spohn H.: 1978, Transport properties of the Lorentz gas: Fourier's law. *J. Stat. Phys.*, **19**, 633–654

Lebowitz J.L. and Spohn H.: 1999, A GallavottiCohen-type symmetry in the large deviation functional for stochastic synamics. *J. Stat. Phys.*, **95**, 333–365

Lepri S., Livi R. and Politi A.: 1998, Energy transport in anharmonic lattices close to and far from equilibrium. *Physica D*, **119**, 140–147

Maes C.: 1999, The Fluctuation theorem as a Gibbs property. *J. Stat. Phys.*, **95**, 367–392

Mittag E. and Evans D.J.: 2003, Time-dependent fluctuation theorem. *Phys. Rev. E*, **67**, 026113

Sano M.M.: 2000, Equilibrium and stationary nonequilibrium states in a chain of colliding harmonic oscillators. *Phys. Rev. E*, **61**, 1144–1151

# STOCHASTIC LANDAU–LIFSHITZ–GILBERT EQUATION WITH DELAYED FEEDBACK FIELD: EFFICIENCY FOR MAINTAINING A UPO

H. Tutu
*Department of Applied Analysis and Complex Dynamical Systems, Graduate School of Informatics, Kyoto University; tutu@acs.i.kyoto-u.ac.jp*

**Abstract.** A control method to cause a stability-change between the Ising and the XY-like states in a single-domain magnetic system is studied. It is assumed that the magnetization prefers the Ising state without control due to the uniaxial anisotropy. Then, using a time-delayed feedback method and the AC field in parallel to the anisotropy axis, we consider to stabilize an oscillation of the magnetization across the equatorial plane perpendicular to the anisotropy axis (swinging motion), which possesses the XY-like symmetry. Employing a stochastic Landau–Lifshitz–Gilbert equation, we investigate effects of thermal fluctuation on the controlled state. It is suggested that the method of the delayed feedback control can maintain the controlled state with a low energy consumption.

**Key words:** Stochastic Landau–Lifshitz–Gilbert equation; Delayed-feedback

## 1. Motivation

Delayed feedback control (DFC) (Pyragas 1992) is one of the powerful methods for stabilizing an unstable periodic orbit (UPO). A brief introduction of DFC is as follows. Suppose a system in which its dynamical variables $\boldsymbol{X}(t)$ obey $\dot{\boldsymbol{X}} = \boldsymbol{F}(\boldsymbol{X})$, and which has a $T$-periodic UPO satisfying $\boldsymbol{X}_{\mathrm{u}}(t) = \boldsymbol{X}_{\mathrm{u}}(t-T)$. Then, the UPO can be stabilized with an delayed-feedback input $\boldsymbol{u} = K[\boldsymbol{g}\{\boldsymbol{X}(t-T)\} - \boldsymbol{g}\{\boldsymbol{X}(t)\}]$ under the controlled system $\dot{\boldsymbol{X}} = \boldsymbol{F}(\boldsymbol{X}) + \boldsymbol{u}$, where $\boldsymbol{g}\{\cdot\}$ is a function represent a conversion from $\boldsymbol{X}$ to the feedback signal, the feedback acts to stabilize the UPO with the period $T$. If the UPO $\boldsymbol{X}_{\mathrm{u}}(t)$ is stabilized, $\dot{\boldsymbol{X}}_{\mathrm{u}} = \boldsymbol{F}(\boldsymbol{X}_{\mathrm{u}})$ and $\boldsymbol{u} \to 0$. Thus, it can maintain the UPO with almost vanishing force.

Thus, in comparison with conventional control methods with some external force, a distinctive feature of DFC is its "non-invasiveness." This feature comes from the capability of this method to stabilize an intrinsically unstable periodic state of the target system and maintain it with a low energy

G. Casati and D. Matrasulov (eds.), *Complex Phenomena in Nanoscale Systems*,
NATO Science for Peace and Security Series B: Physics and Biophysics,

© Springer Science+Business Media B.V. 2009

consumption. The DFC has been studied over the last 15 years, and many applications of the DFC are proposed, see for example Schöll and Schuster (2007) and references cited therein.

There are also a lot of interest for application of the DFC in nano-scale or small scale systems, e.g., to suppress chaos in semiconductor systems [see an article in Schöll and Schuster 2007], or to regulate a motion of the micro-cantilever in AFM (Yamasue and Hikihara 2004). Here, one may ask what are differences in controls between nano- or micro-scopic and macroscopic systems. In feedback controlled systems, relevant problems may be (1) a latency period of feedback loop is not necessarily negligible in comparison with the characteristic relaxation time of the system, (2) some influences of thermal noise, (3) heat production. These problems may be severe on control in a usual sense. Then, our question is how these factors are influence on a stabilization of UPO under DFC (Tutu and Horita 2008). In particular, in the presence of thermal noise, does the DFC has a robustness with non-invasiveness? This study is devoted to a theoretical investigation of the DFC to stabilize a certain unstable state in a classical spin system in a thermal environment, and to clarify the energetic efficiency of the DFC.

## 2. Model

As a physical system, we consider a single domain magnetic system (SDMS). The SDMS is actively studied in various purposes, e.g., application to magnetic storage element. Its typical size is 10–100 nm. It has an anisotropy depending on its shape. Below, we assume SDMS has an anisotropy axis in $z$-direction, and the uneasy-plane is the $xy$-plane.

As shown in Fig. 1, let us consider a physical system to stabilize an swinging oscillation of the magnetic orientation across uneasy-plane with the schematic device consisting of SDMS, DFC, and an applied AC-field. This

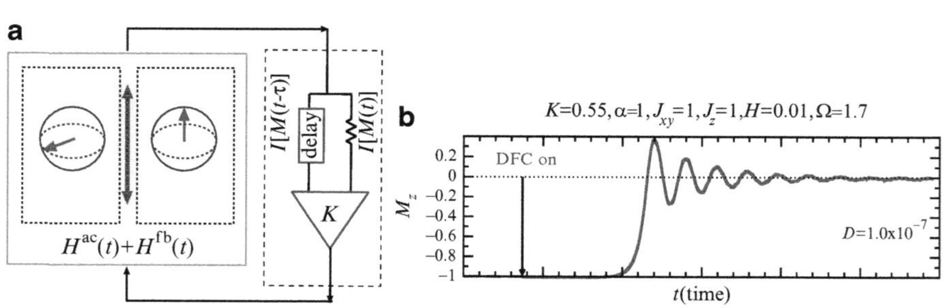


*Figure 1.* (**a**) Schematic representation of the controlled system. By use of the sinusoidal periodic field and the DFC, the XY-like state takes the place of the Ising one. (**b**) The time series of $M_z$ demonstrates the switch from the Ising to the SRO state after the control is applied

can maintain a XY-like spin state with a possibly weak AC-field under appropriate conditions. This system can be regarded as a model controlled system with the DFC in small-scale magnetic systems. Namely this corresponds to the inverted pendulums which often studied in the control of mechanical systems, such a model enables us to examine new control methods or estimate their efficiencies.

For a theoretical argument, let us write down a model with a classical $N$-spin system consisting of a set $\{\boldsymbol{S}_j\}_{j=1}^N$, $\boldsymbol{S}_j = (S_j^x, S_j^y, S_j^z)^T$ ($|\boldsymbol{S}_j| = 1$). The motion of each spin is assumed to obey the Landau–Lifshitz–Gilbert (LLG) equation:

$$\alpha \dot{\boldsymbol{S}}_j = (1+\alpha^2)\boldsymbol{S}_j \times \boldsymbol{H}_j - \alpha^2 \boldsymbol{S}_j \times \dot{\boldsymbol{S}}_j \tag{1}$$

where the first and second terms represent torque and damping. $\alpha$ (= 1) is the damping constant. The equation is normalized in a certain way. The total field reacting on the $j$th spin is $\boldsymbol{H}_j = \boldsymbol{H}_j^{\mathrm{i}} + H^{\mathrm{ac}}\boldsymbol{e}_z + H^{\mathrm{fb}}\boldsymbol{e}_z + \boldsymbol{H}_j^{\mathrm{n}}$, where $\boldsymbol{H}_j^{\mathrm{i}} = -\partial\mathcal{H}/\partial\boldsymbol{S}_j$, i.e., the internal field is generated from the Hamiltonian $\mathcal{H}$:

$$\mathcal{H} = -\frac{1}{2N}\sum_{j,k=1}^{N}\left[J_\perp\left\{S_j^x S_k^x + S_j^y S_k^y\right\} + J_z S_j^z S_k^z\right] - \frac{1}{2}\sum_{j=1}^{N}(S_j^z)^2,$$

$H^{\mathrm{ac}} = h\cos(\Omega t)$: the periodic driving field with the period $T = 2\pi/\Omega$, $H^{\mathrm{fb}} = -K[M_z(t) + M_z(t - T/2)]$: the feedback field with the characteristic strength $K$ using the mean field $M_z(t) = \sum_j S_j^z/N$, and $\boldsymbol{H}_j^{\mathrm{n}}(t)$ represents the thermal noise satisfying $\langle H_j^{\mathrm{n},\beta}(t)\rangle = 0$, $\langle H_j^{\mathrm{n},\beta}(t) H_{j'}^{\mathrm{n},\beta'}(t')\rangle = 2D\delta(t-t')\delta_{j,j'}\delta_{\beta,\beta'}$ ($\beta, \beta' = x, y, z$).

In order to clarify the target state to be stabilized and its mechanism, let us consider a noise-free case with $N = 1$, $J_z = J_\perp = 0$ (a single spin case). This case is described with only the $z$-component of the single spin variable $\boldsymbol{S}$, then, we have

$$\dot{S}_z = (1 - S_z^2)\left[S_z + h\cos(\Omega t) - K\{S_z(t) + S_z(t - T/2)\}\right] \tag{2}$$

There are two characteristic solutions in this system. One is the Ising state, $S_z = \pm 1$, and another is a state which satisfies

$$S_z(t) = -S_z(t - T/2). \tag{3}$$

The latter is the symmetric state with the invariance for the conversion $t \to t + T/2$ and $S_z \to -S_z$, thus we call this "symmetry restored oscillation" (SRO) state. The Ising state is stable in the absence of the feedback, $K = 0$, and it is stabilized while $K$ is in appropriate range, or both state is unstable if $K$ is too large. The latter phase corresponds to quasi-periodic state, which is caused by the emergence of a new oscillatory mode. In the $N$-spin system, the major phases consist of Ising state (all spins up or down), SRO state [the $z$-components of all spins satisfy (3), and their orientations align a certain direction in the $xy$-plane], anti-ferromagnetic state (a half number of spins up and the remains down), and the quasi-periodic state.

Figure 1b shows the switch from the Ising state to the SRO state after the feedback field is applied. This is a time series of $M_z(t) \equiv M_z$, and the Ising state corresponds to $M_z \approx 1$, and the SRO state $M_z \approx 0$.

### 2.1. STABILITY CRITERIA

Let us show the linearly stability conditions (LSC) in which the SRO state is stable. The condition can be obtained from a set of linearized equations for the deviation from the SRO state. Here, the $z$-component of the SRO state satisfies (2) for $K = 0$ with (3), and its $xy$-component takes an orbital orientation in the $xy$-plane. Letting $S_z$ be the $z$-component in the SRO state, we have the LSC as follows $J_\perp > \left\{\overline{(1 - S_z^2)^{-1}}\right\}^{-1}$ and

$$\epsilon < \frac{\cos^{-1}((1 + J_z - J_\perp)/K - 1)}{\tau\sqrt{(1 + J_z - J_\perp)(2K - 1 - J_z + J_\perp)}} \qquad 0 < 1 + J_z - J_\perp < 2K \tag{4}$$

or, $1 + J_z - J_\perp < 0$, where the overline means the temporal average over the period $T$, and $\epsilon \equiv \left(1 - \overline{S_z^2}\right) > 0$ is the Floquet exponent of the SRO state as mentioned above. The derivation of these condition is omitted here because it requires more pages. If the latter inequality is satisfied the SRO state is stable without control. Thus, the relevant conditions are the former two inequalities. The exponent $\epsilon$ is a monotonic function of $h$ and $\Omega$ as $\epsilon = \epsilon(h, \Omega)$. Hence, the inequalities can be rewritten into the equivalent inequalities between $h$ and $K$ with the other parameters fixed, as a consequence, we can obtain the stability diagram as shown in Fig. 2a. The bifurcation diagrams based on Fig. 2a are

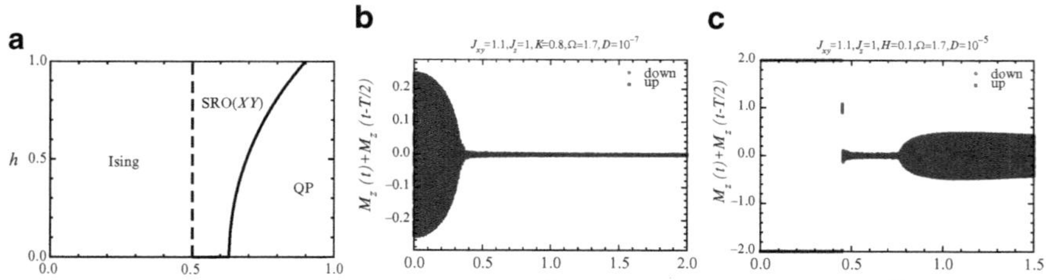


*Figure 2.* (**a**) Stability diagram for the SRO state in the $K$–$h$ plane for $\Omega = 1.7$, $J_\perp = 1.7$, and $J_z = 1$. (**b**) Bifurcation as $h$ increases and decreases. (**c**) Bifurcation as $K$ changes

also shown in Fig. 2b, c: the former exhibits that the SRO (XY) state becomes unstable and gets into the quasi-periodic (QC) state as $h$ decreases, and the latter exhibits that the Ising, the SRO, and the QS states appears in this order as $K$ increases.

## 3. Energetic Efficiency to Maintain the Controlled State

In order to evaluate the energetic efficiency of the DFC, let us investigate the work carried out by applied fields, $H^{\text{tot}}(t) = H^{\text{ac}}(t) + H^{\text{fb}}(t)$, as defined by $W_{\text{x}}(t) = \sum_j \int \mathrm{d}\boldsymbol{S}_j(s) \cdot \boldsymbol{H}^{\text{x}}(s)$, $[\boldsymbol{H}^{\text{x}}(s) = H^{\text{x}}(s)\boldsymbol{e}_z]$, where the suffix x stands for "tot," "ac," and "fb" for the total external field $H^{\text{tot}}(t)$, the AC-field $H^{\text{ac}}(t)$, and the feedback field $H^{\text{fb}}(t)$, respectively. For a relative assessment, let us compare the DFC with another feedback method without delay. Then, for $H^{\text{fb}}$, we examine two methods, the DFC and externally applied anisotropic-field (EAF) as

$$H^{\text{fb}} = \begin{cases} -K[M_z(t) + M_z(t - T/2)] & \text{DFC} \\ -2KM_z(t) & \text{EAF} \end{cases}$$

The EAF corresponds to the case with the vanishing delay period. In comparison, all the other parameters is set to be the same between two methods. Together with the work, we also define power as $P_{\text{x}} = \lim_{t\to\infty} t^{-1} W_{\text{x}}(t)$ with $\text{x} = \text{tot, ac, fb}$.

Figure 3a shows the time series of the work done by the AC-field under the DFC and the EAF methods in the regime in which the SRO state is stable. Since the slope of the work with respect to time corresponds to the power, this exhibits that the power fed by $H_{\text{ac}}$ under the DFC method is smaller than

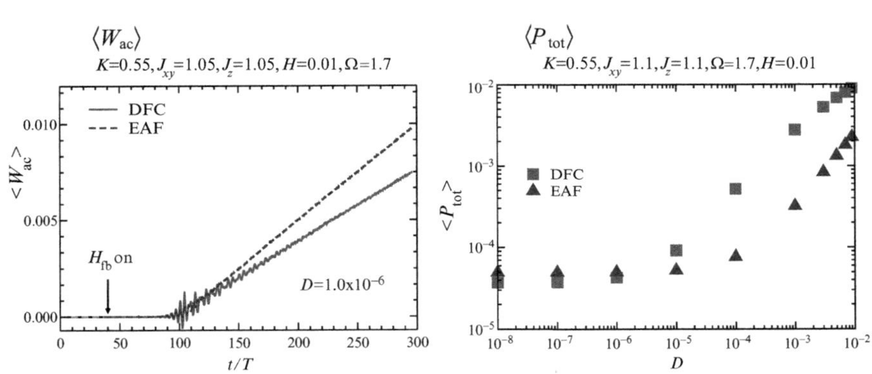


*Figure 3.* (**a**) Time series of the work carried out by the AC-field $W_{\text{ac}}$. The solid and dashed curves correspond to $W_{\text{ac}}$ in the DFC and EAF methods. (**b**) Power fed by the total field $P_{\text{tot}}$ vs. the noise intensity ($D$). The squares and triangles correspond to $P_{\text{tot}}$ in the DFC and EAF methods

that in the EAF. Figure 3b shows the power fed by the total fields under two methods as a function of the strength of noise $D$. This also equivalent to the energy consumption to maintain the SRO state if it is stable. This exhibits that, for a small-$D$ region, the energy consumption $\langle P_{\mathrm{tot}} \rangle$ under the DFC is lower than that under the EAF. In such situation, the fluctuation is neglectable in comparison with the amplitude of the SRO state. Thus, the difference in the energy consumption under two methods results in a property of the realized periodic orbit. The expression of the work is written as

$$\langle W_{\mathrm{tot}} \rangle_t = N \int_0^t (1 - S_z^2)\,(H^{\mathrm{ac}})^2\, \mathrm{d}s + N(\gamma + J_z - J_\perp) \int_0^t (1 - S_z^2) S_z H^{\mathrm{ac}} \mathrm{d}s + O(D).$$

Below, we omit the terms of order $O(D)$, because we concern the small-noise situation. The integrand in the first term is always positive, and the integral linearly increases with $t$. The second term gives the most relevant contribution to the difference in the two methods. Here, the coefficient in the second term is positive, and whether the integral increases or decreases with $t$ crucially depends on the sign of the factor $S_z H^{ac}$ in the integrand. Then, we may extract the essential part of the internal with $\int_0^t S_z H^{\mathrm{ac}} \mathrm{d}s$.

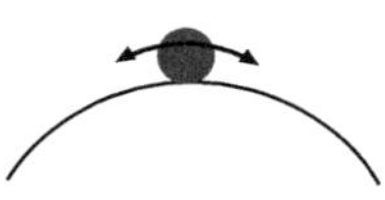

$\dot{S}_z \approx \epsilon S_z + H^{\mathrm{ac}}$

$\lim_{t \to \infty} t^{-1} \int_0^t S_z H^{\mathrm{ac}} \mathrm{d}s < 0$

$\dot{S}_z \approx -|\epsilon| S_z + H^{\mathrm{ac}}$

$\lim_{t \to \infty} t^{-1} \int_0^t S_z H^{\mathrm{ac}} \mathrm{d}s > 0$

As shown in the left figure, the integral decreases in the case that the UPO is stabilized with the DFC. The orbit is approximated as a linear response from $H^{\mathrm{ac}}$. On the other hand, it increases in the case the UPO is stabilized with the EAF, because the EAF makes a potential well structure. The key role of DFC is that it does not change the property of UPO, which is just of non-invasiveness, and this contributes to the decrease of the energy consumption.

## 4. Summary

We considered the DFC for maintaining XY-like state in a globally coupled spin system in the situation where Ising state is stable. We showed that the energy consumption in the DFC is lower than that in the EAF.

## References

Pyragas K.: 1992. *Phys. Lett. A*, **170**, 421
Schöll E. and Schuster H.G.: 2007. *Handbook of Chaos Control*, Wiley, Berlin
Tutu H. and Horita T.: 2008. *Prog. Theor. Phys.*, **120**, 315
Yamasue K. and Hikihara T.: 2004. *Phys. Rev. E*, **69**, 056209

# SPIN DYNAMICS AND QUANTUM TRANSPORT IN QUANTUM SPIN CHAINS UNDER AN OSCILLATING FIELD

K. Kudo*
*Ochadai Academic Production, Ochanomizu University, Tokyo, Japan; kudo.kazue@ocha.ac.jp*

T.S. Monteiro
*Department of Physics and Astronomy, University College London, Gower Street, London WC1E 6BT, UK; t.monteiro@theory.phys.ucl.ac.uk*

**Abstract.** Quantum spin-chains subjected to external oscillating fields exhibit a wide range of dynamical behavior depending on the strength and frequency of the field. We investigate here the particular case of a Heisenberg model with a single spin flip; this maps onto a tight-binding model with an oscillating field, which, in turn, is equivalent to a driven-Harper model with sinusoidal time-driving. The Harper models can be mapped to image classical phase-spaces. We find that peculiar structures called "shearless tori" feature very prominently in the classical phase space of the driven-Harper model. We show that quantum-spin wave packets initially located on a "shearless torus" can travel for a quite long time without broadening. The behavior has potential applications as it may provide a mechanism for non-dispersive transfer of quantum information.

**Key words:** Quantum spin chains; Shearless torus; Quantum information

## 1. Introduction

Quantum dynamics of spin chains are, these days, frequently studied in the context of quantum information (Bose 2003; Amico et al. 2008). Applications of nonlinear dynamics in quantum information have also been attracting considerable attention. For example, it has been realized that a many-body Hamiltonian can be analyzed in terms of the dynamics of one-body image quantum and classical Hamiltonians (Prosen 1998, 1999, 2002; Boness 2006; Kudo and Monteiro 2008).

In this work, we have investigated quantum spin dynamics and quantum transport for a ferromagnetic Heisenberg system, with a single excitation, subjected to an additional magnetic field oscillating in time and position (Kudo and Monteiro 2008). Although we solve for the full quantum spin Hamiltonian, we show that the observed behavior in this case is closely

G. Casati and D. Matrasulov (eds.), *Complex Phenomena in Nanoscale Systems*,
NATO Science for Peace and Security Series B: Physics and Biophysics,

© Springer Science+Business Media B.V. 2009

related to the structure of an underlying classical image phase-space which is a variant of the well-known Harper map. Because of the sinusoidal driving, we term it the driven-Harper map. A remarkable feature of the classical phase-space of the driven Harper map (and which are much less prominent in the better known kicked-Harper map) are classical barriers termed "shearless tori." We find that there is non-dispersive quantum transport for a spin wave packet initially localized on or close to a shearless torus. In contrast, when the initial wave packet is located on a normal torus, it rapidly delocalizes along the length of the torus.

We show that the non-spreading of quantum wave packets moving on shearless tori is due to the harmonic character of the underlying Floquet spectrum. We also calculate the time evolution of entanglement corresponding to the dynamics of the non-dispersive spin wave packets.

## 2. Models

We consider a ferromagnetic Heisenberg spin chain under an oscillating magnetic field. In a spin-$1/2$ and one-spin-flip case, the model corresponds to a tight-binding model with an oscillating field. The Hamiltonian is written as

$$H(t) = \frac{J}{2}\sum_{j=1}^{N}(c_j^\dagger c_{j+1} + c_{j+1}^\dagger c_j) + B_0 \sin\omega t \sum_{j=1}^{N} \cos\left(\frac{2\pi}{N}j\right) c_j^\dagger c_j, \tag{1}$$

where $c_j^\dagger$ and $c_j$ are the creation and annihilation operators of a fermion at the $j$th site, respectively. The amplitude and frequency of the field are denoted by $B_0$ and $\omega$, respectively. Equation (1) maps onto the one-body classical "image" Hamiltonian,

$$H(x, p) = J \cos p + B_0 \sin\omega t \cos[(2\pi/N)x]. \tag{2}$$

We call this model "driven Harper model" (Kudo and Monteiro 2008). The corresponding classical equations of motion are given by

$$\begin{cases} \dot{x} = -J \sin p & (0 < x \le N), \\ \dot{p} = (2\pi/N)B_0 \sin\omega t \sin[(2\pi/N)x] & (-\pi < p \le \pi). \end{cases} \tag{3}$$

We set $J = -1$, $B_0 = 2$ and $N = 100$ in the numerical calculations.

## 3. Quantum Dynamics and Classical Phase Space

We calculate the time evolution of a spin wave packet, using the quantum spin Hamiltonian equation (1), and thus obtain the quantum spin distribution

$$P(j, t) = |\langle j|\psi(t)\rangle|^2, \tag{4}$$

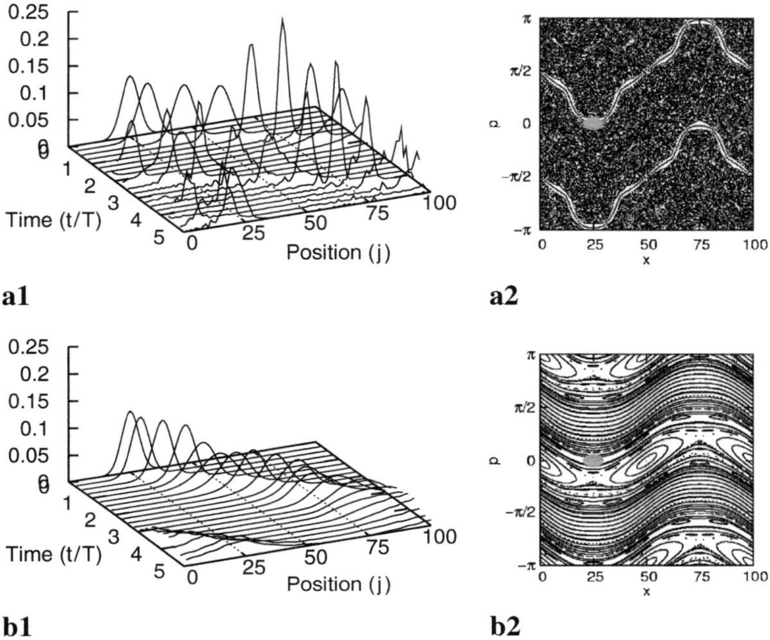


*Figure 1.* Dynamics of quantum wave packets ((**a1**) and (**b1**)) and the corresponding classical surfaces of section ((**a2**) and (**b2**)) (Kudo and Monteiro 2008). The frequency of the field is $\omega = 0.12$ for (**a1**) and (**a2**), and $\omega = 0.2$ for (**b1**) and (**b2**)

where $|j\rangle$ is the state where the $j$th spin is down, and $|\psi(t)\rangle$ is calculated by the Schrödinger equation. Here, the initial state is given as a Gaussian spin wave packet centered at $j = 25$ with width $\Delta_j = 5$. Figure 1a1, b1 show the time evolution of $P(j, t)$ for $\omega = 0.12$ and 0.2, respectively.

Classical surfaces of section are obtained by plotting values of $(x, p)$ at integer multiples of the period. They are shown in Fig. 1b1, b2 for $\omega = 0.12$ and 0.2, respectively. The patched circles in the figures correspond to the initial Gaussian spin wave packets. The regular channels in the chaotic sea in Fig. 1a2 contain shearless tori.

The dynamics of the spin wave packet shows non-dispersive traveling when it is initially localized at the region near a shearless torus (Fig. 1a1). In contrast, the spin distributions show delocalization for the spin wave packet whose initial condition corresponds to normal tori (Fig. 1b1).

The difference of the quantum dynamics of Fig. 1a1, b1 can be also understood from their quasi-energy spectra. Since the system is temporally periodic, the stationary states of the system are represented by the eigenstates of the one-period unitary evolution operator (Floquet states) $\phi_m(x, t)$ and associated eigenphases (quasi-energies) $\varepsilon_m$.

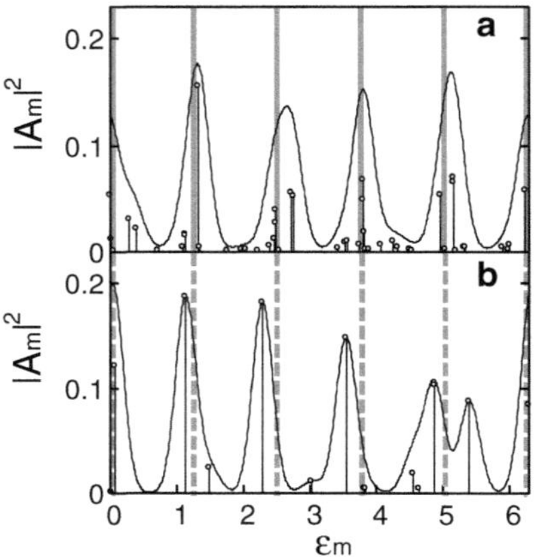


*Figure 2.* Low-resolution quasi-energy spectra for (**a**) $\omega$ = 0.12 and (**b**) $\omega$ = 0.2 (Kudo and Monteiro 2008)

Figure 2 shows the local Floquet spectra for the initial Gaussian wave packets. It is obtained by calculating the overlap between initial wave packet and the Floquet states, $A_m = \langle \phi_m(x, t = 0)|\psi_0\rangle$. The probability $|A_m|^2$ and its low-resolution smoothed spectra are plotted in Fig. 2. Figure 2a, which corresponds to Fig. 1a1, is obtained for the wave packet initially localized near a shearless torus. The low-resolution spectrum has equally-spaced peaks. The interval of the peaks is $2\pi/5$; this corresponds to the fact that the wave packet traveling along the spin chain returns at the initial position after about five cycles. On the contrary, the spectrum of Fig. 2b, which is for the wave packet on normal tori, has no equal spacings.

## 4. Entanglement

The entanglement is also transported when the spin wave packet travels. The concurrence $C_{i,j}$ is a measure of the bipartite entanglement of two sites $i$ and $j$ (Wootters 1998). Figure 3 shows the time evolution of $C_{25,26}$. Figure 3a corresponds to the spin dynamics for the spin wave packet initially located in the region near a shearless torus (Fig. 1a1). It shows periodic sharp peaks corresponding to the non-dispersive transport of the spin wave packet. In contrast, Fig. 3b, which corresponds to the dynamics on normal tori (Fig. 1b1), shows no sharp peaks.

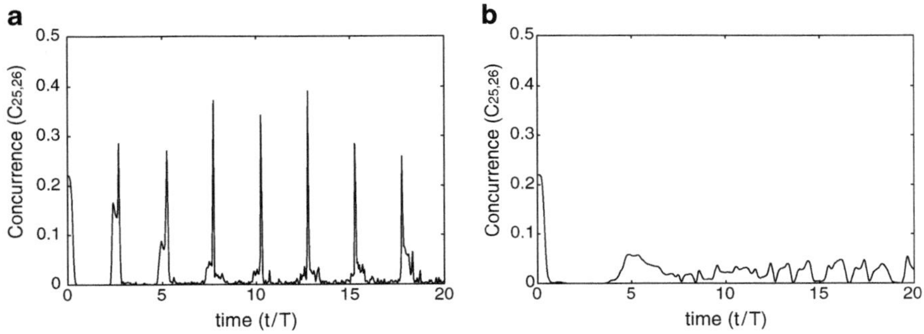


*Figure 3.* Time evolution of the concurrence $C_{25,26}$ for (**a**) $\omega = 0.12$ and (**b**) $\omega = 0.2$ (Kudo and Monteiro 2008)

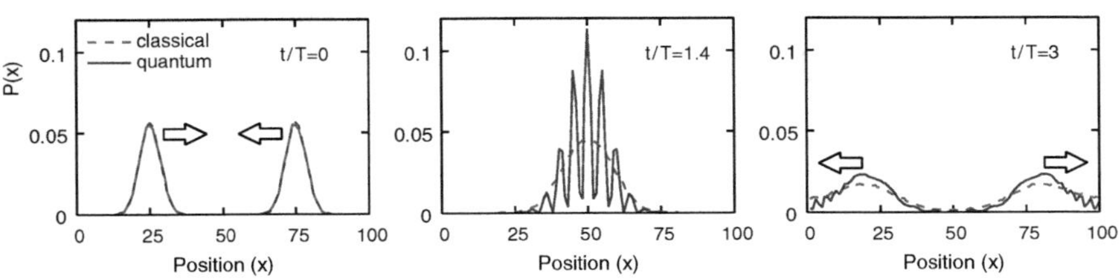


*Figure 4.* Quantum and classical distributions for two spin wave packets

## 5. Quantum and Classical Distributions

So far we have considered the dynamics of one wave packet. In this section, we consider the situation that the initial state is the superposition of two wave packets. Namely, the initial state is given as

$$
\begin{aligned}
|\psi(0)\rangle &= A_{\text{norm}} \sum_{j=1}^{N} [g(j-j_1) + g(j-j_2)]|j\rangle, \qquad (5)\\
g(x) &= \exp[-x^2/2\Delta_j^2],
\end{aligned}
$$

where $j_1 = 25$, $j_2 = 75$ and $A_{\text{norm}}$ is a normalization factor.

Figure 4 shows the time evolution of the quantum and classical distributions. The classical distributions show the classical dynamics of the particles whose distribution of starting points consists of two Gaussian ones. At the initial time, both the quantum and classical distributions coincide. The two wave packets move approaching each other. When they meet around $x = 50$, the quantum distribution shows interference, while the classical distribution looks like just one large wave packet. After the collision, the two wave packet continues to travel. The centers of mass of quantum and classical distributions seem to coincide. The essential difference between quantum and classical dynamics is the appearance of interference.

## 6. Conclusions

The dynamics of a quantum wave packet in a Heisenberg system with a single spin flip can be explained well by the corresponding classical phase space. The spin wave packet initially located in the region near a shearless torus shows approximately non-dispersive behavior. The shearless tori can also provide an advantage in transporting entanglement around the spin chain.

## Acknowledgements

The authors thank K. Nakamura and S. Bose for helpful discussions.

## References

Amico L., Fazio R., Osterloh A. and Vedral V.: 2008. *Rev. Mod. Phys.*, **80**, 517
Boness T., Bose S. and Monteiro T.S.: 2006. *Phys. Rev. Lett.*, **96**, 187201
Bose S.: 2003. *Phys. Rev. Lett.*, **91**, 207901
Kudo K. and Monteiro T.S.: 2008. *Phys. Rev. E*, **77**, 055203
Prosen T.: 1998. *Phys. Rev. Lett.*, **80**, 1808
Prosen T.: 1999. *Phys. Rev. E*, **60**, 1658
Prosen T.: 2002. *Phys. Rev. E*, **65**, 036208
Wootters W.K.: 1998. *Phys. Rev. Lett.*, **80**, 2245

# EXTRACTION OF PARTIAL WAVES AND FLUXES OF THE HYDROGEN ATOM IN A STRONG MAGNETIC FIELD USING THE COMPLEX ROTATION METHOD

S.K. Avazbaev*
*Heat Physics Department of the Uzbek Academy of Sciences, 28 Katartal Street, 100135 Tashkent, Uzbekistan; sanat_avazbaev@yahoo.com*

D. Delande
*Laboratoire Kastler Brossel, Universitŭ Pierre et Marie Curie, 4, place Jussieu 75252 Paris Cedex 05, France*

**Abstract.** The preliminary results to extract partial waves and fluxes of the resonances using the complex rotation method are presented in the example of the hydrogen atom in a magnetic field.

**Key words:** Atomic resonances; Complex rotation method

## 1. Introduction

Apart from the fundamental question of quantum chaos – classical-quantum correspondence, rapid development in semiconductor-microstructure devices, based on quantum dots (Nakamura and Harayama 2004) or tunneling structures (Fromhold et al. 1994) demand a deeper knowledge about the behavior of quantum chaotic systems (Blümel and Reinhardt 1997; Stöckmann 1999; Haake 2001).

Resonances are very common phenomena in atomic physics. For example, resonances are observed in the hydrogen atom in a static electric field (Alvarez et al. 1991), in a static magnetic field (Delande et al. 1991; Friedrich and Wintgen 1989; Hasegawa et al. 1989). And the physical properties of the system are dominated by its resonances. This is true especially for highly excited Rydberg states of atoms, where the density of states is very large. That is why the ability to compute the energies and lifetimes of resonances is an important step in describing a number of physical phenomena.

For determining resonance parameters, direct methods are efficient. These direct methods aim at calculating the complex pole of the Green function, or equivalently, of the $S$-matrix, the real part of which is identified as the

G. Casati and D. Matrasulov (eds.), *Complex Phenomena in Nanoscale Systems*,
NATO Science for Peace and Security Series B: Physics and Biophysics,

© Springer Science+Business Media B.V. 2009

energy of the metastable state and the imaginary part as its width (Isaacson et al. 1978; Ho 1983). The character of direct methods is that they don't require approximations inherent to golden rule expressions and Feshbach type projection operators (Feshbach 1962; Hickmann et al. 1976). A number of authors used $R$-matrix theory to investigate the structure of the scattering matrix (Schneider 1981; Morgan and Burke 1988; McCartney et al. 1990).

One of such direct approaches is the method of complex rotation (Ho 1983). When the complex rotation is applied to the resonance state (a solution of the Schrödinger equation obtained under outgoing boundary conditions), the exponentially diverging wavefunctions become convergent. And bound state methods can be used to compute resonances. In this way the resonances, which are unphysical solutions of the Schrödinger equation, can be mapped onto square integrable eigenfunctions of the complex rotated Hamiltonian (Ho 1983). The exterior complex scaling has been applied to study three-body breakup problem (Rescigno et al. 1999).

The helium atom is the simplest atom for which an analytic solution of the Schrödinger equation is impossible. A major open problem is to understand the consequences of the classically chaotic dynamics (Richter et al. 1993; Tanner et al. 2000) on the quantum properties of the system when the two electrons are simultaneously highly excited (Jiang et al. 2008).

The method of complex rotation was applied to obtain the positive energy spectrum of the hydrogen atom in a magnetic field, the total photoionization cross-section was obtained (Delande et al. 1991) and found a perfect agreement with experimental results (Iu et al. 1991). Currently the method provides only the total decay rate, but is unable to compute partial decay rates corresponding to different open channels. This isn't fundamental limitation, as the complex rotation gives the Green function of the system, which contains all the information on the dynamics (Buchleitner et al. 1994). The practical problem is to extract the partial decay rates from the knowledge of the Green function. In this work we present preliminary results on the calculation of partial fluxes of the hydrogen atom in a magnetic field by the complex rotation method.

## 2. Extraction of Partial Waves and Fluxes

The hydrogen atom in a magnetic field is an ideal system to construct and test new computational techniques because of the availability of highly accurate calculations (Delande et al. 1991; Friedrich and Wintgen 1989). The Hamiltonian of the hydrogen atom in a magnetic field (which is directed along $z$ axis) is (in atomic units)

$$H = \frac{p^2}{2} - \frac{1}{r} + \frac{\gamma}{2}L_z + \frac{\gamma^2}{8}\rho^2, \tag{1}$$

where $\gamma$ is the magnetic field and it is given by $\gamma = \frac{B}{B_c}$, $B_c = 2.35 \times 10^5$ T. $L_z$ is the $z$ component of the angular momentum and $L_z = 0$ case is considered. At very strong magnetic field, diamagnetic field dominates and shrinks the atom in the $\rho$ plane. That is why the rapid motion in the $\rho$ direction can be adiabatically separated from the slow one along $z$ axis. Eventually, an infinite set of Landau thresholds appear each supporting one-dimensional Rydberg series. All these series (except the lowest one) are composed of resonances. The application of the complex rotation $r = re^{i\theta}$, $p = pe^{-i\theta}$ to the Hamiltonian equation (1) gives the rotated one

$$H(\theta) = \frac{p^2}{2}e^{-2i\theta} - \frac{e^{-i\theta}}{r} + \frac{\gamma^2}{8}\rho^2 e^{2i\theta}. \tag{2}$$

The rotated Hamiltonian $H(\theta)$ is a non-Hermitian operator, whose spectrum is complex and depends on the rotation angle $\theta$. The great property of the complex rotation method is that the spectrum of $H(\theta)$ can be related to the resonances of the Hamiltonian $H$ (Ho 1983). The spectrum of $H(\theta)$ has the following properties (Fig. 1): (1) The continua are rotated by the angle $2\theta$ around the Landau thresholds; (2) the discrete spectra of $H(\theta)$ and $H$ coincide below the first Landau threshold; (3) The resonances of $H$ coincide with the complex eigenvalues of $H(\theta)$. The real part (the energy) and the imaginary part (negative of the half width) of the resonance are $\theta$ independent, if the rotation of the continua has uncovered the resonances.

$H(\theta)$ is diagonalized in Sturmian functions basis (Buchleitner et al. 1994; Halley et al. 1993). Because of the non-orthogonality of the Sturmian functions, the Schrödinger equation written in this basis corresponds to a generalized eigenvalue problem. The advantage of the Sturmian basis is that

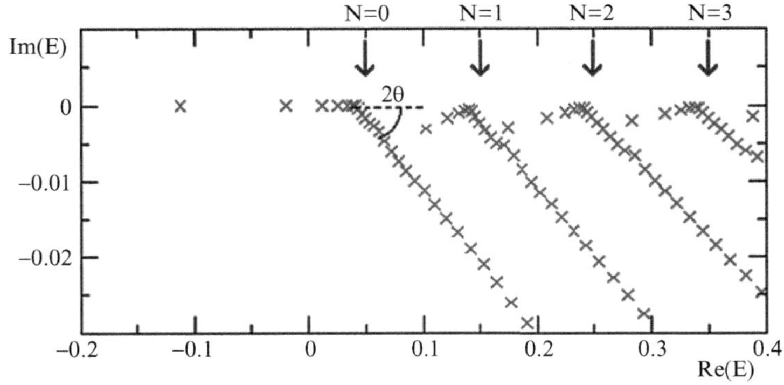


*Figure 1.* Odd parity spectrum of the rotated Hamiltonian $H(\theta)$, $\gamma = 0.1$, $\theta = 0.03$, $\alpha = 0.5$ (the scaling parameter of the Sturmian functions) (Delande et al. 1991). The arrows indicate the Landau thresholds

the strong selection rules of the operators in the Hamiltonian on the quantum numbers give sparse matrices, allowing efficient Lanczos diagonalization algorithm (Delande et al. 1991; Halley et al. 1993).

At sufficiently large distances from the nucleus the wavefunction is purely outgoing and decay properties are contained in the wavefunction. In the case of the hydrogen atom in a magnetic field, the transverse to the field (along $\rho$) behavior of the wavefunction is defined by Landau eigenfunctions

$$\varphi_N(\rho) = L_N^0\left(\frac{\gamma\rho^2}{2}\right)\exp\left(-\frac{\gamma\rho^2}{4}\right), \tag{3}$$

where $L_N^0$ are associated Laguerre polynomials. Projection of the total wavefunction ($\psi(\rho, z)$) on the corresponding Landau eigenfunctions gives partial waves

$$\phi_N(z) = \int_{-\infty}^{+\infty} \psi(\rho, z)\varphi_N(\rho)\rho d\rho. \tag{4}$$

From the knowledge of partial waves we are able to compute partial fluxes

$$F_N(z) = Im\left(\phi_N(z)\frac{d\phi_N(z)}{dz}\right). \tag{5}$$

Let's take a resonance with energy $E = 0.3381 - i0.237\times10^{-3}$ for $\gamma = 0.1$, which has three open channels. The partial waves for this resonance is shown in Fig. 2.

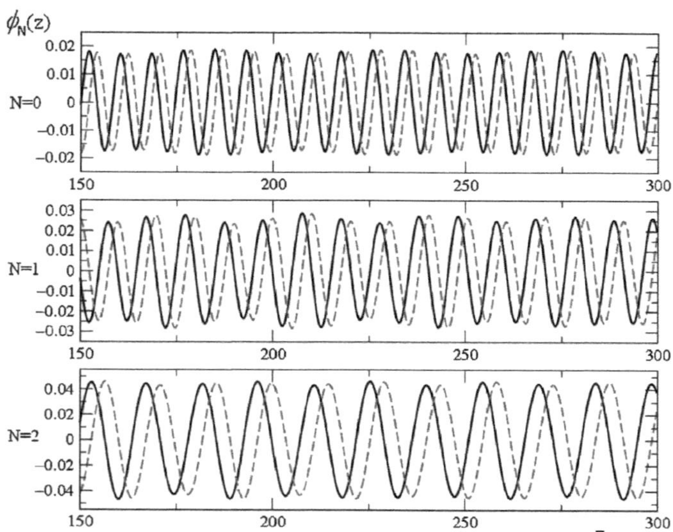


*Figure 2.* Partial waves for the resonance with energy $E = 0.3381 - i0.237 \times 10^{-3}$, solid and dashed lines denote the real and imaginary parts of the wavefunction, respectively

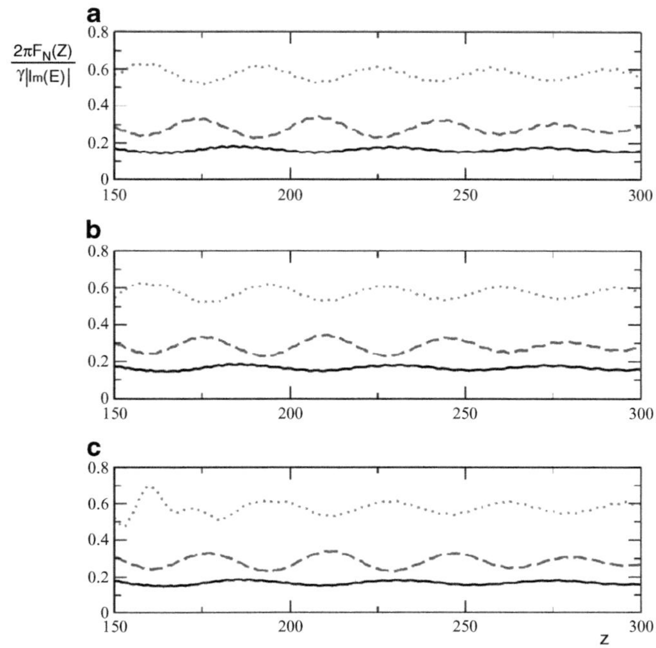


*Figure 3.* Scaled fluxes for three consecutive resonances $E = 0.3381 - i0.237 \times 10^{-3}$ (**a**), $E = 0.3410 - i0.159 \times 10^{-3}$ (**b**) and $E = 0.3430 - i0.112 \times 10^{-3}$ (**c**). Solid, dashed and dot lines denote partial fluxes corresponding to $N = 0$, $N = 1$, $N = 2$ channels respectively

To estimate the accuracy of the calculations we apply the above described technique to three consecutive resonances with energies $E = 0.3381 - i0.237 \times 10^{-3}$, $E = 0.3410 - i0.159 \times 10^{-3}$ and $E = 0.3430 - i0.112 \times 10^{-3}$. In Fig. 3 the normalized partial fluxes (i.e. the resonance and $\gamma$ dependencies are removed) are given for these resonances. The sum of the normalized partial fluxes shows that the inaccuracy of the calculations is a few percent.

## 3. Conclusion and Perspectives

We have described the direct technique to extract partial waves and to compute partial fluxes from the complex rotation method. The accuracy of partial fluxes signals about the efficiency of the complex rotation for determining resonance parameters. Currently the studies to compute partial cross-sections for this system is in progress. Future application of the above described technique to compute partial cross-sections for the doubly excited states of helium (for which there are some experimental data and theoretical calculations Jiang 2006; Czasch et al. 2005) is expected.

## Acknowledgements

The work of S.K. Avazbaev is supported in part by Marie Curie Training Fellowship of the European Union (MEST-CT-2004-503847), INTAS Young Scientist Collaborative Fellowship with Uzbekistan (Nr. 06-1000023-6422) and by a grant of the Uzbek Academy of Sciences (ФA-Ф2-Ф082). S.K. Avazbaev acknowledges the Laboratoire Kastler-Brossel for hospitality during his several visits.

## References

Alvarez G., Damburg R.J. and Silvestone H.J.: 1991. *Phys. Rev. A*, **44**, 3060
Blümel R. and Reinhardt W.P.: 1997. *Chaos in Atomic Physics*, Cambridge University Press, Cambridge, MA
Buchleitner A., Grémaud B. and Delande D.: 1994. *J. Phys. B: At. Mol. Opt. Phys.*, **27**, 2663
Czasch A. et al: 2005. *Phys. Rev. Lett.*, **95**, 243003
Delande D., Bommier A. and Gay J.C.: 1991. *Phys. Rev. Lett.*, **66**, 141
Feshbach H.: 1962. *Ann. Phys.*, **5**, 287
Friedrich H. and Wintgen D.: 1989. *Phys. Rep.*, **183**, 37
Fromhold T.M. et al: 1994. *Phys. Rev. Lett.*, **72**, 2604
Haake F.: 2001. *Quantum Signatures of Chaos*, Springer, Berlin
Halley M.H., Delande D. and Taylor K.T.: 1993. *J. Phys. B: At. Mol. Opt. Phys.*, **26**, 1775
Hasegawa H., Robnik M. and Wunner G.: 1989. *Prog. Theor. Phys.*, **98**, 198
Hickmann A.P., Isaacson A.D. and Miller W.H.: 1976. *Chem. Phys. Lett.*, **37**, 63
Ho Y.K.: 1983. *Phys. Rep.*, **99**, 1
Isaacson A.D., McCurdy C.W. and Miller W.H.: 1978. *Chem. Phys.*, **34**, 311
Iu C.H., Welch G.R., Kash M.M., Kleppner D., Delande D. and Gay J.C.: 1991. *Phys. Rev. Lett.*, **66**, 145
Jiang Y.H.: 2006. PhD Dissertation, Freie Universität, Berlin, http://www.diss.fu-berlin.de/2006/269
Jiang Y.H., Püttner R., Delande D., Martins M. and Kaindl G.: 2008. *Phys. Rev. A*, **78**, 021401(R)
McCartney M., Burke P.G., Morgan L.A. and Gillan C.J.: 1990. *J. Phys. B: At. Mol. Opt. Phys.*, **23**, L415
Morgan L.A. and Burke P.G.: 1988. *J. Phys. B: At. Mol. Opt. Phys.*, **21**, 2091
Nakamura K. and Harayama T.: 2004. *Quantum Chaos and Quantum Dots*, Oxford University Press, Oxford
Rescigno T.N., Baertschy M., Isaacs W.A. and McCurdy C.W.: 1999. *Science*, **286**, 2474
Richter K., Tanner G. and Wintgen D.: 1993. *Phys. Rev. A*, **48**, 4182
Schneider B.I.: 1981. *Phys. Rev. A*, **24**, 1
Stöckmann H.-J.: 1999. *Quantum Chaos: An Introduction*, Cambridge University Press, Cambridge
Tanner G., Richter K. and Rost J.-M.: 2000. *Rev. Mod. Phys.*, **72**, 497

# MARKOVIAN LIMIT OF A SPATIO-TEMPORAL CORRELATED OPEN SYSTEMS

T. Monnai
*Department of Applied Physics, Waseda University, Tokyo, Japan; monnai@aoni.waseda.jp*

**Abstract.** Large fluctuation of Brownian particles is affected by the finiteness of the correlation length of the background noise field. Indeed a Fokker–Planck equation is derived in a Markovian limit of a spatio-temporal short correlated noise. Corresponding kinetic quantities are renormalized due to the spatio-temporal memory. We also investigate the case of open system by connecting a thermostat to the system.

**Key words:** Spatio-temporal colored noise

## 1. Introduction

Largely fluctuating Brownian particle can experience locality or decorrelation of the background noise (Golubovic et al. 1991; Deutsch 1985; Arvedson et al. 2006; Bezuglyy et al. 2006; Monnai et al. 2008; Rosenbluth 1992; Monnai 2008). Even in equilibrium, the background noise field fluctuates both spatially and temporally. For a Markovian limit of the diffusion caused by a spatio-temporal short correlated noise and a systematic force, a well-defined Fokker–Planck description has been derived (Monnai et al. 2008; Rosenbluth 1992; Monnai 2008). Indeed, so-obtained Fokker–Planck equation shows subtle correction to that of the thermal diffusion induced by a temporally stochastic noise.

In this paper, a thermostat is connected to the system, and the equilibrium condition is explored. Especially, we shall derive a Fokker–Planck equation for the total system, and show that a nontrivial potential condition uniquely determines the temperature which guarantees the equilibrium canonical distribution of the total system.

This paper is organized as follows. In Sect. 2, we briefly review the derivation of the Fokker–Planck equation in a Markovian limit for a thermally isolated system. In Sect. 3, similar short correlation limit of a spatio-temporal correlated open system is discussed as well.

G. Casati and D. Matrasulov (eds.), *Complex Phenomena in Nanoscale Systems*,
NATO Science for Peace and Security Series B: Physics and Biophysics,

© Springer Science+Business Media B.V. 2009

## 2. Fokker–Planck Equation of a Spatio-Temporal Short Correlated Noise

In this section, we review the Markovian limit of the spatio-temporal correlated noise (Monnai et al. 2008; Rosenbluth 1992).

### 2.1. MODEL

Let us consider a one-dimensional overdamped Langevin equation

$$\eta \dot{x}(t) = -U'(x(t)) + f(x(t), t), \tag{1}$$

where $\eta$ is the friction coefficient, and $U$ is the potential of external perturbation. The noise $f$ depends on the present position, as well as the time. In Fourier representation, $f(x,t) = \int dk d\omega c(k,\omega) e^{i(kx-\omega t)}$, the coefficient $c(k,\omega)$ is a stochastic variable. As a noise statistics, we assume that $f(x,t)$ is Gaussian for both $x$ and $t$ with the mean and variance

$$\langle f(x,t) \rangle = 0, \langle f(x,t) f(x',t') \rangle = C(x-x', t-t'). \tag{2}$$

Here, we assume that the noise is invariant under the uniform translation. Also, the correlation function is sufficiently smooth, and has a characteristic correlation length $\xi$ and a correlation time $\tau$. For concreteness, we consider the case of Gaussian kernel,

$$C(x,t) = C_0 e^{-\frac{x^2}{\xi^2} - \frac{t^2}{\tau^2}}, \; C_0 = \frac{2\eta k_B T}{\sqrt{\pi}\tau}. \tag{3}$$

The noise strength $T$ equals the thermodynamic temperature in the absence of the spatial randomness $\xi \to \infty$.

### 2.2. POPULATION DYNAMICS

It is convenient to formulate the time evolution of the probability density $p(x,t)$ to find the particle at a position $x$ at a time $t$. This is achieved with the use of stochastic Liouville equation (Zwanzig 2001) $\frac{\partial}{\partial t} p(x,t) = -Lp(x,t) - \frac{1}{\eta}\frac{\partial}{\partial x} f(x,t)p(x,t)$. $L = \frac{1}{\eta}\frac{\partial}{\partial x} U'(x)$ is the Liouville operator. Then with the aid of the formula $p(x,t) = e^{-tL}p(x,0) - \frac{1}{\eta}\int_0^t ds e^{-(t-s)L} \frac{\partial}{\partial x} f(x,s)p(x,s)$, the stochastic Liouville equation is given as

$$\frac{\partial}{\partial t} p(x,t) = -Lp(x,t) - \frac{1}{\eta} e^{-tL} p(x,0)$$
$$+ \frac{1}{\eta^2}\frac{\partial}{\partial x} f(x,t) \int_0^t ds \frac{\partial}{\partial x} f(e^{(t-s)L}x, s) p(e^{(t-s)L}x, s). \tag{4}$$

Note that the time evolution operator acting on the noise and the population is time reversed as $e^{-tL}f(x,s) = f(e^{tL}x,s)$. Averaging over the noise, $P(x,t) = \langle p(x,t)\rangle$, and neglecting the coupling between the noise $f$ and the population at earlier time $p$, one has

$$\frac{\partial}{\partial t}P(x,t) = -LP(x,t) + \frac{1}{\eta^2}\frac{\partial}{\partial x}\langle f(x,t)\int_0^t dse^{-(t-s)L}\frac{\partial}{\partial x}f(x,s)\rangle\langle p(x,s)\rangle. \quad (5)$$

The neglect of the $O(\tau)$ correlation between the explicit noise and implicit noise in the population is justified in the Markovian limit $\tau \to 0$. In terms of the correlation function, the corresponding Fokker–Planck equation is

$$\frac{\partial}{\partial t}P(x,t) = \frac{1}{\eta}\frac{\partial}{\partial x}(U'(x) - \frac{1}{\eta}\int_0^\infty ds\frac{\partial C}{\partial x}(\frac{U'(x)}{\eta}s,s) + \frac{1}{\eta}\int_0^\infty dsC(\frac{U'(x)}{\eta}s,s)\frac{\partial}{\partial x})P(x,t)$$

## 2.3. RENORMALIZED FOKKER–PLANCK EQUATION

For the Gaussian noise correlation (3), the lowest order truncation of the hierarchy of the stochastic Liouville equation is available in the Markovian limit defined as $\tau \to 0$, $\xi \to 0$, and $\kappa \equiv \frac{2\eta\tau k_B T}{\sqrt{\pi}\eta\xi^2}$ kept finite. In this case, the Fokker–Planck equation is

$$\frac{\partial}{\partial t}P(x,t) = \frac{1}{\eta}\frac{\partial}{\partial x}(U'(x)(1-\kappa) + k_B T\frac{\partial}{\partial x})P(x,t). \quad (6)$$

Intuitively, this limit can capture the finiteness of the spatial correlation $\xi$ of the background noise expressed by the important parameter $\kappa$. Note that unlike slow diffusion in disordered systems, the spatio-temporal correlated system can show accelerated diffusion since the kinetic barriers are lowered. In Fig. 1, for a double well potential $U(x) = x^4/4 - x^2/2$, the renormalized canonical distribution is compared with the steady state distribution calculated from the stochastic simulation.

# 3. Spatio-Temporal Correlated Open System

In this section, we generalize the Markovian limit approach to the case of open system. As far as the system is driven only by the external perturbation $-U'(x)$, the drift velocity seems to be renormalized. In contact with a spatially uncorrelated heat bath, however, it is unknown how does the interaction with the reservoir is affected. We explore this issue with the use of the Fokker–Planck equation for total system.

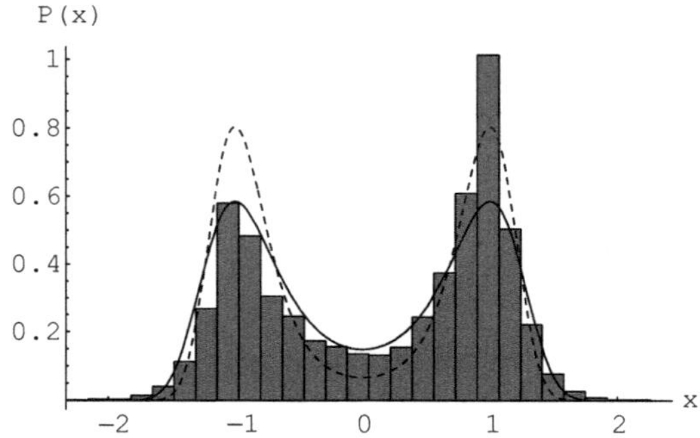


*Figure 1.* The histogram shows the steady state distribution calculated from the colored Langevin equation (1). The solid-line indicates the renormalized canonical distribution $\frac{1}{Z}e^{-U(x)(1-\kappa)/T}$. The canonical distribution without potential renormalization $\kappa = 0$ is plotted as the broken-line

### 3.1. MODEL OF AN OPEN SYSTEM

The spatio-temporal correlated system is connected to a thermostat at a temperature $T_0$. The thermostat is composed of a Brownian particle suspended to a Langevin heat bath. Let us denote the position of this particle as $y(t)$. Another particle introduced in the previous section $x(t)$ is linearly couples with the particle $y(t)$

$$\eta_1 \dot{x} = -U'(x) + f(x,t) - k(x-y)$$
$$\eta_2 \dot{y} = -V'(y) + k(x-y) + f_y(t), \tag{7}$$

where $f(x,t)$ is the spatio-temporal noise and $f_y(t)$ is the thermal white noise. We assume that $f(x,t)$ and $f_y(t)$ are independent Gaussian processes. The spring constant $k$ is small-enough, i.e. coupling is weak-enough. As a concrete example, we consider the case of the Gaussian memory $\langle f(x,t)f(x',t')\rangle = \frac{2\eta k_B T}{\sqrt{\pi}\tau} e^{-\frac{(x-x')^2}{\xi^2} - \frac{(t-t')^2}{\tau^2}}$, and the delta correlation, $\langle f_y(t) f_y(s)\rangle = 2\eta_2 k_B T_0 \delta(t-s)$.

### 3.2. FOKKER–PLANCK EQUATION

The stochastic Liouville equation for the distribution of $x$ and $y$ coordinates is

$$\frac{\partial}{\partial t} p(x,y,t) = -L_2 p(x,y,t) - \frac{\partial}{\partial x}\frac{1}{\eta_1} f(x,t) p(x,y,t) - \frac{\partial}{\partial y}\frac{1}{\eta_2} f_y(t) p(x,y,t). \tag{8}$$

Here $L_2 = \frac{1}{\eta_1}\frac{\partial}{\partial x}U'(x) + \frac{1}{\eta_2}\frac{\partial}{\partial y}V'(y)$. With the use of the truncation for the hierarchy of the stochastic Liouville equation as in the previous section, the Fokker–Planck equation is given as

$$\begin{aligned}\frac{\partial}{\partial t}P(x,y,t) &= \frac{1}{\eta_1}\frac{\partial}{\partial x}((U'(x)+k(x-y))(1-\kappa)+k_BT\frac{\partial}{\partial x})P(x,y,t)\\ &\quad+\frac{1}{\eta_2}\frac{\partial}{\partial y}(V'(y)+k(y-x)+k_BT_0\frac{\partial}{\partial y})P(x,y,t),\end{aligned} \tag{9}$$

where $\kappa \equiv \frac{2\tau k_B T}{\sqrt{\pi}\xi^2\eta_1}$. The correlation between $f$ and $p$ is $O(\tau)$ and is neglected in the Markovian limit.

It is remarkable that the interaction force acting on the $x$ particle is $k(x-y)(1-\kappa)$, while that on the $y$ particle is just $k(x-y)$. This is one of the main result of this paper. With the linear transformation of the coordinates $x' = \sqrt{\eta_1}x$, $y' = \sqrt{\frac{\eta_2 T}{T_0}}y$, the Fokker–Planck equation is rewritten as $\frac{\partial}{\partial t}P = -\nabla\cdot\vec{J}$, $-\vec{J} = \vec{D}P + k_BT\nabla P$.

Then the potential condition $\nabla\times\vec{D} = 0$ is rewritten as

$$\frac{\partial}{\partial y'}\left(\frac{\partial U}{\partial x'} + k(\frac{x'}{\eta_1} - \sqrt{\frac{T_0}{\eta_1\eta_2 T}}y')\right)(1-\kappa) = \frac{\partial}{\partial x'}\left(\frac{T}{T_0}\frac{\partial V}{\partial y'} + k(\frac{y'}{\eta_2} - \sqrt{\frac{T}{\eta_1\eta_2 T_0}}x')\right), \tag{10}$$

or equivalently $T_0 = T/(1-\kappa)$. This condition guarantees vanishing current $\vec{J} = 0$ at the stationary state,

$$P(x,y) = \frac{1}{Z}e^{-(U(x)+V(y)+k(x-y)^2/2)/T_0}. \tag{11}$$

Therefore, there is a unique thermodynamic temperature specified by the noise strength and the spatial correlation $T_0 = T/(1-\kappa)$.

### 3.3. NUMERICAL RESULTS

Numerical simulation shows that for $T_0 = T/(1-\kappa)$, (1) the canonical distribution (11) is achieved, and (2) energy dissipated to the reservoir is zero. We have used the harmonic trapping potentials $U(x) = 0.5(x+3)^2$, $V(y) = 0.5y^2$, $k = 0.5$, and the parameters $\tau = 0.01$, $\xi = 0.05$, $T = 0.1$, $T_0 = T/(1-\kappa)$, $\eta_1 = \eta_2 = 1$, and the time step $\delta t = 0.08$ (Fig. 2).

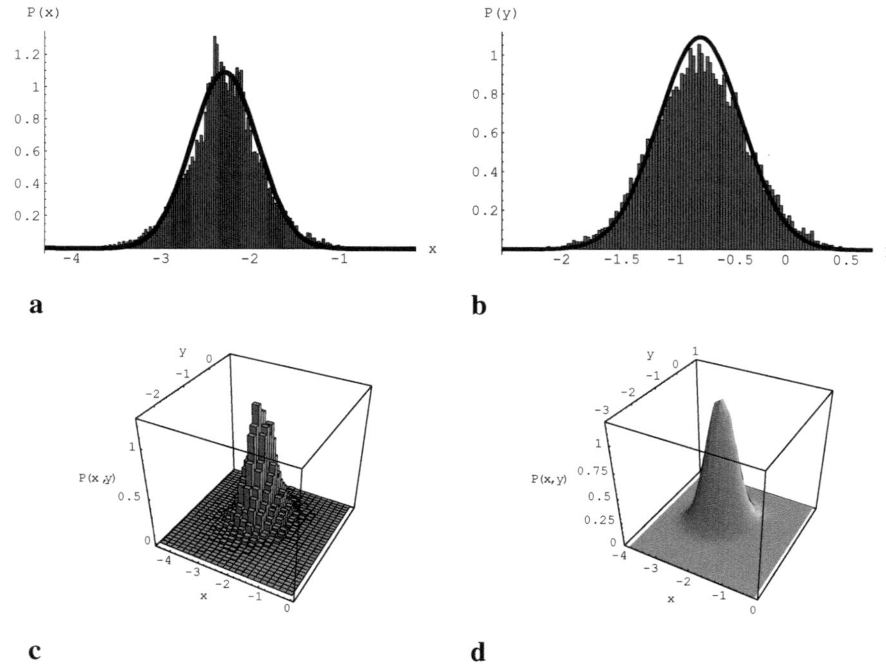


*Figure 2.* The stationary position distributions of (**a**) $x$-particle $P(x)$, and (**b**) $y$-particle $P(y)$. (**c**) shows stationary distribution on the $x$–$y$ plane $P(x, y)$. The histograms are obtained by the stochastic simulation of the Langevin equations. Fifty-one trajectories are sampled for 300 time steps. Parameters are $\tau = 0.01$, $\xi = 0.05$, $T = 0.1$, $T_0 = T/(1-\kappa)$, $\eta_1 = \eta_2 = 1$, and time step is 0.08. The histograms well agrees with the canonical distribution indicated as a solid-line (**a, b**), and as a smooth surface (**d**)

## 4. Summary

In summary, for the spatio-temporal short correlated noise field, a largely fluctuating particle experiences the decorrelation, which amounts to the renormalization of the drift velocity appearing in the Fokker–Planck equation. In the presence of the thermostat, the renormalization of the interaction yields a nontrivial potential condition, which determines a well-defined temperature $T_0 = T/(1-\kappa)$.

## Acknowledgements

The author is grateful to Professor K. Nakamura and Dr. Sugita for fruitful discussions. This work is partially supported by Waseda University Grant for Special Research Projects.

## References

Arvedson E., Wilkinson M., Mehlig B. and Nakamura K.: 2006. *Phys. Rev. Lett.*, **96**, 030601

Bezuglyy V., Mehlig B., Wilkinson M., Nakamura K. and Arvedson E.: 2006. *J. Math. Phys.*, **47**, 073301

Deutsch J.M.: 1985. *J. Phys. A*, **18**, 1457

Golubovic L., Feng S. and Zeng F.A.: 1991. *Phys. Rev. Lett.*, **67**, 2115

Monnai T.: 2008. *J. Stat. Mech.*, 01055

Monnai T., Sugita A. and Nakamura K.: 2005. *Europhys. Lett.*, **84**, 20005

Rosenbluth M.N.: 1992. *Phys. Rev. Lett.*, **69**, 1831

Zwanzig R.: 2001. *Nonequilibrium Statistical Mechanics*. Oxford University Press, Oxford

# INDEX